Multifrequency Electromagnetic Data Interpretation for Subsurface Characterization

Multifrequency Electromagnetic Data Interpretation for Subsurface Characterization

Siddharth Misra
Harold Vance Department of Petroleum Engineering,
College of Engineering, Texas A&M University, TX, United States;
Department of Geology and Geophysics, College of Geoscience,
Texas A&M University, TX, United States

Yifu Han
Schlumberger Beijing Geoscience Center, Beijing, China

Yuteng Jin
Harold Vance Department of Petroleum Engineering,
Texas A&M University, TX, United States

Pratiksha Tathed
Petrophysicist at BP, Houston, TX, United States

ELSEVIER

Elsevier
Radarweg 29, PO Box 211, 1000 AE Amsterdam, Netherlands
The Boulevard, Langford Lane, Kidlington, Oxford OX5 1GB, United Kingdom
50 Hampshire Street, 5th Floor, Cambridge, MA 02139, United States

Library of Congress Cataloging-in-Publication Data
A catalog record for this book is available from the Library of Congress

British Library Cataloguing-in-Publication Data
A catalogue record for this book is available from the British Library

ISBN: 978-0-12-821439-8

For information on all Elsevier publications visit our website at https://www.elsevier.com/books-and-journals

Publisher: Candice Janco
Acquisitions Editor: Amy Shapiro
Editorial Project Manager: Pat Gonzalez
Production Project Manager: Paul Prasad Chandramohan
Cover Designer: Christian J. Bilbow

Typeset by TNQ Technologies

To Swati

For your guidance, encouragement,
patience, and care.
For being there from the conceptualization
to completion of the book and beyond.
For being a wonderful person who is always
ready to laugh and make others laugh.

Contents

Section 4
Applications of multifrequency electromagnetic data and interpretation

Preface

The ideas, concepts, and formulations developed in this book can be applied to several disciplines requiring material/process characterization, such as colloid and interfacial sciences, bioscience, electrochemistry, geophysics, water resources, geothermal resources, fossil energy resources, and mining exploration. Multifrequency electromagnetic (EM) sensing and its interpretation is a crucial tool to characterize the subsurface and materials. Characterization of materials enables efficient and effective engineering of systems and processes. Multifrequency laboratory measurement of a dispersive EM property, such as electrical conductivity, dielectric permittivity, or magnetic permeability, is commonly analyzed for purposes of material characterization. Such an analysis requires inversion of the multifrequency measurement based on a forward mechanistic/empirical model. In doing so, the estimated parameters of the model serve as characteristics/identifiers of the material.

Chapters 1 provides a comprehensive review of current practices in multifrequency EM data acquisition and interpretation in the laboratory and in the subsurface. The chapter also provides information about various interpretation models and workflows for geological characterization and formation evaluation. The chapter presents the use of various EM logging tools for subsurface multifrequency EM data acquisition. Several case studies of subsurface EM log acquisition and interpretation are elaborated and analyzed in the chapter. Different EM log-interpretation workflows are discussed for various formations, including conventional sandstone, shaly-sand, carbonate, turbidities, and unconventional formations. At the end, Chapter 1 discusses the integration of EM measurements with other physical measurements, such as nuclear magnetic resonance and formation testing measurements, for purposes of reducing the uncertainties in estimation and to estimate new formation properties that can only be computed using joint-interpretation methods.

Chapters 2 provides a brief review of mechanistic and empirical models used for multifrequency EM data interpretation. Chapter 2 first describes popular empirical models that are used for the interpretation of resistivity measurements acquired in the laboratory and subsurface environments. Next, the chapter provides information about various multifrequency mechanistic models that can be used for the interpretation of high-frequency conductivity and permittivity frequency-based measurements. Following that, several mechanistic models developed using effective-medium and percolation approximations are presented that are used for the interpretation of low-frequency and mid-frequency conductivity and permittivity measurements. Mechanistic models implementing effective-medium or percolation approximations can handle dispersive behavior of EM properties that makes them suited for the interpretation of multifrequency EM logs for geological characterization and formation evaluation.

Chapters 3, 4, and 5 develop and validate mechanistic models for the interpretation of multifrequency EM data. Chapter 3 describes the development of a mechanistic model of multifrequency conductivity and permittivity for a homogeneous porous media containing uniformly distributed surface charge–bearing nonconductive particles (e.g., clays and sand). This model is referred as the SCAIP model. The mixtures under investigation are assumed to be saturated with hydrocarbon and brine/water. The model is useful in the frequency range of 100 Hz to 10 MHz. The model accounts for the frequency-dependent interfacial polarization effects on the conductivity and permittivity of fluid-filled porous materials because of the surface conductance of nonconductive particles. The chapter focused on the interfacial polarization effects of clays on the effective conductivity in the frequency range of 100 Hz to 100 kHz and on the effective permittivity in the frequency range of 0.5 MHz to 1 GHz. Overall, porosity, surface conductance of particles, and size of the particles play an important role in determining the conductivity and permittivity of mixtures containing nonconductive particles possessing surface charges.

Chapter 4 describes the development of a mechanistic model of multifrequency conductivity and permittivity for a homogeneous porous media containing uniformly distributed conductive particles (e.g., pyrite). This model is referred as the PPIP model that is combined with the SCAIP model from Chapter 3. The mixtures under investigation are assumed to be saturated with hydrocarbon and brine/water. The later part of the chapter focuses on the interfacial polarization effects of both conductive and nonconductive minerals on the effective conductivity in the frequency range of 100 Hz to 100 kHz and on the effective permittivity in the frequency range of 0.5 MHz to 1 GHz. The presence of conductive particles increases the effective permittivity and decreases the effective conductivity at lower frequencies. Effective conductivity reduces and effective permittivity increases for mixtures containing conductive particles of smaller characteristic lengths. In the frequency windows mentioned above, the frequency dispersion of complex conductivity because of the IFP effects for clays is negligible compared to conductive particles.

In Chapter 5, we develop a new mechanistic model to quantify the effects of wettability of conductive particles (e.g., pyrite) and nonconductive particles (e.g., clay) on the multifrequency complex conductivity/permittivity of fluid-filled porous geological materials. The model allows the wettability to range from strongly wet to intermediate wet with respect to the two pore-filling fluids. The model considers the fluid saturation and the operating frequency of the externally applied EM field. The model reveals that low-frequency conductivity, unlike high-frequency conductivity, of the mixture is more sensitive to contact angle (wettability) of the conductive particle as compared to the oil saturation. Furthermore, as compared to low-frequency permittivity, high-frequency permittivity is much less sensitive to oil saturation. As the volume fraction of water-wet conductive particles decreases with the simultaneous increase in volume fraction of oil-wet conductive particles, the frequency dispersion reduces. Counterintuitively, low-frequency conductivity decreases with increase in water wetness of conductive particles. Overall, the effect of oil saturation is less than the effect of contact angle on the frequency dependence of conductivity and permittivity (i.e., the contact angle has a primary effect and oil saturation has a secondary effect).

Chapters 6 and 7 demonstrate simple deterministic inversion-based interpretation of multifrequency EM data. EM-based characterization generally involves estimation of the empirical/physical parameters of a relaxation model that are unique to the material with respect to the measurement and the measured physical process. To process a dispersive EM measurement, a deterministic or stochastic inversion algorithm can be applied to iteratively minimize the difference between the measurements and the numerical predictions of a relaxation model for a set of unknown relaxation model parameters. Chapter 6 presents a unified inversion scheme coupled with relaxation model that was developed to process any multifrequency complex conductivity, complex resistivity, complex permittivity, and complex impedance measurements. The proposed inversion scheme was successfully coupled to various type of relaxation models, namely, Havriliak–Negami model, Pelton complex impedance Cole–Cole model, and complex conductivity Cole–Cole model, to name a few. This unified inversion scheme can be coupled to any type of relaxation models to independently process the multifrequency measurement of any EM property for purposes of improved EM-based material characterization. The proposed deterministic-inversion method can estimate up to seven relaxation-model parameters while exhibiting convergence for random initializations of the relaxation-model parameters within three orders of magnitude variations. The inversion method implements a bounded Levenberg–Marquardt algorithm with tunable damping parameter and its iterative adjustment factor. Notably, jump-out step and jump-back-in step are implemented as automated methods in the inversion scheme to prevent the inversion from getting trapped around local minima and to honor physical bounds of model parameters. The unified inversion scheme will facilitate consistent and robust material characterization based on multifrequency measurements of various dispersive EM properties.

Borehole-based subsurface EM logs, namely galvanic resistivity (laterolog), induction resistivity, propagation resistivity, and dielectric dispersion logs, are commonly used for water saturation estimation in hydrocarbon-bearing formations. EM logs exhibit frequency dependence primarily due to the interfacial polarization effects arising from clay grain surfaces, conductive minerals, and charge blockage in pore throats. To limit operational expenses in near-wellbore subsurface characterization, operators commonly employ only one of the four EM logging tools operating at a specific frequency. Also, it is not a common practice to measure permittivity along with resistivity when using the laterolog and induction logging tools. Deployment of all the four EM tools in a single well followed by a joint interpretation of the multifrequency electrical conductivity and relative permittivity acquired over a broadband frequency range for water saturation estimation is not a common practice. Chapter 7 proposes joint deterministic interpretation method to invert relative permittivity and electrical conductivity logs derived from a combination of the four aforementioned EM logs. In real world, a joint inversion similar to that proposed in the chapter is challenging because of the limited data points corresponding to the discrete log-acquisition frequencies, which ranges from four to nine frequencies depending on the available EM log combinations. Added to that, in most cases, relative permittivity logs are not generated when using galvanic resistivity and induction logs; consequently, the proposed joint interpretation will most likely lack permittivity data at laterolog and induction log frequencies.

Chapters 8 and 9 present the applications of multifrequency EM data and its interpretation. Chapter 8 describes a stochastic method for joint interpretation of EM induction resistivity, logging-while-drilling propagation resistivity, and dielectric dispersion logging tools in a well drilled European organic-rich shale formation. Multifrequency EM measurements (also referred as the broadband EM dispersion measurements) were acquired at seven EM-log-acquisition frequencies continuously across a depth interval of 1500 m. This was the first ever acquisition of broadband EM dispersion logs, comprising relative dielectric permittivity, and electrical conductivity, measured at seven EM-log-acquisition frequencies in the range of 1 kHz to 1 GHz. The interpretation of the multifrequency EM logs was done using Markov-chain Monte-Carlo—based stochastic inversion scheme coupled with the mechanistic clay-pyrite interfacial polarization model (referred as the PPIP-SCAIP model or the PS model, as described in Chapters 3 and 4). In doing so, multifrequency EM logs can be interpreted to estimate the hydrocarbon saturation, connate water conductivity, and surface conductance of clay in the shale formation.

Chapter 9 presents a method to jointly interpret eight dielectric permittivity and conductivity dispersion logs (10 MHz to 1 GHz) and one laterolog resistivity ($\sim$1 kHz) acquired in 520-feet depth interval of the clay- and carbonate-rich upper Wolfcamp formation. The proposed joint interpretation technique also processed eight dielectric permittivity and conductivity dispersion logs (22 to 960 MHz) and one induction resistivity ($\sim$20 kHz) logs acquired in 350-feet depth interval of the Bakken Petroleum System (BPS). In doing so, a continuous estimation of water saturation, brine conductivity, cementation exponent (textural index), and saturation exponent with ranges of possible values was obtained for the upper Wolfcamp formation and BPS. The proposed log processing technique is robust compared to processing only the dielectric dispersion logs, especially in pyrite-rich, low-porosity, and hydrocarbon-bearing formations. These estimates have higher certainty and better convergence at lower water saturation, which is the desired feature of the processing technique. Compared to the estimates for saturation exponent and brine conductivity, water saturation and cementation exponent estimates exhibit higher certainty. Estimates of cementation exponent obtained using the proposed inversion exhibit higher variation with the increase in depth indicating an increase in heterogeneity/layering with depth.

Chapters 10 and 11 demonstrate the use of advanced inversion-based interpretation of multifrequency EM data. In Chapter 10, water saturation, water conductivity, surface conductance of clay, and contact angle of graphite are simultaneously estimated by processing multifrequency EM measurements using the Markov-chain Monte Carlo (MCMC) inversion and the simulated annealing (SA) inversion. MCMC inversion generates a range of values for the desired estimate denoting the uncertainty in the estimate. Multifrequency EM log responses from 25 distinct layers in a synthetic formation were processed by the two inversion methodologies for purposes of comparing their estimation accuracies and computational times. The determination of data covariance matrix for the MCMC inversion can be a challenge. On the other hand, the SA inversion provides a single value of the desired estimate, which generally represents the optimum value of the estimate. Both the inversion methods provide reliable estimates. MCMC inversion takes about 100 min to perform inversion for all the 25 synthetic layers, while the SA inversion takes about 40 min. MCMC needs more iteration steps to ensure chain convergence as it explores the range of possible values of the desired estimate, representing the uncertainty in the estimation. The computation time of SA can be further reduced by selecting proper values of initial temperature and temperature-decline rate.

Chapter 11 applies and compares parameter estimation methods based on swarm intelligence and probabilistic multiagent algorithms for purposes of multifrequency EM data interpretation. To that end, particle swarm optimization (PSO) and ant colony optimization (ACO) are used to process the multifrequency EM data. In this chapter, water saturation, water conductivity, surface conductance of clay, and contact angle of graphite are simultaneously estimated by processing multifrequency EM measurements using the PSO and ACO-based inversion methods. PSO uses a linearly decreasing inertia weight, and the ACO uses a max-min ant system. Multifrequency EM log responses from 25 distinct layers in a synthetic formation were processed by the two inversion methodologies for purposes of comparing their estimation accuracies and computational times. Water saturation estimates are least accurate, while contact angle estimates are the most accurate when the oil saturation is high. In order of computational speed for the inversion of multifrequency EM data, PSO is the fastest followed by SA and then ACO with Monte Carlo Markov Chain (MCMC) inversion being the slowest. PSO-based inversion is one order of magnitude faster than MCMC-based inversion. The parameters are relatively easy to tune in PSO method.

In summary, a detailed knowledge of reservoir rock properties is crucial for efficient hydrocarbon production, geothermal resource development, and carbon geo-sequestration, to name a few. Geoscientists and petrophysicists have developed a variety of reservoir characterization techniques, such as seismic surveys, well logging, and core analysis, to evaluate the storage potential, resource potential, and producibility of subsurface reservoirs. Currently, a vast majority of exploration and production activities are in geologically complex reservoirs, such as shaly-sands, sand-shale laminations, igneous rocks, and organic-rich mudrocks. New reservoir characterization techniques are required to accurately estimate the reservoir properties.

Existing interpretation methods for subsurface EM measurements in formations containing dispersed and/or laminated clay minerals, clay-sized grains, and conductive minerals rely on empirical models and lack reliable mechanistic models. Electrical conductivity anisotropy, dielectric permittivity anisotropy, and interfacial polarization of these formations significantly influence the EM measurements. Hence, the accuracy of estimation of petrophysical properties based on conventional resistivity interpretation of the EM measurements is generally improved by correlating the subsurface measurements with laboratory measurements on core plugs. Subsurface EM measurements in shaly-sands, sand-shale laminations, and organic-rich mudrocks, to name a few examples, exhibit directional and frequency dispersive characteristics primarily due to the effects of electrical conductivity anisotropy, dielectric permittivity anisotropy, and interfacial polarization phenomena. Conventional resistivity interpretation or high-frequency permittivity interpretation techniques for laboratory and subsurface EM measurements do not account for the frequency-dependent effects of interfacial polarization phenomena. Consequently, the mechanistic models, inversion-based multifrequency EM data interpretation techniques, and subsurface data-processing workflows proposed in this book will be crucial for improved subsurface characterization using multifrequency EM data.

Siddharth Misra
Texas, United States
04/02/2021

Acknowledgments

First and foremost, I want to express my gratitude to my mother, Surekha Mohapatra, who nurtured my fascination with the natural world, and my father, Rabindra Nath Misra, who instilled in me the value of discipline and dedication. As wonderful parents, they held my hand and guided me as I embarked on the journey of life. Next, I thank Dr. Carlos Torres-Verdin, who introduced me to the world of research and taught me to strive for perfection. Sincere thanks to Dr. Dean Homan and John Rasmus for introducing me to the latest and greatest in electromagnetic sensing for applications in subsurface characterization and engineering. This book is made possible by the hard and innovative work being done by the electromagnetic research and engineering community focused on subsurface sensing and monitoring.

Siddharth Misra

Writing a book is harder than I thought and more rewarding than I could have ever imagined. I cannot express enough thanks to my father, Qi Han, and mother, Wenli Yu, for their continuous support and encouragement. A very special thanks to Xiaoyu Zhou, who is my best friend and stands by me during every struggle and all my successes. I would also like to express my eternally gratefulness to Dr. Siddharth Misra, who introduced me to the world of rock physics and inverse theory research and taught me to strive for perfection.

Yifu Han

To my beloved mom, dad, and fiancée. Thanks for the endless love and support. Thanks for the continuous understanding and encouragement. Thanks for the sacrifices and your trust.

Yuteng Jin

Dedicated to my late mother with love and eternal appreciation. Grateful to my family for their unconditional support. Thank you to my advisor, Dr. Siddharth Misra, and friend, Yifu Han, for constant support encouragement when writing this book.

Pratiksha Tathed

Section 1

Multifrequency electromagnetic measurements and interpretation models

Chapter 1

Multifrequency electromagnetic data acquisition and interpretation in the laboratory and in the subsurface: A comprehensive review

Yifu Han[1], Siddharth Misra[2,3], Yuteng Jin[4]

[1]*Schlumberger Technology Corporation, Beijing, China;* [2]*Harold Vance Department of Petroleum Engineering, College of Engineering, Texas A&M University, TX, United States;* [3]*Department of Geology and Geophysics, Texas A&M University, TX, United States;* [4]*Harold Vance Department of Petroleum Engineering, Texas A&M University, TX, United States*

1. Electromagnetic measurements in the laboratory

This section covers the multifrequency electromagnetic (EM) measurements acquired in the laboratory to better understand the dispersive and directional nature of EM properties of geological materials. We will also introduce commonly used instruments for EM data acquisition in the laboratory environment. Although few instruments are presented in this chapter, as authors we do not endorse any company, brand, or product.

1.1 Devices and materials

Keysight's E5071C ENA vector network analyzer (VNA) can perform wide-frequency EM measurement from 9 kHz to 20 GHz with high-measurement speed and low noise. Keysight's E5071C network analyzer allows user to select EM data acquisition frequency either at linear or logarithmic spacing. Calibration process is mandatory before any measurement, which removes systematic errors but not drift errors. Keysight Technologies provides coaxial probe which is a terminated section of rigid coaxial line for the nondestructive relative dielectric-permittivity measurements. There are various coaxial probes

Multifrequency Electromagnetic Data Interpretation for Subsurface Characterization
https://doi.org/10.1016/B978-0-12-821439-8.00005-7

FIGURE 1.1 Keysight E5071C ENA vector network analyzer (on left) which can be used to measure electromagnetic properties of porous materials, such as core samples; Keysight N1501A coaxial probe (on right), which is placed on top of core samples for ensuring contact and nondestructive measurements. *From Garcia AP, Han Y, Heidari Z, October 2018. Integrated workflow to estimate permeability through quantification of rock fabric using joint interpretation of nuclear magnetic resonance and electric measurements. Petrophysics 59 (5), 672–693.*

such as high-performance probe, high-temperature probe, and slim probe. The coaxial probe measurements assume the material is isotropic, homogeneous, and nonmagnetic with flat surfaces (North, 2017). Fig. 1.1 shows the Keysight E5071C ENA VNA and the Keysight N1501A coaxial probe used for multi-frequency relative dielectric-permittivity measurements (Garcia et al., 2018).

Multifrequency impedance analyzer (MFIA) from Zurich Instruments can measure the spectral induced polarization (SIP) measurements or complex impedances at frequencies below 1 MHz (Garcia et al., 2018). Zurich Instruments MFIA is a digital complex impedance analyzer that has two fully configurable demodulators, and it can provide 0.05% basic accuracy at a rate of 20 ms per data. MFIA has high-measurement repeatability with small temperature drift, and can also support voltage and current measurements. The measurement range spanning from 1 mΩ to 1 TΩ. Zurich Instruments MFIA comes with control software and can run on web server to provide graphical user interface for frequency, bias voltage, and test signal amplitude response measurements. The front panel of MFIA contains one current signal input, differential voltage input and differential signal output, two auxiliary inputs, and four auxiliary outputs (Zurich Instruments, 2019).

Another tool called EpsiMu contains a coaxial cell using propagating guided waves and associated software compatible with VNA (Sabouroux and Boschi, 2005). EpsiMu is a dielectric permittivity measurement system that can provide real-time complex permittivity, complex permeability, conductivity and dielectric losses for any type of material including solid, liquid, gels, powder, and granular materials (Sabouroux and Ba, 2011). This instrument adopts a reflection/transmission technique in a coaxial line and de-embedding technique for broadband frequency measurements (Sabouroux and Boschi, 2005). EpsiMu can perform wide-band EM measurements between 1 MHz and 18 GHz. The accuracies of real and imaginary parts of dielectric permittivity measurement vary with the shape of the samples and generally range between

2% and 10%. The minimum thickness required for precise measurements of solid samples is 1.5 mm. EpsiMu only requires the calibration of the VNA and does not need to be recalibrated before measuring different samples.

The experimental setup by Leroy et al. (2008) performed the SIP or multifrequency complex conductivity measurements on water-saturated packs of glass beads by using a four-electrode measurement method, which reduces the electrode polarization. Fig. 1.2A illustrates the schematic experimental setup for SIP experiments of water-saturated packs of glass beads. The electrical current was injected through the outer electrodes A and B, and the electrical field induced by the water-saturated packs of glass beads can be measured by the inner electrodes M and N. The size of the sandbox in Fig. 1.2A is 30 × 20 cm. The experimental setup in Fig. 1.2A has high signal-to-noise ratio, and the drift is negligible. Fig. 1.2B shows the secondary experimental setup for SIP measurements. There is a four-electrode array which includes two porous bronze outer current electrodes and two silver rings inner potential electrodes. The cylinder sample holder has 30 cm length with the inner diameter of 6 cm (Leroy et al., 2008). This experimental setup was tested and validated on electrical networks with phase responses of various sedimentary rocks.

1.2 Data acquisition and data type

A sample measured by Keysight E5071C network analyzer must be thicker than 2 cm to be considered as infinite sample. North (2017) used Keysight E5071C network analyzer with Agilent 85070E high temperature probe to measure the frequency-dependent permittivity and conductivity of fully water

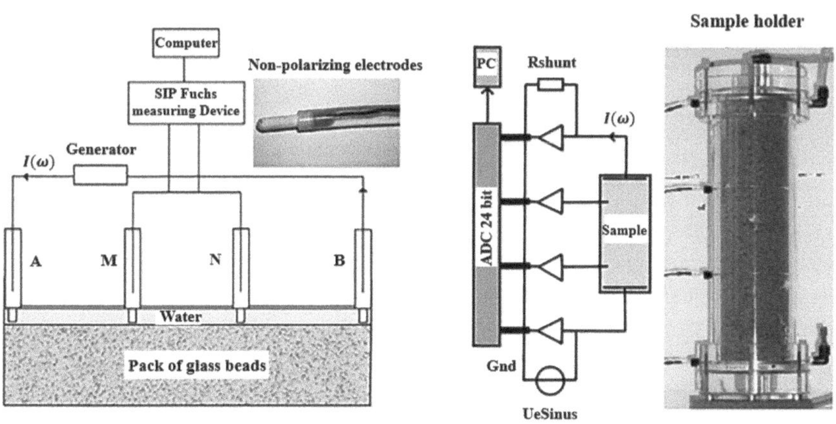

FIGURE 1.2 (A) Sandbox for multifrequency complex conductivity measurement on water-saturated packs of glass beads. (B) Four-electrode array measurement setup and sample holder for the spectral induced polarization experiments on packs of glass beads. *From Leroy P, Revil A, Kemna A, Cosenza P, Ghorbani A, 2008. Complex conductivity of water-saturated packs of glass beads. J. Colloid Interface Sci. 321, 103–117.*

saturated and dry soil samples. The measurement system was set in frequency range of 700 kHz to 7 GHz to measure 401 logarithmically spaced points. Garcia et al. (2018) used the Keysight E5071C network analyzer and placed Keysight N1501A coaxial probe on the top of fully water saturated core samples to measure dielectric permittivity of these samples in the frequency range of 20 MHz to 1 GHz. After the calibration process, Garcia et al. (2018) placed carbonate core samples (dimension of $1\frac{1}{2}$ inch × $1\frac{1}{2}$ inch or $1\frac{1}{2}$ inch × 2 inch) in contact with the coaxial probe with a constant pressure, which confirms the same contact quality between the core sample and the coaxial probe for all the samples. They repeated multifrequency dielectric permittivity measurements including the real and imaginary parts several times in both sides of the samples, which eliminates the effects of surface roughness of the core samples on data acquisition quality.

Fig. 1.3 illustrates the experimental setup of a Zurich Instruments MFIA by Garcia et al. (2018) to measure the complex impedance of the rock samples in the frequency range of 10 Hz–500 kHz. They placed the core sample between two metallic electrodes, and then tightened the core along a threaded rod through the wing nuts. The two electrodes were in contact with core sample and Zurich Instruments MFIA. A sinusoidal voltage excitation with an amplitude of 300 mV was applied on the sample, and the resulting current was recorded to obtain the complex impedance of the core samples. They repeated the measurements and then reliably calculated electrical conductivity of rock samples based on resistance at the minimum phase.

Sabouroux and Ba (2011) used microwave EpsiMu tool to measure complex dielectric permittivity of Fontainebleau sand sample in central France, Dune, of Pilat sand sample in Arcachon Bay area of France, and homogeneous mixture of Fontainebleau sand and 10%–20% clay in the frequency range of 600 MHz to 1 GHz. They used two identical low-loss PVC plugs on each side

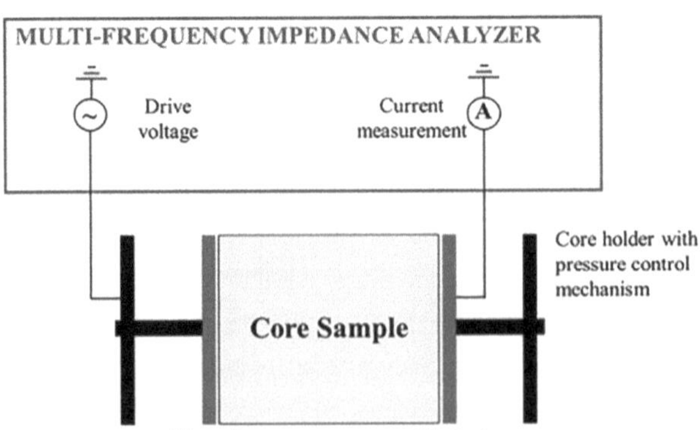

FIGURE 1.3 Schematic of two-terminal experimental setup for complex impedance spectroscopy measurements of core samples. *From Garcia AP, Han Y, Heidari Z, October 2018. Integrated workflow to estimate permeability through quantification of rock fabric using joint interpretation of nuclear magnetic resonance and electric measurements. Petrophysics 59 (5), 672–693.*

of the compacted sand sample to keep it in the coaxial part. The cable effects were removed after VNA calibration process, and the measurement error of real and imaginary parts of the complex dielectric permittivity were less than 1% and 1.5%, respectively. The drift error was not taken into account because the duration of measurement was comparatively short and instrument setup was relatively stable (Sabouroux and Ba, 2011).

Leroy et al. (2008) performed SIP experiments on four samples of glass beads pack with different particle-size distributions and roughness using the four-electrode measurement method. The operation frequency of SIP Fuchs measuring device (setup illustrated in Fig. 1.2A) which is connected to computer is between 10^{-1}–10^{4} Hz. Compared with the setup in Fig. 1.2A, the secondary experimental setup in Fig. 1.2B, which is developed by the Central Laboratory for Electronics in Germany, provides complex conductivity measurements in the frequency range of 1 mHz–45 kHz and higher phase accuracy which is less than 0.1 mrad below 1 kHz. The glass beads were put with deionized water for several weeks before measurements. They also used a laser granulometer (Laser Scattering Mastersizer, MALVERN) which operates in the range of 50 nm to 2 mm to determine the particle-size distribution of samples of glass beads pack under investigation (Leroy et al., 2008).

2. Case studies on the use of core-scale and laboratory-scale multifrequency electromagnetic measurements

2.1 Case study I

This case study focuses on the effect of grain size and grain roughness on the multifrequency complex conductivity. Deionized water was added to glass-bead pack. The water phase conductivity increased with waiting time before SIP measurements because of the dissolution of sodium cations from the glass beads. The measured phase of complex conductivity above 1 kHz is likely to be affected by the wire layout of the device and Maxwell–Wagner polarization effects. Fig. 1.4 illustrates the phase of low- and wide-frequency complex-conductivity measurements on two glass beads pack samples with different surface roughness. All properties of the two glass beads pack samples were the same expect the surface roughness. Surface roughness of the pack was changed by exposing the beads to acid (Leroy et al., 2008). For both smooth and rough glass beads pack, the measured porosity, assumed cementation exponent, measured pH, and the electrical conductivity of the pore-space filled water phase (at 25°C) were 0.4, 1.5, 9.3, and 2.6×10^{-2} S/m, respectively. Leroy et al. (2008) proposed a low-frequency complex conductivity dispersion model to combine Maxwell–Wagner polarization effects and triple-layer electrochemical model, which takes the grain-interface triple-layer electrochemical processes into account. This proposed model also accounts for the grain size distribution. The model predictions agree well with the laboratory-scale multifrequency complex conductivity measurements. Fig. 1.4A shows

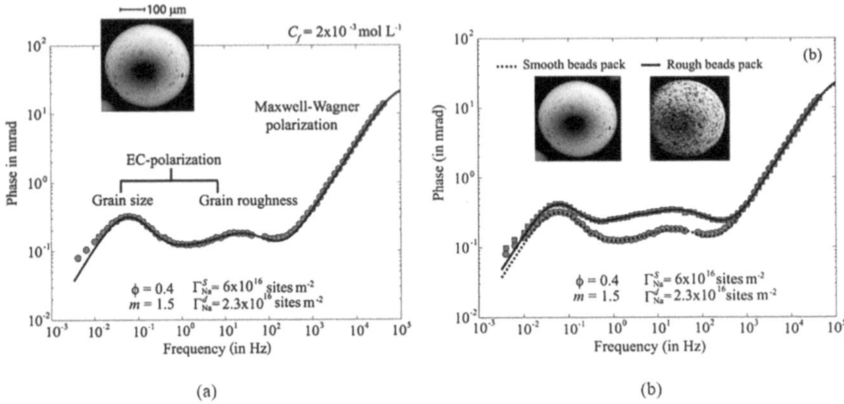

FIGURE 1.4 The comparisons between the measured and modeled phase spectra of complex conductivity of (A) the smooth glass-bead pack sample and those of (B) both the smooth and rough glass-bead pack samples. *From Leroy P, Revil A, Kemna A, Cosenza P, Ghorbani A, 2008. Complex conductivity of water-saturated packs of glass beads. J. Colloid Interface Sci. 321, 103–117.*

the measured phase (dots) and modeled phase (solid line) of the complex conductivity of smooth glass-bead pack sample under investigation. The first peak of the phase of SIP measurements in Fig. 1.4 is around 75 mHz. The first peak corresponds to the grain size distribution of the glass-bead pack under investigation. Fig. 1.4B compares the measured (dots) and modeled (line) phase spectra of the smooth and rough glass beads pack sample. The second peak of the phase of complex conductivity at around 20 Hz is caused by the roughness of the grain surface (Leroy et al., 2008).

The rock physical parameters in the triple-layer electrochemical model for smooth glass-bead sample and rough glass-bead sample were obtained by matching the model predictions with the measurements. The peak of grain size distribution, density of the counterions in the Stern layer, density of the counterions in the diffusion layer, and Dukhin number, which is the ratio of particle surface conductivity to the bulk fluid conductivity multiplied by the particle size, were estimated. The reported values are 187.5 μm, 6.0×10^{16} counterions/m^2, 2.3×10^{16} counterions/m^2, and 0.06, respectively (Leroy et al., 2008). Fig. 1.5 presents the measured and estimated grain size distribution of rough glass-bead pack. The complex conductivity measurements were performed at three different conductivity of the pore-filling water phase: 1.7×10^{-2} S/m, 2.3×10^{-2} S/m, and 4.1×10^{-2} S/m. The estimated grain size distribution agrees well with the measurements, and can capture the dominant grain size (Leroy et al., 2008).

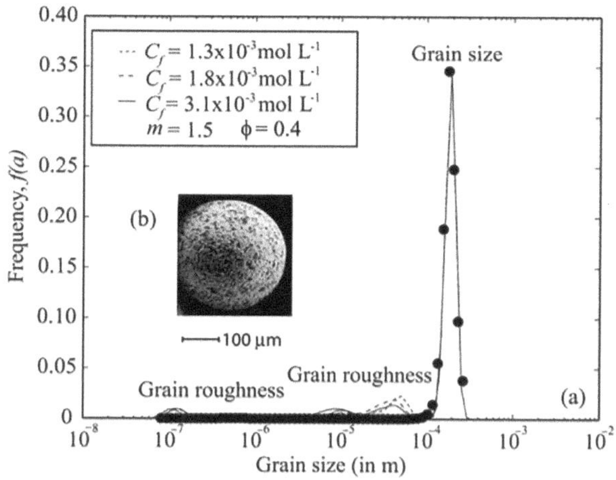

FIGURE 1.5 The comparison of measured (*dots*) and estimated (*lines*) grain size of a rough glass-bead pack sample at various water conductivity. *From Leroy P, Revil A, Kemna A, Cosenza P, Ghorbani A, 2008. Complex conductivity of water-saturated packs of glass beads. J. Colloid Interface Sci. 321, 103–117.*

2.2 Case study II

This case study analyzes the possibility of simultaneously estimating water saturation, salinity, cation exchange capacity (CEC), dominant grain size, and cementation exponent by wideband EM measurements interpretation. Seleznev et al. (2017) estimated water-filled porosity, water conductivity or salinity, CEC, dominant grain size, and cementation exponent by using their proposed differential effective-medium based model to analyze wideband frequency EM measurements. More information about effective medium models are provided in Chapters 2, 3, and 4. The high-frequency EM measurements have higher sensitivity to grain shape, such as the aspect ratio of the spheroids, whereas the low-frequency EM measurements give more insights on grain size distribution (Seleznev et al., 2017). First of all, they estimated those rock physics parameters by inverting model-predicted low-frequency SIP spectra and high-frequency dielectric dispersion data with added Gaussian noise. The high-frequency dielectric dispersion data include the modeled relative permittivity and conductivity at four dielectric-dispersion logging tool operation frequencies ranging from 10 MHz to 1 GHz, and the low-frequency SIP spectra contain 12 logarithmically spaced amplitude and phase angle of complex conductivity between 1 mHz and 12.6 kHz. They validated the proposed inversion method in synthetic cases, and all the estimated petrophysical parameters agree well with their actual values.

Seleznev et al. (2017) measured wideband SIP spectra and dielectric dispersion on clean sandstones with negligible clay minerals, such as Berea sandstone and Massillon sandstone. The SIP spectra experimental setup was tested on NaCl brines before measurements, and they used WTW-3310 conductivity meter to measure water conductivity and temperature before and after SIP spectra measurements. They measured dielectric dispersion data on the twin core samples from the same rock type. The core samples were initially saturated with 0.1 ppk NaCl brine for few days before SIP spectra measurements, then were cleaned and resaturated with 5 ppk NaCl brine for another SIP measurements. Firstly, they inverted SIP spectra from 1 mHz to 100 Hz to estimate the cementation exponent, dominant grain radius, and CEC by coupling the proposed differential effective-medium model. The water-filled porosity, salinity, and volumetric fraction of the uncharged ellipsoidal inclusions were input into the inversion scheme. The estimated dominant grain radius by inversion of only low-frequency SIP spectra for both 0.1 ppk and 5 ppk brine saturated cases are illustrated in Fig. 1.6. The inversion estimates are comparable to the volume-weighted grain size obtained from the image analysis.

Following that, the SIP spectra and dielectric dispersion measurements were jointly inverted to obtain water-filled porosity, salinity, cementation exponent, dominant grain radius, and CEC. The estimated dominant grain size for low-salinity brine (0.1 ppk) saturated measurements agree well with effective volume-weighted grain size obtained by scanning electron microscope images analysis, and were slightly underestimated but still reasonable for high-salinity brine (5 ppk) saturated measurements. The inverted CEC is

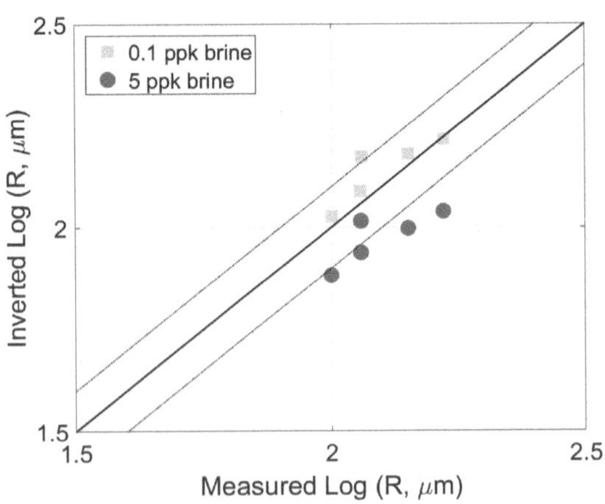

FIGURE 1.6 The comparison of the volume-weighted grain size obtained from the image analysis against the inversion-derived estimate of dominant grain size (R) by processing the spectral induced polarization spectra measurements on 0.1 ppk and 5 ppk brine saturated core, respectively. *Modified from Seleznev N, Hou CY, Freed D, Habashy T, Feng L, Fellah K, June 17–21, 2017. Coherent interpretation of wideband electromagnetic measurements in the millihertz to gigahertz frequency range. SPWLA 58th Ann. Logg. Symp.; Source: Authors.*

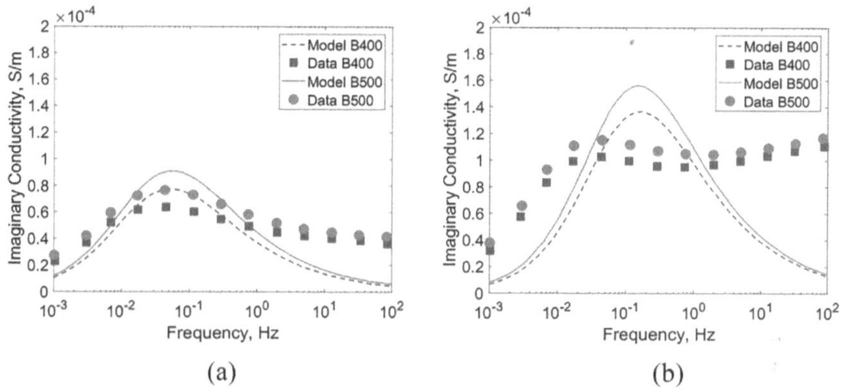

FIGURE 1.7 The comparison of the modeled imaginary part of complex conductivity based on the wideband electromagnetic data inversion estimates against the measured imaginary conductivity for brine saturated core at salinity of (A) 0.1 ppk and (B) 5 ppk, respectively (Seleznev et al., 2017). *Source: Authors.*

low since the samples are clean sandstone. Even though the sensitivity of cementation exponent to SIP spectra varies with brine salinity, the inverted cementation exponent is in the typical range of clean sandstone. Fig. 1.7A shows that the modeled imaginary part of complex conductivity can characterize the peak position of measured imaginary conductivity at low frequencies when the salinity is low (0.1 ppk). However, Fig. 1.7B indicates that the modeled imaginary conductivity cannot characterize the flattening of the measured imaginary conductivity dispersion at low frequencies for the higher salinity case (5 ppk). Even though few physical parameters can be reasonably estimated by separately inverting the low-frequency SIP spectra or high-frequency dielectric dispersion measurements, Seleznev et al. (2017) mentioned that the joint inversion of the low-frequency SIP spectra and the high-frequency dielectric dispersion measurements can more accurately and reliably estimate more petrophysical parameters.

2.3 Case study III

This case study introduces a newly proposed multiphysics method to estimate permeability and rock fabric parameters by processing electromagnetic measurements. The electrical resistivity and dielectric permittivity measurements can be used to interpret tortuosity, and the constriction factor which is the function of the ratio of effective pore-body and pore-throat size. Garcia et al. (2018) combined multi-physics methods including electrical resistivity, dielectric permittivity, and nuclear magnetic resonance (NMR) in both core-scale and pore-scale domain to estimate directional permeability and pore-throat size distribution. They measured permeability, electrical resistivity, dielectric permittivity, and NMR measurements on core samples in core-scale domain. The core samples include three cores from Texas Cream limestone (E_TC), one Estaillâdes limestone core (E_ES), one Indiana limestone core (E_IN), one Austin Chalk core (E_AC), and one core obtained from SACROC formation (E_SR). Fig. 1.8 shows these carbonate core samples from left to right. They estimated rock fabric parameters such as tortuosity, effective pore-body size, and constriction factor by the joint interpretation of multi-physics measurements. Then they proposed and validated the workflow which incorporates these rock fabric parameters for permeability and pore-throat size distribution assessment.

Garcia et al. (2018) used the capacitive formation factor proposed by Bitterlich and Wobking (1970) to characterize tortuosity (Clennell, 1997). The capacitive formation factor was defined as the ratio of relative dielectric permittivity of fluid and rock, and can be expressed as

$$F_C = \frac{\varepsilon_f}{\varepsilon_R} = \frac{\tau_{eW}^2}{\phi_T} \tag{2.1}$$

where ε_f and ε_R are the relative dielectric permittivity of fluid and rock, ϕ_T is the total porosity, τ_{eW} is the dielectric tortuosity. They used the rock relative dielectric permittivity measurement at 960 MHz to calculate the dielectric tortuosity by inputting the fluid permittivity and total porosity. Fig. 1.9 contained the multifrequency relative permittivity measurements on above-mentioned carbonate core samples in the frequency range of 20 MHz to 1 GHz. The measured multifrequency relative permittivity on various carbonate samples which are fully saturated with brine are distinct because of different porosity and rock fabric parameters. They applied the analytical

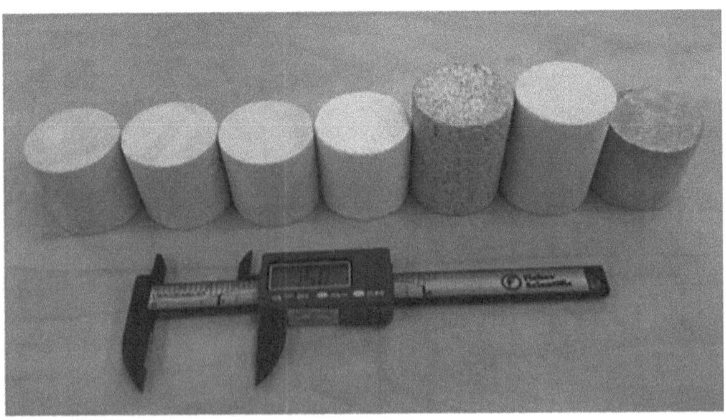

FIGURE 1.8 Carbonate core samples used for experimental measurements. The core samples include three cores from Texas Cream limestone, one Estaillades limestone core, one Indiana limestone core, one Austin Chalk core, and one core obtained from SACROC formation. *From Garcia AP, Han Y, Heidari Z, October 2018. Integrated workflow to estimate permeability through quantification of rock fabric using joint interpretation of nuclear magnetic resonance and electric measurements. Petrophysics 59 (5), 672–693.*

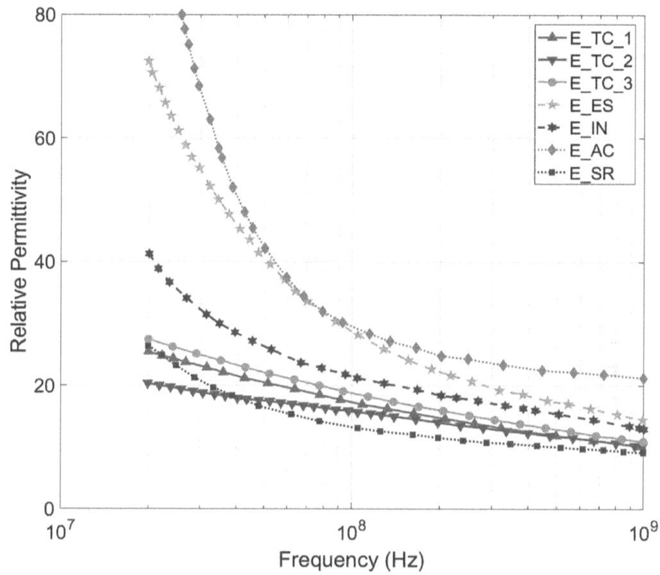

FIGURE 1.9 Multifrequency relative permittivity measurements on various types of carbonate core samples with different porosity and pore-size distribution. *Modified from Garcia AP, Han Y, Heidari Z, October 2018. Integrated workflow to estimate permeability through quantification of rock fabric using joint interpretation of nuclear magnetic resonance and electric measurements. Petrophysics 59 (5), 672–693; Source: Authors.*

equation (Berg, 2012) to relate the electrical conductivity of water-solid two-phase porous media, which can be expressed as

$$\frac{\sigma_R}{\sigma_w} = \frac{\phi_c}{C_E \tau_E} \tag{2.2}$$

where σ_w and σ_R are the conductivity of brine and rock, ϕ_c is the connected porosity, C_E is the electrical constriction factor, and τ_E is the electrical tortuosity. They assumed that the dielectric tortuosity obtained from the interpretation of relative dielectric permittivity at 960 MHz is roughly equal to both the electrical and hydraulic tortuosity, and connected porosity is close to the total porosity. Then, they calculated the electrical constriction factor based on the measured electrical resistivity of rock. NMR measurements were integrated to estimate the pore-body size distribution and finally estimate the permeability and pore-throat size distribution, which will be discussed in the coming section about case studies on the uses of integrating EM measurements with other physics measurements. Table 1.1 includes the measured permeability (k), total porosity (ϕ), and electrical conductivity of carbonate core samples (σ_R), and the estimated permeability (k_{est}), and calculated rock fabric parameters of core samples such as effective pore-throat (r_T) and effective

TABLE 1.1 The measured permeability, total porosity, and electrical conductivity of carbonate core samples, and the estimated permeability and calculated rock fabric parameters of those core samples using the proposed multi-physics workflow.

Sample	k(mD)	ϕ (%)	σ_R (S/m)	τ_{eW}	$r_P(\mu m)$	C_E	$r_T(\mu m)$	k_{est} (mD)
E_TC_1	5.2	26.9	0.186	1.37	1.04	2.46	0.66	7.8
E_TC_2	7.5	27.1	0.173	1.37	1.03	2.66	0.63	7.2
E_TC_3	9.9	27.9	0.202	1.35	1.04	2.44	0.67	8.5
E_ES	43.8	23.3	0.1	1.08	3.76	6.40	1.49	55.1
E_IN	129.5	17.6	0.089	1.00	5.28	6.38	2.09	96.2
E_AC	19.0	31.0	0.3	1.02	1.16	3.20	0.65	15.8
E_SR	0.2	2.2	0.005	1.00	0.94	13.41	0.26	0.2

Data reported by Garcia AP, Han Y, Heidari Z. Integrated workflow to estimate permeability through quantification of rock fabric using joint interpretation of nuclear magnetic resonance and electric measurements. Petrophysics. October 2018;59(5):672-693.

pore-body size (r_P), tortuosity (τ_{eW}), and constriction factor (C_E), which were reported by Garcia et al. (2018). This case study indicates that more rock fabric parameters can be estimated when combined with multi-physics measurements.

2.4 Case study IV

This case study discusses an inversion-based interpretation method coupled with a new proposed rock physics model to simultaneously estimate rock physics parameters such as water saturation and wettability. Garcia et al. (2019) introduced a rock-physics model for the interpretation of wideband electrical resistivity and dielectric permittivity to estimate the porosity, saturation, and wettability of core samples. The proposed model incorporates grain size, porosity, water saturation, wettability, and considers the water-solid interfacial polarization. Various polarization mechanisms co-exist for various frequency ranges. For certain frequency range, one polarization mechanism of the many coexisting mechanisms becomes the dominant mechanism. Similar, in-situ wettability estimation technique has been developed by Jin et al. (2020) and elaborated in Chapter 5. Fig. 1.10 indicates dominant polarization mechanisms at different frequency ranges (Revil, 2012, 2013). The relative dielectric permittivity dispersion effects from 1 Hz to 1 MHz are because of water-solid interfacial polarization and the Maxwell–Wagner effects. Bona et al. (2001) demonstrated that the relative dielectric permittivity measurements are also

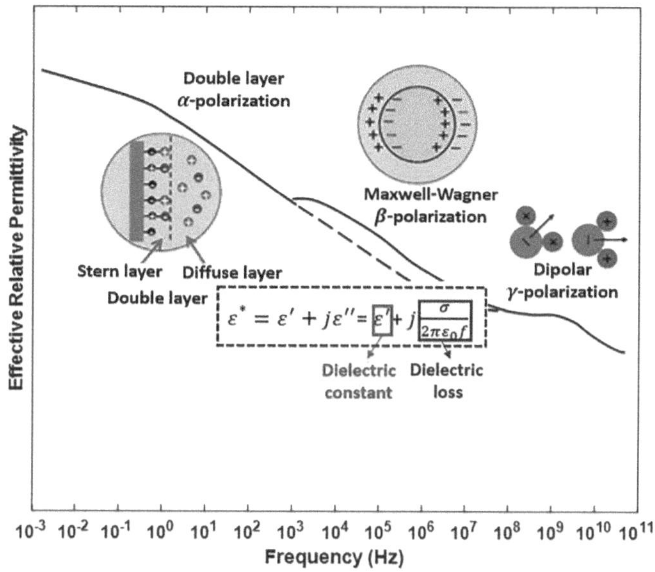

FIGURE 1.10 Dominant frequency-dependent polarization mechanisms including double layer polarization, Maxwell—Wagner polarization, and dipolar polarization in porous media. *Modified from Garcia AP, Heidari Z, Torres-Verdin C, June 17—19, 2019. Multifrequency interpretation of electric resistivity and dielectric permittivity measurements for simultaneous assessment of porosity, water saturation, and wettability. SPWLA 60th Ann. Logg. Symp.*

affected by wettability of rock. The interpretation method developed by Garcia et al. (2019) considers the effect of wettability on the fluid distribution in porous media; hence, the fractional wettability index (X_w) can be estimated from the EM interpretation. They defined the fractional wettability index as the fraction of water-wet grain surface and the change of water-wet rock to hydrocarbon-wet rock happens at X_w equaling to 0.5. Fig. 1.11 illustrates the proposed forward-modeling workflow to calculate the effective permittivity of fluid phase including water and hydrocarbon and effective permittivity of rock.

Garcia et al. (2019) performed inversion of synthetic complex permittivity containing 10% Gaussian noise to validate the proposed inversion scheme. They added 10% Gaussian noise to the modeled real and imaginary parts of multifrequency permittivity computed using the predefined rock-physics parameters, and then inverted the multifrequency data to estimate porosity (ϕ), water saturation (S_w), water salinity (N), dominant grain size (a_G), depolarization factor (N_{HC}), fractional wettability index (X_w), and effective conductivity of hydrocarbon phase ($f_{\sigma w}$). Fig. 1.12A presents the modeled relative permittivity and electrical conductivity (square dot) based on inversion estimates of rock physics parameters and synthetic data with 10% Gaussian noise (circle dot). Table 1.2 lists the inversion results for the synthetic data for

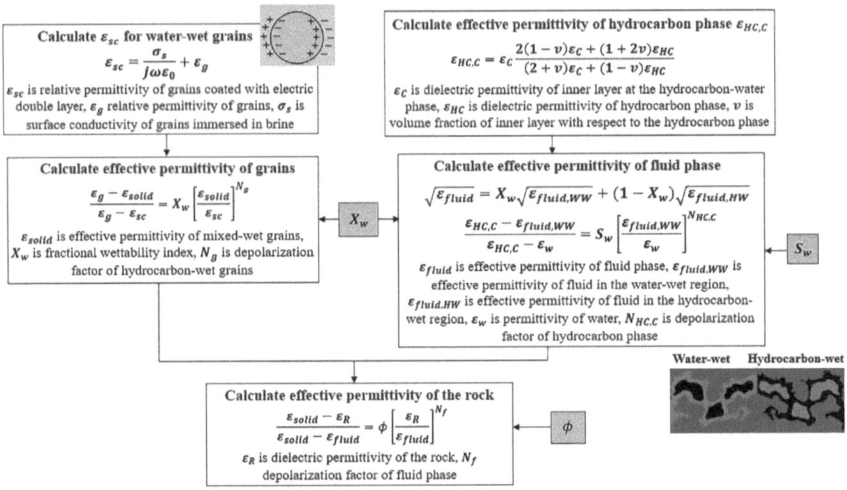

FIGURE 1.11 The proposed forward model workflow integrating total porosity, water saturation, and fractional wettability index to calculate the rock effective permittivity. *According to Garcia AP, Heidari Z, Torres-Verdin C, June 17–19, 2019. Multifrequency interpretation of electric resistivity and dielectric permittivity measurements for simultaneous assessment of porosity, water saturation, and wettability. SPWLA 60th Ann. Logg. Symp.; Source: Authors.*

purposes of validation. The dominant grain size estimate has the largest error compared to its actual value. Finally, they applied the inversion scheme coupled with their proposed forward model workflow to process multifrequency complex dielectric laboratory measurements (such as circle dots in Fig. 1.12B) on core samples documented by Bona et al. (2001). Table 1.3 presents inversion results for complex permittivity of partial water-filled core samples. These core samples are all sandstone with different wettability and levels of water saturation (0.09–0.18). They only estimated porosity, water saturation, and fractional wettability index, which have higher sensitivity or lower relative error as compared to the synthetic inversion case. The results in this case study illustrate that water saturation and wettability can be simultaneously estimated or inverted from multifrequency EM measurements. Similar mechanistic models and log-inversion methods have been developed by Jin et al. (2020).

3. Multifrequency electromagnetic measurements in the subsurface

3.1 Devices and materials

There are four widely used EM logging tools deployed in the borehole environment for formation evaluation and hydrocarbon volume assessment.

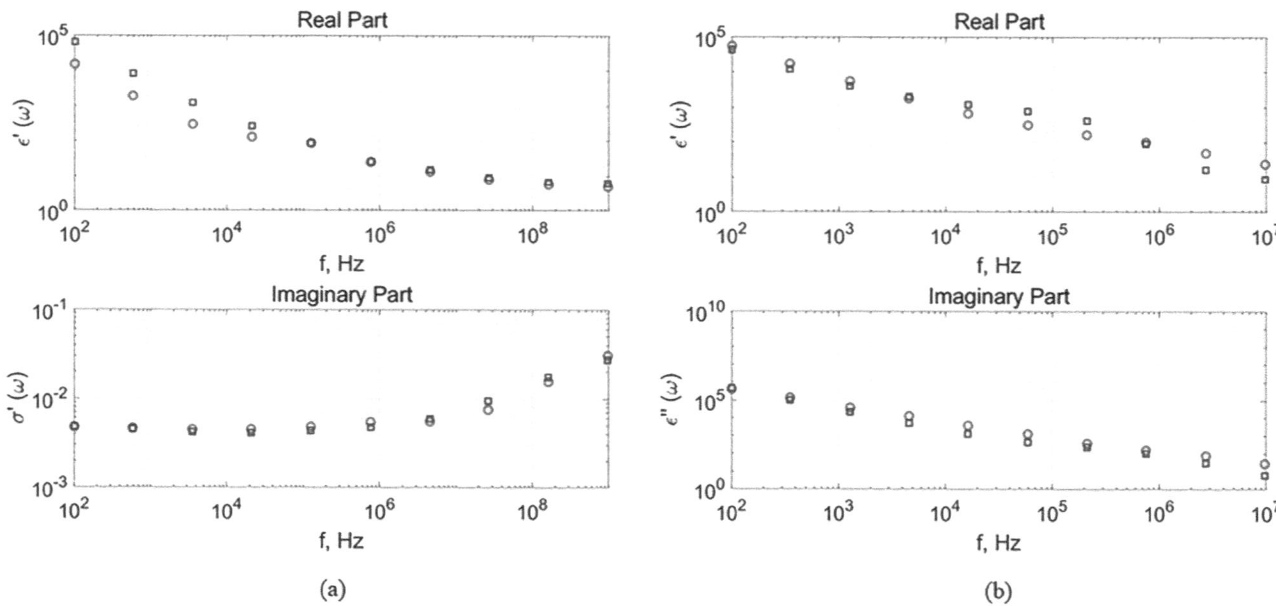

FIGURE 1.12 (A) The comparsion of the model predictions of relative permittivity and electrical conductivity based on the inversion estimates (square *dots*) and the synthetic data containing 10% Gaussian noise (circle *dots*); (B) the comparison of model predictions of real and imaginary parts of multifrequency complex permittivity based on the inversion estimates (square *dots*) and the measured real and imaginary parts of dielectric permittivity of water-wet core sample K13 (circle *dots*). *Modified from Garcia AP, Heidari Z, Torres-Verdin C, June 17–19, 2019. Multifrequency interpretation of electric resistivity and dielectric permittivity measurements for simultaneous assessment of porosity, water saturation, and wettability. SPWLA 60th Ann. Logg. Symp.*

TABLE 1.2 The inversion-derived estimates of rock-physics parameters, their actual values, and estimation error for the synthetic case containing 10% Gaussian noise.

Parameter	Actual	Estimated	Relative error (%)
a_G	150 μm	250 μm	66.7
N	50.0 normality	36.6 normality	26.8
S_w	0.15	0.14	6.7
ϕ	0.25	0.28	12.0
N_{HC}	0.5	0.492	1.6
X_w	0.5	0.48	4.0
$f_{\sigma w}$	0.01	0.009	10.0

Data source is reported by Garcia AP, Han Y, Heidari Z. Integrated workflow to estimate permeability through quantification of rock fabric using joint interpretation of nuclear magnetic resonance and electric measurements. Petrophysics. October 2018;59(5):672–693.

TABLE 1.3 The inversion-derived estimates of porosity (ϕ), water saturation (S_w), and fractional wettability index (X_w) against the measured values for four sandstone core samples.

Sample	ϕ_{actual}	$\phi_{estimated}$	S_w_actual	S_w_estimated	X_w_actual	X_w_estimated
K3	0.23	0.22	0.18	0.14	<0.5	0.29
K98	0.22	0.27	0.17	0.13	<0.5	0.34
K13	0.23	0.24	0.08	0.06	>0.5	0.78
K99	0.22	0.27	0.09	0.07	>0.5	0.7

Data source is reported by Garcia AP, Han Y, Heidari Z. Integrated workflow to estimate permeability through quantification of rock fabric using joint interpretation of nuclear magnetic resonance and electric measurements. Petrophysics. October 2018;59(5):672–693.

These include laterolog/galvanic resistivity, induction resistivity, EM propagation, and dielectric dispersion tools. Laterolog/galvanic resistivity logging tool operates at frequencies lower than 10 kHz, whereas induction logging tool works around 10–80 kHz. EM propagation logging tool is deployed in 400 kHz–10 MHz, whereas dielectric dispersion tool is operational in the frequency range of 10 MHz to 1 GHz. More information about these logging tools and methods to jointly process these logs will be provided in Chapters 7, 8 and 9. Although some of the widely used logging tools are introduced, the introduction of these brand names does not imply any institution's endorsement. Fig. 1.13 shows a schematic of dispersion of relative dielectric permittivity and electrical conductivity along with the four aforementioned

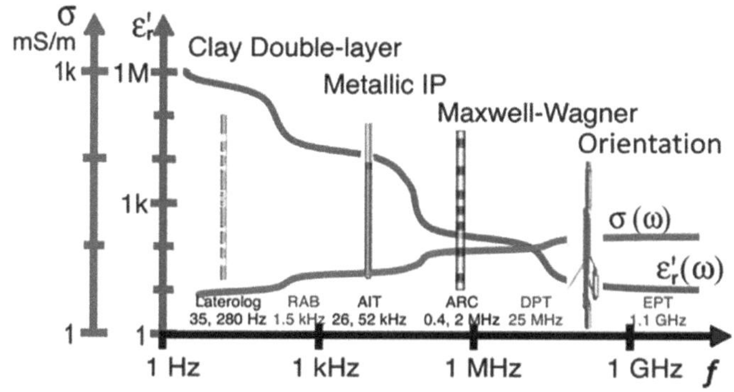

FIGURE 1.13 The dispersion of relative dielectric permittivity and electrical conductivity in the broadband-frequency ranging from 1 Hz to 1 GHz. The four EM logging tools, including laterolog resistivity, induction resistivity, propagation resistivity, and dielectric dispersion tools, are shown approximately at their operating frequency ranges. *Modified from Anderson BI, Barber TD, Luling MG, Rasmus J, Sen PN, Tabanou JR, Haugland M, June 3–6, 2007. Observations of large dielectric effects on LWD propagation-resistivity logs. SPWLA 48th Ann. Logg. Symp.*

EM logging tools at their corresponding operating frequencies (Anderson et al., 2007; Misra and Han, 2016). The frequency dispersion effects on the multifrequency EM measurements are because of the electronic polarization, dipolar polarization, and interfacial polarization effects (Hizem et al., 2008). Electronic polarization is active over a large range of frequencies from low to high, whereas interfacial polarization is active over much smaller range in the lower frequencies. At very high frequencies, electronic polarization remains, whereas at low frequencies, the interfacial polarization dominates. Around 1 GHz, dipolar/orientational polarization tends to dominate in the subsurface because of the presence of dipoles in the pore-filling water.

Laterolog tools are used to determine formation resistivity in conductive borehole or mud environments and operated at discrete frequencies lower than 10 kHz, with a depth of investigation ranging from 30 to 90 in (Han and Misra, 2018a) into the formation. Laterolog resistivity logging tools are commonly operated in the range of 10 Hz to 2 kHz (Anderson, 2001). The Schlumberger high-resolution laterolog array (HRLA) tool has symmetric tool string design and optimal array focusing. The sonde in HRLA tool injects a focused current into the formation and measures the voltage potential at electrodes providing the true formation resistivity. The sonde actually measures the intensity for a constant voltage drop across fixed interval, and the array has one source electrode in the central and six electrodes above and below it. The length of HRLA tool string is 24.1 ft with the depth of investigation of 50 in (median response at the ratio of true formation resistivity and invaded zone resistivity equaling to 0.1), and the vertical resolution is 12 in (Donadille et al., 2017). Laterolog tool implements specific design of measure, buck, and source electrodes to allow multiple depths of investigation at the same depth.

Wireline array induction tool (AIT) uses multiple induction coil arrays and is operated at multiple frequencies. AIT was originally designed for resistivity measurements in wells drilled with oil-based mud or air-drilled boreholes. AIT consists of one transmitter and receiver coil arrays, and measures the secondary voltages induced in the receiver coil because of the eddy currents induced in the formation. The Halliburton "ACRt" AIT (Xiao et al., 2006) has a pair of receivers for each array, one main receiver and one bucking receiver, and all receiver coil arrays are in one side of transmitter. AIT provides robust multichannel signal processing with high radial and vertical resolution, and it corrects thermal, skin, and borehole effects. EM induction tools typically operate at discrete frequencies between 10 and 80 kHz. The length of AITs vary from 16 to 33.5 ft, with five radial depth of investigation: 10, 20, 30, 60, 90 in, and the vertical resolution is 1 ft, 2 ft, or 4 ft (Xiao et al., 2006). Induction logging tool is sensitive to conductivity of the formation.

The logging-while-drilling (LWD) propagation resistivity tools such as Schlumberger arcVISION tool are used for real-time formation evaluation and geosteering while drilling, and they also provide multiple depths of investigation even in high dogleg severity wells and high-resistivity formations. The arcVISION tool provides real-time resistivity, gamma ray, inclination, and annular pressure measurements. There are five transmitters and two receivers in the antenna array of LWD propagation resistivity tools which are typically operated at 400 kHz and 2 MHz. The borehole compensated attenuation and phase shift resistivity measurements at operating frequencies can provide up to 20 depths of investigation. The combination of inclination measurements near the bit, annular pressure, and the multiple depths of investigation resistivity measurements can help to keep the wellbore in the target formation. The length of arcVISION tool string is 25.18 ft, and the depth of investigation of 2 MHz phase shift propagation resistivity measurement is around 12–17 in at formation resistivity equaling to 1 Ωm for different spacings, with the vertical resolution of 0.7 ft (Schlumberger, 2007).

The Schlumberger Dielectric Scanner tool operates in the frequency range of tens of megahertz to a gigahertz. The tool performs dielectric dispersion measurements, including relative dielectric permittivity and electrical conductivity. The typical operating frequencies of Dielectric Scanner are 22, 100, 350, and 960 MHz (Claverie, 2012; Han and Misra, 2018b), and the depth of investigation is 4 inches (10 cm). Due to high operating frequency, these tools have shallow depth of investigation. Figs. 1.13 and 1.14 show the tool string and the short and multispacing antenna array pads of dielectric dispersion logging tool. The blue (dark gray in printed version) and red (gray in printed version) dipoles in Fig. 1.14 are the longitudinal and transverse polarization, respectively (Mosse et al. 2009). There are two coaxial electrical probes in pad antenna, which are used for assessing and quality control of mud and mudcake dielectric properties. The transmitters are in the center with receivers placed symmetrically around transmitters, and the antenna pad is controlled to ensure the best contact against the borehole wall. As the logging-tool operating frequency increases, the depth of investigation decreases. The length of the tool string is 11.27 ft, and it can be operated in both oil and water-based mud conditions (Claverie, 2012).

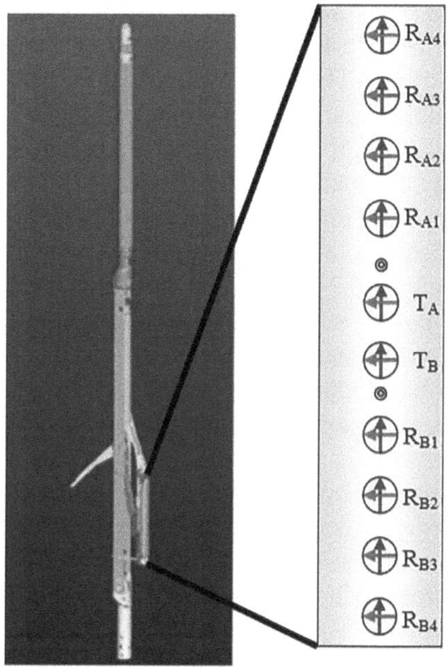

FIGURE 1.14 Dielectric dispersion logging tool sketch and pad antenna. *From Mosse L, Carmona R, Decoster E, Faivre O, Hizem M, June 21–24, 2009. Dielectric dispersion logging in heavy oil: a case study from the Orinoco Belt. SPWLA 50th Ann. Logg. Symp.*

3.2 Data acquisition and data type

Donadille et al. (2017) reported that the HRLA laterolog tool samples every 2 inches at logging speed of 3600 ft/h. HRLA provides five independent and focused formation resistivity measurements with increasing depths of investigation and one mud resistivity measurement. The tool string design of HRLA ensures that the various depths of investigation are simultaneously measured at any given depth. The range of resistivity measurement is 0.2–100,000 Ωm at a mud resistivity of 1 Ωm, and 0.2 to 20,000 Ωm at a mud resistivity of 0.02 Ωm (Donadille et al., 2017). The measurement accuracy is ±5% for the measurement range of 1 to 2000 Ωm, and the measurement accuracy decreases for the larger measurement range. The shallowest formation resistivity measurement is mostly sensitive to borehole wash out and eccentricity effects. HRLA can provide both one-dimensional (1D) and two-dimensional (2D) real-time formation true resistivity, invaded zone resistivity, depth of invasion estimates by 1D radial inversion which assumes radial variation of resistivity and an infinite thick formation, and advanced 2D inversion which accounts for thin-bed or shoulder-bed layer effects, respectively (Donadille et al., 2017).

The secondary voltages induced in the receiver coil is amplified, digitized, and then sent to the surface. Finally, the signals are transformed to apparent conductivity after applying tool calibration and offsets. The apparent conductivity is corrected for the temperature, skin effect because of high-conductivity material, and borehole/environmental effects. The shallowest measurement from shortest-spacing receiver array of AIT is severely affected by borehole effects. The borehole correction method was developed to estimate environmental variables including tool eccentricity (Xiao et al., 2006). The radial inversion of apparent conductivity is applied to obtain true formation conductivity, invaded zone conductivity, and depth of invasion. The advanced quantitative 2D imaging of formation resistivity improves the characterization of bedding and invasion. For the AIT tool, the range of resistivity measurement is 0.1–2000 Ωm, and the measurement accuracy is $\pm 0.75\ \mu$S/m conductivity. In high-contrast thinly bedded formations, the presence of low-resistivity shoulders (<2 Ωm) can be detected by AIT log response (Griffiths et al., 2000). Misra et al. (2015a,b), and Misra (2015) lay out various concepts and design elements essential for triaxial inductive conductivity measurements in the laboratory environment.

The arcVISION array propagation resistivity measurements are compensated to remove rugosity effects and electronics drift. The 2 MHz phase shift resistivity measurement has the shallowest depth of investigation, while 400 kHz attenuation resistivity measurement has the deepest depth of investigation, which is around 24–36 inches with a vertical resolution of 4 ft (Schlumberger, 2007). Generally, with increase in frequency of measurement, the depth of investigation decreases. The measurement range of 2 MHz phase shift resistivity is 0.2 to 60 Ωm with ±2% accuracy and 60 to 3000 Ωm with ±0.3 mS/m conductivity accuracy. On the other hand, the measurement range of 400 kHz attenuation resistivity is 0.1 to 3 Ωm with ±3% accuracy and 3 to 10 Ωm with ±0.3 mS/m conductivity accuracy (Schlumberger, 2007). The tool eccentricity severely affects 2 MHz resistivity measurements in oil-based mud drilled low resistivity formations but not that much for 400 kHz resistivity measurements. The interpretation of LWD propagation resistivity measurements can estimate horizontal and vertical resistivities by using anisotropy model, and formation resistivity and dielectric constant can be estimated by using the dielectric model (Li et al., 2004).

The Dielectric Scanner tool can measure multifrequency formation dielectric properties, and the different array spacings with both longitudinal and transverse polarization ensure high resolution of measurements. The vertical resolution of Dielectric Scanner is only 1 inch, and it directly measures attenuation and phase-shift of complex-valued EM signal. The measurements are processed or transformed to obtain relative permittivity and conductivity at each operating frequency by assuming that the formation is homogeneous (Hizem et al., 2008). The relative permittivity and conductivity measurement accuracy is ±1% and correspond 0.002 water-filled porosity unit (Claverie, 2012). The range of relative permittivity and conductivity measurements of Dielectric Scanner is 1 to 100 and 0.1 to 3000 mS/m, respectively. It can measure both water-filled porosity and water salinity, even in low-water salinity shaly and laminated sand formations. The multifrequency dielectric dispersion measurements can be also used to estimate continuous Archie *m-n* exponent log and CEC (Claverie, 2012).

4. Case studies on the use of near-wellbore and field-scale multifrequency electromagnetic measurements in the subsurface

4.1 Case study I

This case study compares wireline and LWD laterolog resistivity logs acquired in a vertical well, and LWD laterolog resistivity, and propagation logs acquired in a horizontal well. Compared to induction resistivity logs, the laterolog galvanic resistivity measurements are less affected by the presence of thin-bed layer or shoulder-bed effects. Ortenzi et al. (2011) mentioned a new LWD laterolog tool, which uses two monitored button electrodes and five toroidal antennas to provide laterolog resistivity and borehole electrical image logs measurements. The true formation resistivity can be inverted from the laterolog resistivity measurements. Fig. 1.15 shows the schematic of the monitored buttons in the LWD laterolog tool string which can be operated to measure both formation resistivity with azimuthal sensitivity and mud resistivity.

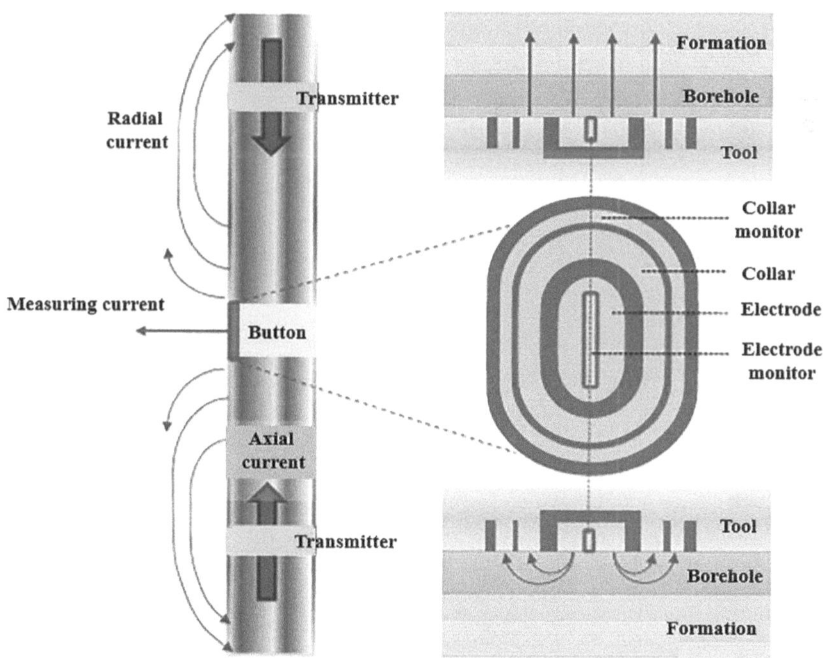

FIGURE 1.15 Schematic of the cylindrical focusing methodology and the monitored buttons in the logging-while-drilling laterolog tool string. *From Ortenzi L, Dubourg I, Os RV, Han SY, Koepsell R, Chow S, Ha Y, May 14–18, 2011. New azimuthal resistivity and high-resolution imager facilitates formation evaluation and well placement of horizontal slim boreholes. SPWLA 52nd Ann. Logg. Symp.*

The toroidal antennas can also be operated as transmitters and receivers without the monitored buttons to measure resistivity. The axially focused toroidal resistivity measurements cannot provide azimuthal sensitivity and have lower vertical resolution.

Fig. 1.16 indicates that logging-while-drilling laterolog resistivity measurements acquired in a vertical well are comparable to wireline resistivity measurements (Ortenzi et al., 2011). The button resistivity measurements, namely shallow (RES_BS), medium (RES_BM), deep (RES_BD), and extra deep (RES_BX), have higher resolution than toroidal resistivity (RES_TRD). Interestingly, RES_TRD is consistent with wireline laterolog resistivity (RLA1, ..., RLA5). The button resistivity can be acquired in azimuthal bins to cover full borehole circumference. The real-time transmission of resistivity images, which contains information about the thin conductive layers, using the MWD tool facilitates real-time structural analysis and fracture characterization while drilling.

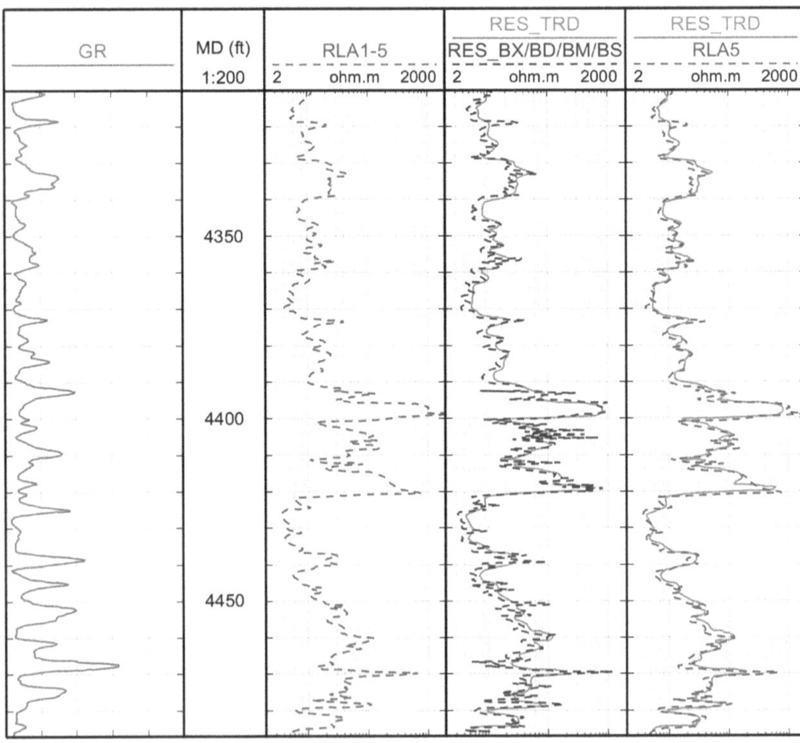

FIGURE 1.16 The comparison between wireline and logging-while-drilling laterolog resistivity logs acquired in a vertical well. *According to Ortenzi L, Dubourg I, Os RV, Han SY, Koepsell R, Chow S, Ha Y, May 14–18, 2011. New azimuthal resistivity and high-resolution imager facilitates formation evaluation and well placement of horizontal slim boreholes. SPWLA 52nd Ann. Logg. Symp.; Source: Authors.*

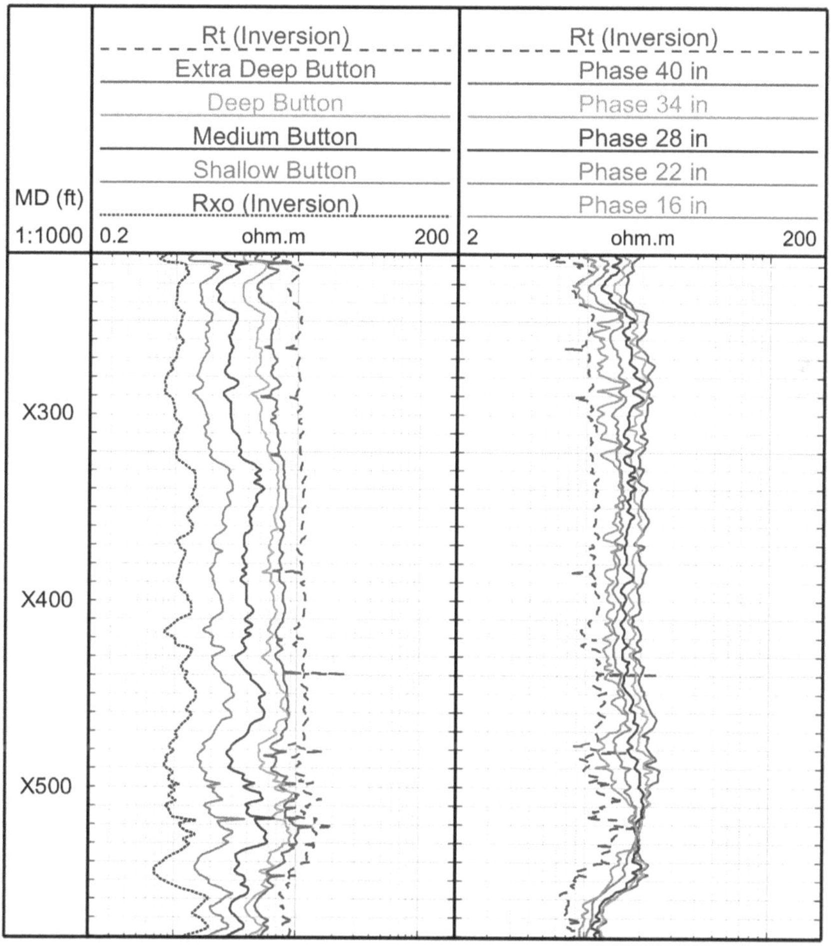

FIGURE 1.17 The comparison between logging-while-drilling laterolog resistivity with multiple depth of investigation and propagation logs with multiple spacing in a horizontal well in laminated turbiditic formation. *According to Ortenzi L, Dubourg I, Os RV, Han SY, Koepsell R, Chow S, Ha Y, May 14–18, 2011. New azimuthal resistivity and high-resolution imager facilitates formation evaluation and well placement of horizontal slim boreholes. SPWLA 52nd Ann. Logg. Symp.; Source: Authors.*

Fig. 1.17 illustrates resistivity logs in a horizontal well in laminated turbiditic formation. Both the LWD propagation and laterolog tools were included in bottom hole assembly (BHA) to measure the true formation resistivity. The first resistivity track in Fig. 1.17 shows the separation of multispacing laterolog resistivity logs because of the invasion of mud with low salinity. The second resistivity track indicates that the inverted true formation resistivity by laterolog resistivity is close to the LWD propagation phase shift resistivity with the shortest spacing (phase 16 in) which is less affected by laminations. The real-time transmission of resistivity images and other measurements helps to confirm whether the trajectory of borehole is within the target formation. The interpretation of propagation resistivity logs is problematic in horizontal well (Ortenzi et al., 2011), the new LWD laterolog tool with multiple spacing can be used to interpret true formation resistivity in such condition.

4.2 Case study II

This case study investigates the impact of dielectric effect on propagation resistivity measurements, and the results indicate that the relative dielectric permittivity and conductivity need to be directly inverted rather than assuming a correlation between dielectric permittivity and resistivity. The responses of LWD propagation resistivity logs including phase shift and attenuation resistivity are controlled by formation dielectric permittivity and conductivity, mud, and invasion. The formation dielectric permittivity has small effect on the EM propagation resistivity log for conventional sedimentary rocks. An empirical relationship between dielectric permittivity and resistivity is generally assumed to calculate formation conductivity. The correlation between relative dielectric constant and resistivity was developed using 286 carbonate and sandstone core samples (Wu et al., 1999), and is demonstrated in Fig. 1.18 (Altman et al., 2008). However, Altman et al. (2008) pointed out that the dielectric assumption was no longer valid even in several documented clastic formations. They reported the anomalous behavior of measured dielectric permittivity and resistivity on volcanic rocks that does not agree with the conventional correlation because of the interfacial polarization effects on grain surface. Fig. 1.18 illustrates that the blue (gray in printed version) data points of volcanic rocks with large dielectric permittivity because of charge buildup. The relative dielectric permittivity and conductivity should be directly inverted from the measured LWD propagation phase shifts and attenuation resistivity when the dielectric assumption is not valid.

Fig. 1.19 shows the wireline induction resistivity and LWD propagation resistivity logs acquired in a vertical well drilled with oil-based mud in an extrusive igneous rock formation (Altman et al., 2008). The oil-based mud invasion was clearly detected from the separation of induction resistivity logs. The track 1 indicates that the shallower 2 MHz phase shift resistivity has lower calculated resistivity than 400 kHz phase shift resistivity, which can be explained by electrical conductivity dispersion with frequency. Generally, conductivity increases with frequency because of various polarization phenomena. However, the track 3 illustrates that 2 MHz phase shift resistivity with lower depth of investigation has lower calculated resistivity than 2 MHz

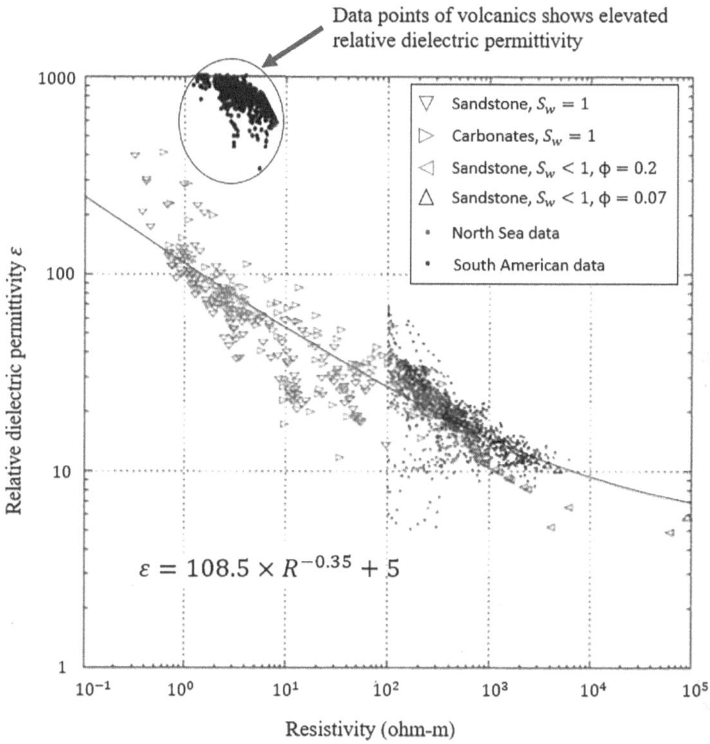

FIGURE 1.18 The dielectric permittivity and resistivity of 286 carbonate and sandstone samples measured at 2 MHz, as reported by Wu et al. (1999) and volcanic rocks. *Modified from Altman R, Anderson B, Rasmus J, Luling MG, May 25–28, 2008. Dielectric effects and resistivity dispersion on induction and propagation-resistivity logs in complex volcanic lithologies: a case study. SPWLA 49th Ann. Logg. Symp.*

attenuation resistivity, which indicates the violation of dielectric assumption. The resistivity logs separation is not obvious or visible in the lower part of formation, which is a shaly zone. The propagation resistivity measurements at 2 MHz were then processed and corrected for dielectric effects, while the relative dielectric permittivity and corrected conductivity were obtained through inversion. The track 1 in Fig. 1.20 demonstrates log interpretation results of assumed and computed relative dielectric permittivity. The track 2 includes the dielectric corrected and uncorrected 2 MHz propagation resistivity, 400 kHz phase shift resistivity, and induction resistivity in the volcanic interval (Altman et al., 2008). This case study illustrates frequency dispersion effects of resistivity measurements. The case shows that the traditional correlation between relative permittivity and resistivity cannot be assumed for resistivity log interpretation.

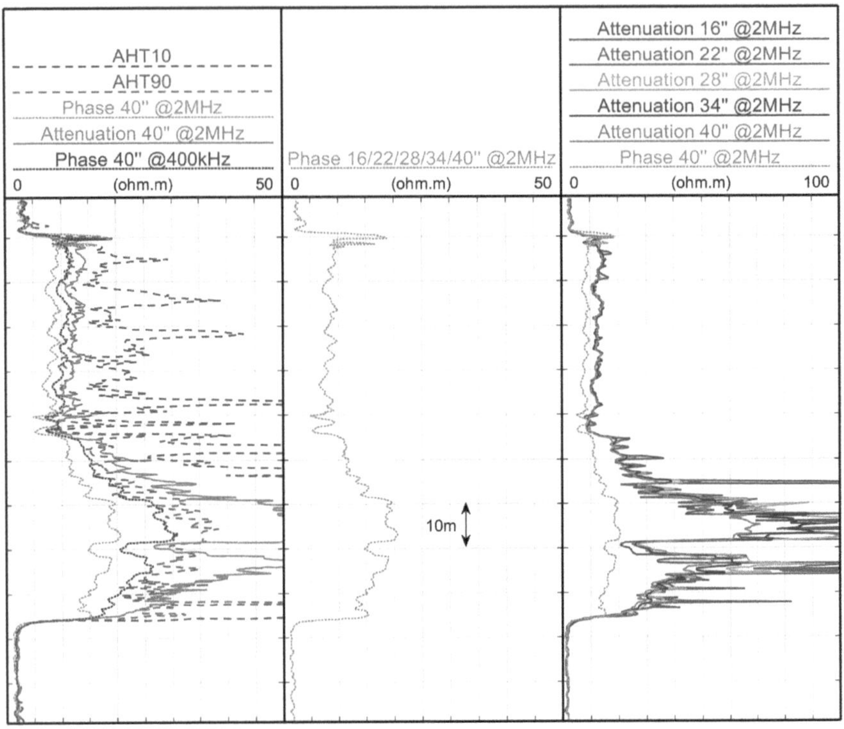

FIGURE 1.19 The dielectric effects and dispersion of induction resistivity and logging-while-drilling propagation resistivity logs acquired in an extrusive igneous rock formation. *According to Altman R, Anderson B, Rasmus J, Luling MG, May 25–28, 2008. Dielectric effects and resistivity dispersion on induction and propagation-resistivity logs in complex volcanic lithologies: a case study. SPWLA 49th Ann. Logg. Symp.; Source: Authors.*

4.3 Case study III

This case study shows that the new-generation dielectric dispersion logging tool (Hizem et al., 2008) can provide more accurate petrophysical formation evaluation in thin-beds laminated sand and carbonate formations. Hizem et al. (2008) applied the widely used Stroud-Milton-De (SMD) model, which was developed in clean-sand, to process dielectric dispersion logs to simultaneously estimate water-filled porosity, salinity, and cementation exponent, which is a rock texture parameter. The dielectric dispersion effects because of rock texture or fabric are significant in high-salinity formation, and are more obvious for higher tortuosity formation (Han et al., 2012). Fig. 1.21 shows the application of dielectric dispersion logs in thin-beds shale laminated heavy-oil sand reservoir formation evaluation. The track 1 in Fig. 1.21 illustrates the comparison water saturation estimates derived from dielectric dispersion logs and those from deep-induction resistivity log. In the upper laminated thin-beds

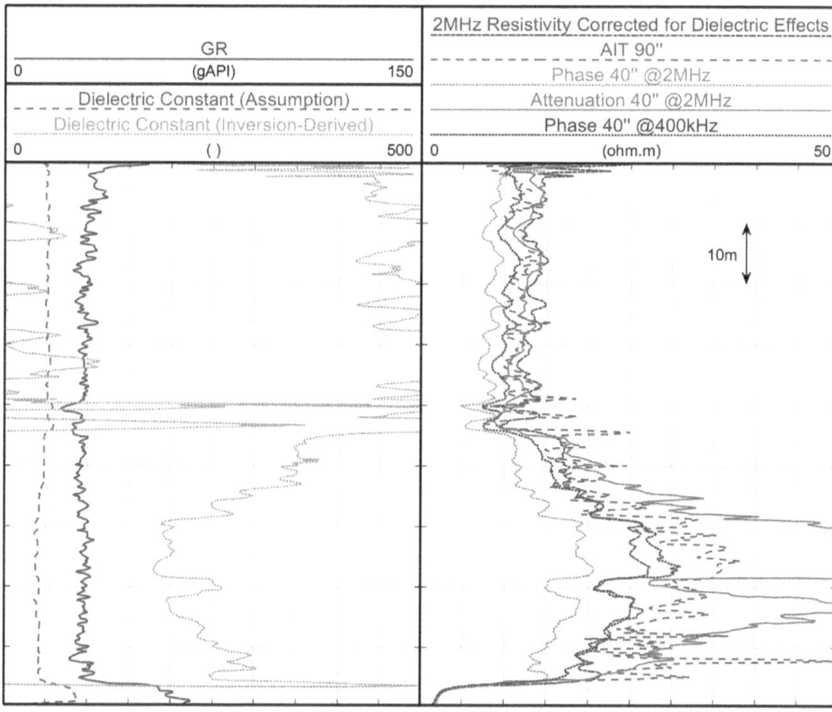

FIGURE 1.20 The assumed and interpreted relative dielectric permittivity (track 1); the resistivity dispersion of the dielectric corrected and uncorrected 2 MHz propagation resistivity, 400 kHz phase shift resistivity, and induction resistivity in the volcanic interval (track 2). *According to Altman R, Anderson B, Rasmus J, Luling MG, May 25–28, 2008. Dielectric effects and resistivity dispersion on induction and propagation-resistivity logs in complex volcanic lithologies: a case study. SPWLA 49th Ann. Logg. Symp.; Source: Authors.*

section (XX00–XX17) of the interval under investigation, the dielectric-dispersion log interpretation shows the presence of more hydrocarbon as compared to the interpretation of induction-resistivity log. Water-filled porosity derived from dielectric-dispersion logs shows there is no hydrocarbon presence in the shale laminations. The laminated thin shale beds are as thin as 2 inches and can be detected in the borehole image logs (Hizem et al., 2008). Water saturation estimates derived from deep-induction resistivity log is overestimated because of the lower vertical resolution of induction log than dielectric dispersion logs, which lead to lower induction-resistivity measurements as a consequence of water-bearing laminated thin-bed shale. The ratio of horizontal and vertical resistivity from tri-axial induction also indicates higher anisotropy, because of laminations, in the upper section (Hizem et al., 2008). Water saturation estimates derived from dielectric dispersion logs and from induction resistivity log are similar in the lower sand formation (XX17–XX25), where there are no laminations.

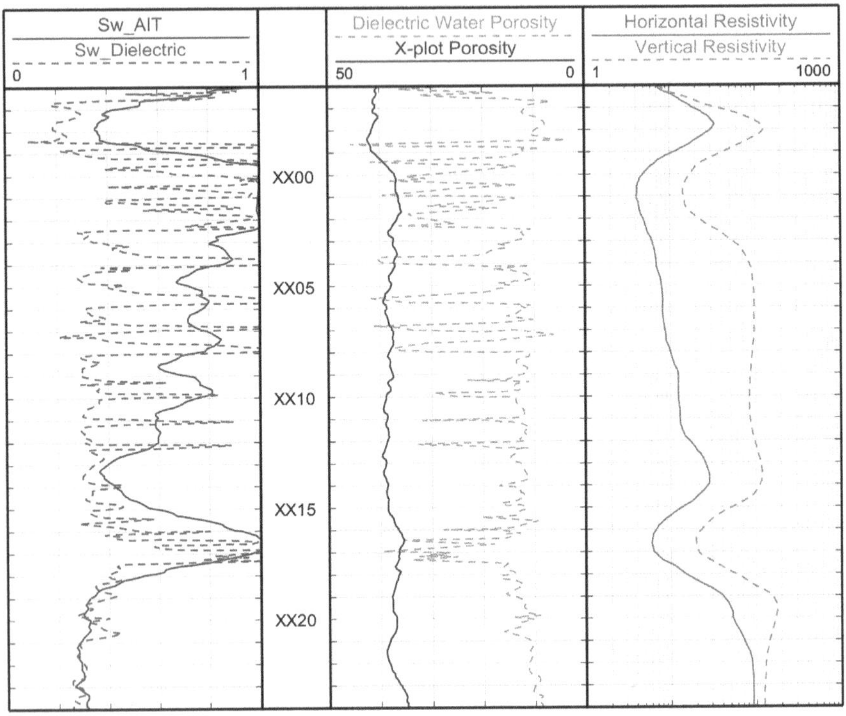

FIGURE 1.21 Dielectric dispersion log-interpretation in thinly bedded and laminated heavy-oil sand reservoir. *According to Hizem M, Budan H, Deville B, Faivre O, Mosse L, Simon M, September 21–24, 2008. Dielectric dispersion: a new wireline petrophysical measurement. SPE Ann. Tech. Confer. Exhibit.; Source: Authors.*

Interpretation of triple combo logs in a fully water-saturated zone of clean carbonate formation is illustrated in Fig. 1.22. The calculated porosity by standard log interpretation is around 0.15 to 0.2, and an unexpected high resistivity zone (XX14–XX22) can be detected, where the water saturation estimate derived from resistivity log using Archie equation is around 0.5. However, the water saturation estimates derived from dielectric dispersion logs using CRI model (CRIM) and SMD model indicate fully water-filled porosity (Fig. 1.23). The salinity estimates in track 1 agree with the measured mud-filtrate salinity, and the inverted cementation exponent in track 2 is around 2.7 in the unexpected high resistivity zone and is 2 in other zones (Hizem et al., 2008). The case demonstrates that an interpretation of dielectric dispersion logs can provide more accurate and consistent formation evaluation results in complex formations, such as thinly bedded and laminated heavy-oil sands and carbonate reservoirs.

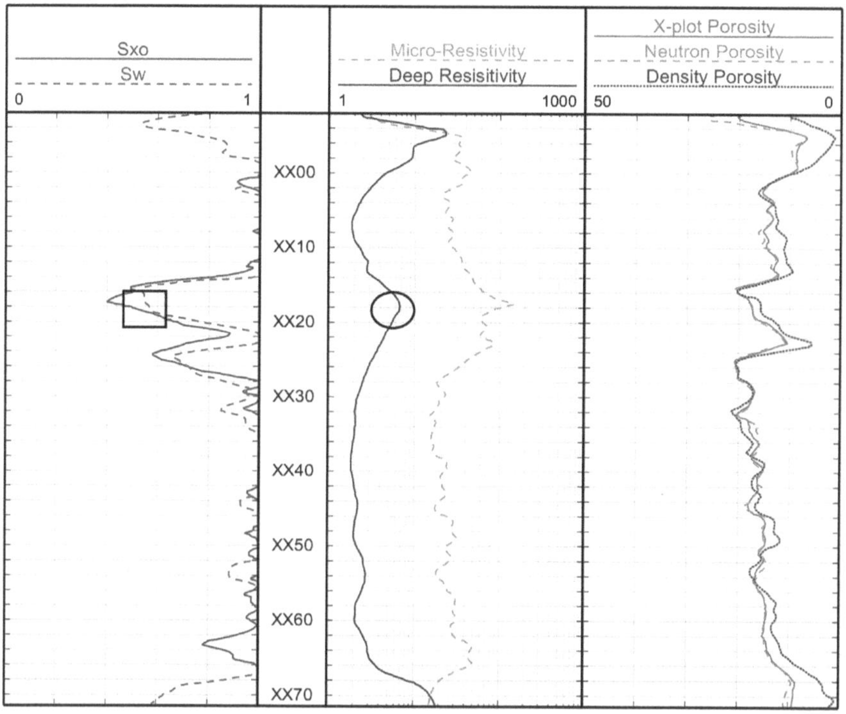

FIGURE 1.22 Triple combo log-interpretation in carbonate reservoir. *According to Hizem M, Budan H, Deville B, Faivre O, Mosse L, Simon M, September 21–24, 2008. Dielectric dispersion: a new wireline petrophysical measurement. SPE Ann. Tech. Confer. Exhibit.; Source: Authors.*

4.4 Case study IV

This case study focuses on documenting the effects on clay content on resistivity logs. Moreover, the resistivity log-interpretation model should consider clay effects to accurately estimate water saturation. A large amount of hydrocarbon is produced from shaly-sand formations, and the formation evaluation in shaly-sand formation especially in high water-saturation zone is challenging because of the clay effects. Wang and Poppitt (2013) elaborated the effects of clay content on the multifrequency relative dielectric permittivity and electrical conductivity logs. Fig. 1.24 compares the multifrequency dielectric constant and conductivity estimated at various frequencies, corresponding to the log-operation frequency of wireline induction, LWD propagation, and dielectric dispersion tools, at two depths with different clay content. Dielectric permittivity decreases while conductivity increases with the increase in frequency.

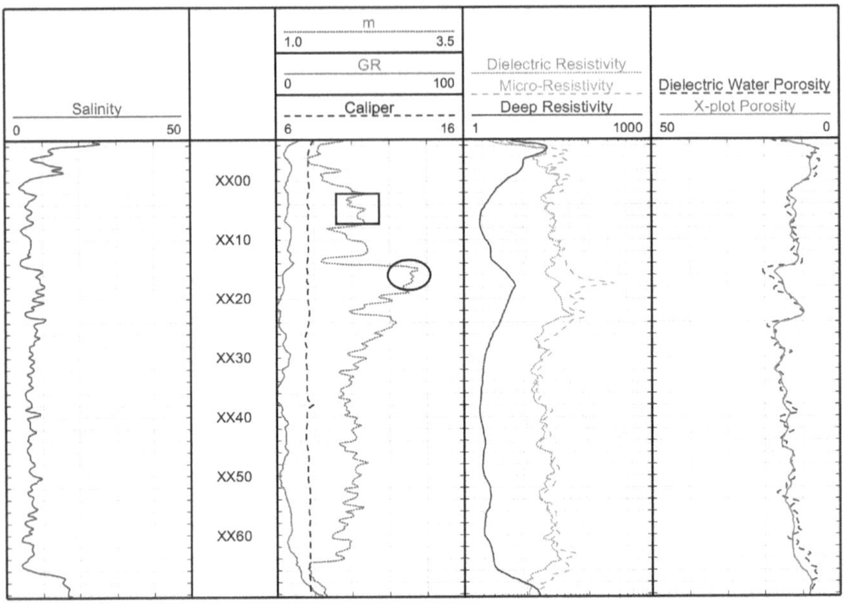

FIGURE 1.23 Dielectric dispersion log-interpretation in carbonate reservoir. *According to Hizem M, Budan H, Deville B, Faivre O, Mosse L, Simon M, September 21–24, 2008. Dielectric dispersion: a new wireline petrophysical measurement. SPE Ann. Tech. Confer. Exhibit.; Source: Authors.*

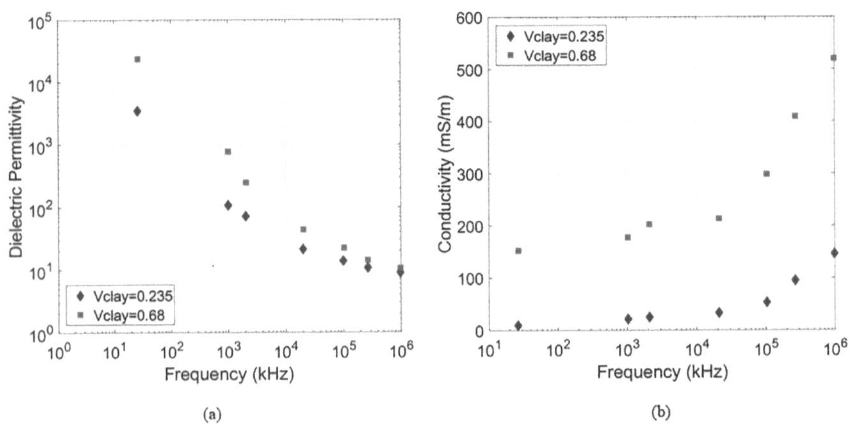

FIGURE 1.24 The effect of clay on the multifrequency relative dielectric permittivity and electrical conductivity measurements. *Modified from Wang H, Poppitt A, June 22–26, 2013. The broadband electromagnetic dispersion logging data in a gas shale formation: a case study. SPWLA 54th Ann. Logg. Symp.*

CEC is the amount of charge per 100 grams of clay. CEC can be approximated by the amount of embedded negative charges of clay particles and varies with clay mineralogy. CEC is an input parameter for the Waxman-Smits model and dual-water model for the resistivity log interpretation. CEC values could not be directly calculated from resistivity logging data by using Waxman-Smits model (Juhasz., 1981). CEC values can be determined from laboratory. Smectite clay mineral has the highest CEC (80−150 meq/100g), and kaolinite clay mineral has the lowest (4−10 meq/100g) (Ellis et al., 2007).

Han et al. (2012) incorporated a highly dispersive phase caused by the clay content to include the CEC effects in the widely used SMD model. The proposed SHSD model defines the conductive phase as a linear combination of clay and water dispersive phases. The dielectric dispersion effects because of clay content is limited at salinity above 30 ppk since the double layer surrounding clay grains is compressed, and is significant in fresh-water formation (Han et al., 2012). Fig. 1.25 shows the log interpretation results by using the SHSD model in a well drilled in a fresh-water formation in the Middle Valley of the Magdalena River in Colombia. The water salinity of the formation (FSXO_ADT in track 2) is low and varies with depth and is approximately 10 to 15 ppk. The mud filtrate is fresh with a salinity around 3 ppk. The inverted formation water salinity is close to the mud filtrate salinity in zones with movable hydrocarbon, which can be confirmed by SP logs in track 1. The total porosity (PHIT) and clay content of the formation is high varying from 0.25 to over 0.3, and 0.1 to above 0.5, respectively. The mineral contents of this formation include dominant quartz, mainly illite and kaolinite type of clay, varying Feldspars, and very little carbonate (Han et al., 2012). The standard core analysis, Dean−Stark extraction, Fourier-transform infrared spectroscopy (FTIR), and CEC measurements were performed on 30 preserved cores taken from this well. The multimineral analysis was performed to calculate the total porosity and matrix permittivity. The last two tracks in Fig. 1.25 show more obvious dielectric dispersion effects in clay-rich zones in the top and bottom of the interval. The water-filled porosity (PWXO_ADT) derived from the dielectric dispersion logs show constant nonmovable hydrocarbon in the entire sand interval except in the top and bottom zones, as shown in Fig. 1.25. The water-filled porosity lies between the total porosity and core measured water-filled porosity (PHIW Core) because of mud filtrate invasion that replaces movable hydrocarbon. However, the resistivity logs in track 5 including the true formation resistivity, invaded zone resistivity, and microresistivity logs significantly vary along depth because of the clay effects. The estimated CEC in track 3 agrees well with CEC measurements on core samples (Han et al., 2012). This case study introduces the application of a shaly-sand model which takes clay effects into account to process the dielectric dispersion logs and estimate water-filled porosity and CEC values.

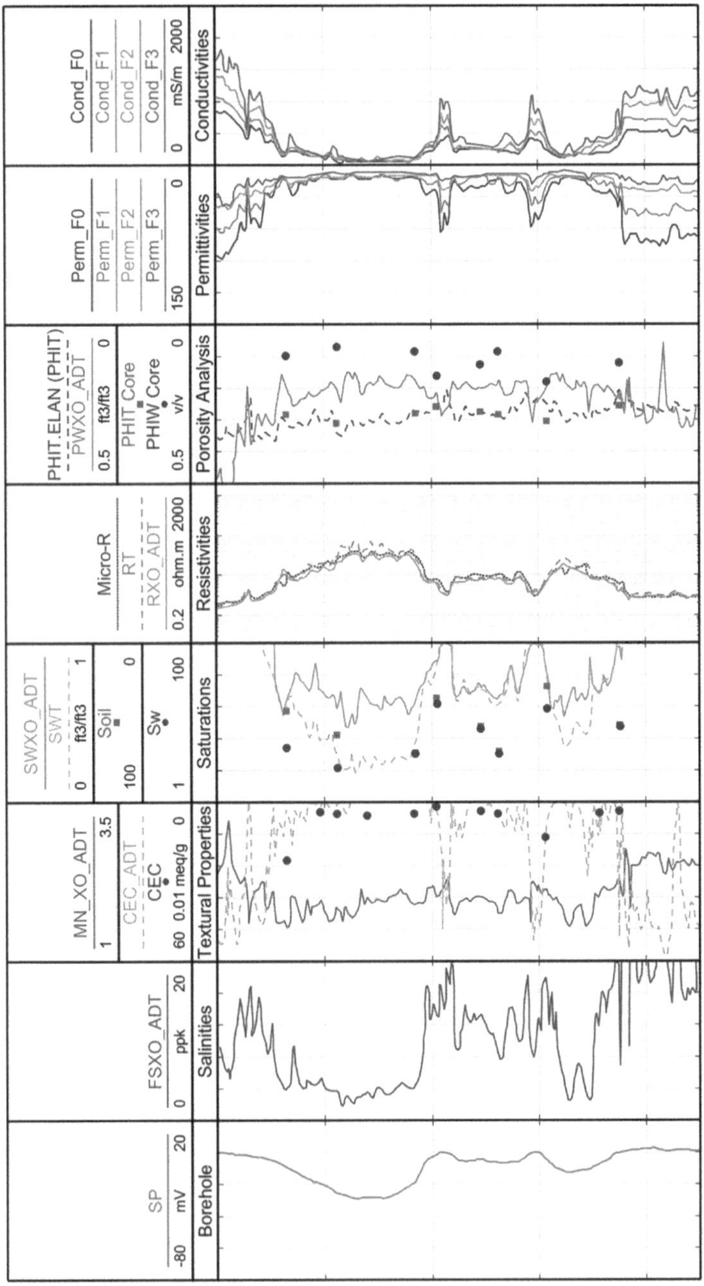

FIGURE 1.25 The water-filled porosity and cation exchange capacity values derived from the dielectric dispersion logs, and the comparison with core measurements for a well drilled in a fresh-water-bearing formation in an alluvial fan deposit. *According to Han M, Cuadros J, Suarez CAP, Decoster E, Faivre O, Mosse L, Seleznev N, June 16–20, 2012. Continuous estimate of cation exchange capacity from log data: a new approach based on dielectric dispersion analysis. SPWLA 53rd Ann. Logg. Symp.; Source: Authors.*

4.5 Case study V

This case study discusses the application of a physics-based rock model to process dielectric dispersion logs and estimate rock physics parameters. The formation EM measurements interpretation is significantly affected by the presence of clays, and the accuracy of water saturation estimates requires the prior information of CEC of clays. Unlike including a dispersive phase in the SMD model to account for the effect of clay content, such as the case study in previous section, Freed et al. (2016) alternatively applied their proposed physics-based model, which is illustrated in Fig. 1.26, for the interpretation/analysis of multifrequency dielectric dispersion logs. The rock grains and hydrocarbon phase are mixed together as the nonconductive phase through CRIM. The texture or rock fabric effects on the dielectric dispersion measurements in shaly-sand formation are assumed to come from clays (Freed et al., 2016). By setting the value of CEC to 0, the proposed shaly-sand model reduces to the bimodal model.

They simultaneously estimated water-filled porosity, salinity, CEC, and water-phase tortuosity in fresh-water formations. They used the averaged rock matrix permittivity for all minerals including clay minerals, and temperature and pressure were used as inputs to calculate the complex permittivity of water. The total porosity which was obtained from standard log analysis, and volume fraction of clay (V_{clay}) which can be obtained from multimineral analysis or geochemical logs were inputs to the inversion scheme. They set the volume fraction of charged spheroidal grains to $0.25V_{clay}$ as the effective clay volume. The clay grains orientation parameter f_a was set to 0.45 considering the effects of noncharged rock matrix texture (Freed et al., 2016). Fig. 1.27 illustrates the log interpretation results by using their proposed rock physics

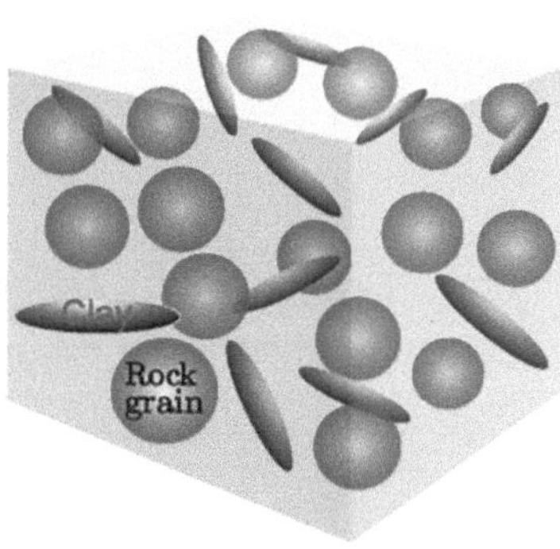

FIGURE 1.26 Schematic of differential effective medium model, which contains spherical rock grains, charged spheroidal grains, and water phase host. *Proposed by Freed D, Seleznev N, Hou CY, Fellah K, Little J, Dumy G, Sen P, June 25–29, 2016. A physics-based model for the dielectric response of shaly sands. SPWLA 57th Ann. Logg. Symp.*

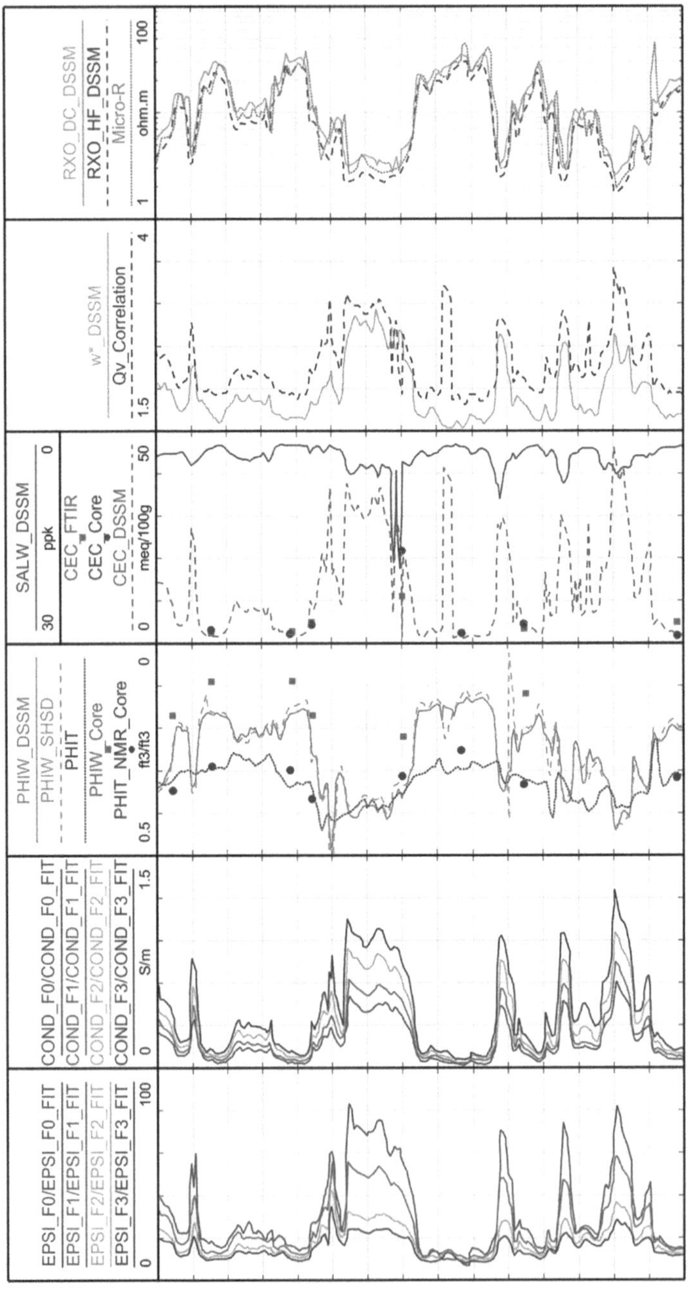

FIGURE 1.27 Log interpretation results including the estimates of water saturation and cation exchange capacity calculated using various methods in a well drilled in Girasol field in Colombia (Freed et al., 2016). *Source: Authors.*

model in the same well as Case Study IV. The tracks 1 and 2 in Fig. 1.27 show the dielectric dispersion logs (EPSI_F0,...,COND_F3) agree well with the modeled dielectric permittivity (EPSI_F0_FIT,...,etc) and conductivity (COND_F0_FIT,...,etc) based on the inversion-derived estimates. The comparisons among the inversion-derived estimates of water-filled porosity calculated using the proposed physics-based model by Freed et al. (2016) (PHIW_DSSM) and those using the shaly-sand model proposed by Han et al. (2012) (PHIW_SHSD), total porosity (PHIT), NMR interpreted total porosity (PHIT_NMR_Core), and water-filled porosity from Dean Stark (PHIW_Core) are illustrated in track 3. The water-filled porosity estimates based on the proposed physics-based model agree well with those based on the shaly-sand model proposed by Han et al. (2012). The water-filled porosity estimates agree with total porosity in some shaly zones, and are between total porosity and Dean Stark water-filled porosity, which was underestimated because of water lost during sample handling in some hydrocarbon-bearing zones. The estimated salinity (SALW_DSSM) along depth was around 2 to 15 ppk, and the CEC estimates (CEC_DSSM) agree well with CEC obtained from the core measurements and the CEC based on FTIR measurements. The estimated water-phase tortuosity (w^*_DSSM) in fully water-saturated zones is comparable to the cementation exponent calculated using the charge per pore volume Q_v and cementation exponent correlation (Q_v_Correlation) (Freed et al., 2016). The estimates of water-phase tortuosity in hydrocarbon-bearing zones are lower than the calculated cementation exponent. The last track in Fig. 1.27 indicates that the calculated shallow resistivity (RXO_DC_DSSM) by the propose model (Freed et al., 2016) is also close to the resistivity measurement by microresistivity tool (Micro-R). This case study demonstrates that rock physics parameters including water saturation and CEC values can be simultaneously and accurately estimated by the newly proposed physics-based model and agree with core measurements.

5. Electromagnetic log-interpretation methodologies for subsurface characterization

5.1 Electromagnetic log-interpretation methodologies for conventional formations

Archie's equation is the most widely used resistivity logs interpretation method to estimate hydrocarbon saturation in conventional formations, such as sandstone. Three important requirements for a valid application of Archie's equation are: (1) water-wet grains, (2) simple intergranular pores, and (3) single conduction mechanism, i.e., absence of clays and conductive minerals. The laterolog or induction resistivity tool can be deployed to obtain the resistivity logs at multiple depth of investigation. Medium and deep resistivity measurements can be combined with the microresistivity measurements to calculate the true resistivity at a specific depth. Cementation index and tortuosity factor need to be determined to use Archie's equation to process the true resistivity for calculating the hydrocarbon saturation at the specific depth. Generally, a higher cementation index m indicates more vuggy porosity and lower m corresponds fracture porosity, provided the tortuosity factor a is assumed as constant. This conclusion is invalid when the m and a are simultaneously determined (Adisoemarta et al., 2000). Adisoemarta et al. (2000) derived that the tortuosity factor is a function of the average angle of electrical movement, and cementation index corresponds to the constriction factor or the ratio of flow area of pore throat and pore body. The theoretical minimum value of a is 1 (Adisoemarta et al., 2000). The cementation index is affected by pore structure, consolidation, grain shape and sorting, and lithology (Bassiouni, 1994). The saturation exponent is determined by wettability, formation fluids composition and distribution, clay minerals, and overburden pressure (Tiab and Donaldson, 2004).

The resistivity anisotropy measurements using multiarray triaxial induction tool is well documented. A typical measurement of resistivity anisotropy requires horizontal and vertical resistivities assuming a transverse isotropic medium. Triaxial induction tool can measure resistivity in the three orthogonal directions. The invading water-based mud into formation reduces anisotropy, and oil-based mud filtrate increases the anisotropy measurements (Barber et al., 2004). Shoulder-bed effects on the nine triaxial couplings measured using the triaxial induction tool can be corrected using inversion method to estimate horizontal and vertical resistivity R_h and R_v at several radial depths of investigation. Fig. 1.28 shows the interpretation results of the multiarray triaxial induction logs acquired in a well drilled in a fluvial deltaic formation. The large R_v and R_h contrast can be observed for very low-resistivity zones measured by the AIT logs (Barber et al., 2004); therefore, triaxial induction and anisotropy measurements can enable the detection of bypassed/undetected hydrocarbon masked because of the presence of conductive shale layers. In Fig. 1.28, majority of depths exhibit anisotropy, especially greater than 3.

The dielectric dispersion logs interpretation including water saturation assessment in clean sandstone formations by using textural model such as SMD model is also well documented. Bean et al. (2013) reported the interpretation of dielectric dispersion logs acquired in a well drilled in a Triassic-aged fluvial formation on the Northwest Shelf in offshore Australia. The sands zones are clean, with a small amount of kaolinite as the dominant clay mineral, and the porosity is typically around 0.15 to 0.25. The well was drilled with oil-based mud, and Fig. 1.29 shows the water saturation estimates obtained by using textural model, such as SMD model, on dielectric dispersion log. These water saturation estimates are consistent with those obtained from standard induction resistivity log calculations and Dean–Stark core measurements in the clean gas sands zone (Bean et al., 2013). The interpretation models, such as

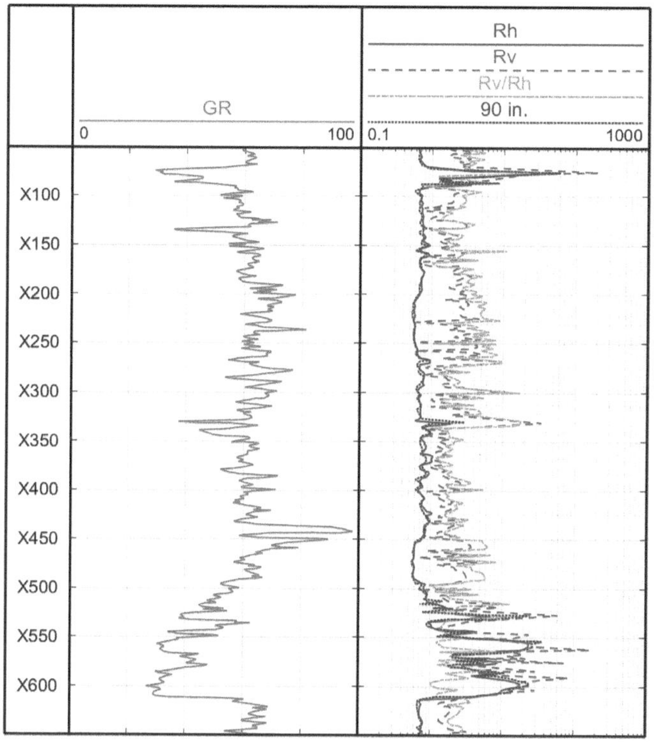

FIGURE 1.28 The multiarray triaxial induction logs from a well drilled in a fluvial deltaic formation. *According to Barber T, Anderson B, Abubakar A, Broussard T, Chen KC, Davydycheva S, Druskin V, Habashy T, Homan D, Minerbo G, Rosthal R, Schlein R, Wang H, September 26-29, 2004. Determining formation resistivity anisotropy in the presence of invasion. SPE Ann. Tech. Confer. Exhibit.; Source: Authors.*

Archie's equation and Waxman-Smits equation, can be used to process laterolog and induction resistivity logs, and SMD model and PS model can be used to process dielectric dispersion logs.

5.2 Electromagnetic log-interpretation methodologies for turbiditic formation

Deep water turbidites formations, which consist mainly of the submarine channels and depositional lobes, can have large amounts of hydrocarbon in-place interbedded with shaly-sand and shale layers. The submarine channels because of gravity-flow often are highly heterogeneous and have thin-bed channels or layers, and the lithology of the associated leveed channels changes rapidly at centimeter scale (Basu et al., 2007). Fig. 1.30 illustrates the typical depositional

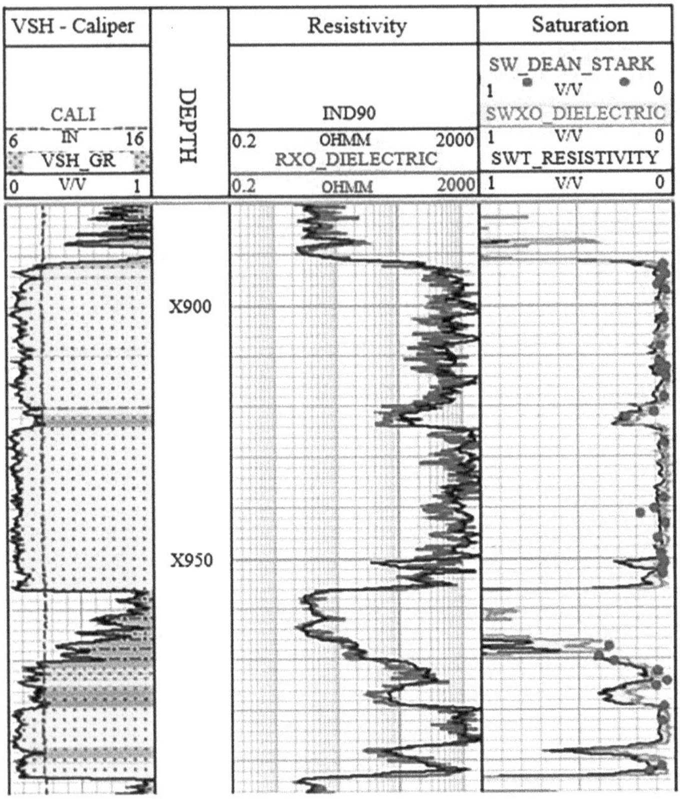

FIGURE 1.29 Log interpretation results including standard induction resistivity and dielectric dispersion logs for a sand zone in a well drilled in a Triassic-aged fluvial formation in offshore Australia. *Modified from Bean C, Cole S, Boyle K, Kho D, Neville TJ, June 22–26, 2013. New wireline dielectric dispersion logging tool result in fluvio-deltaic sands drilled with oil-based mud. SPWLA 54th Ann. Logg. Symp.*

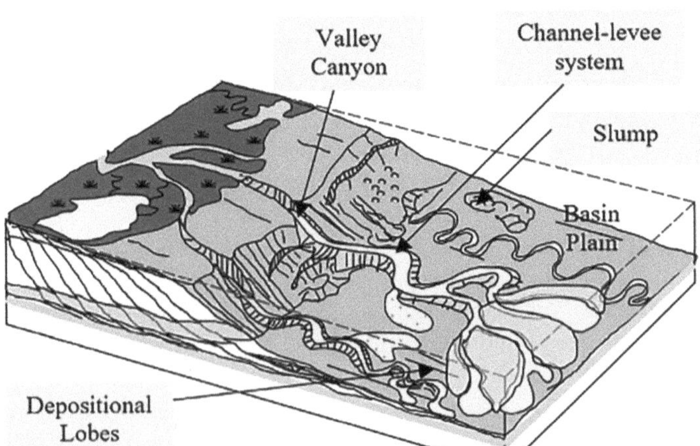

FIGURE 1.30 Schematic depositional model for deep water turbidites formations. *From Basu I, Machin N, Tyagi A, Saxena K, Altman R, Mathew A, Kumar R, June 3–6, 2007. Enhanced reservoir characterization in a deep water turbidite system using borehole images and spectroscopy logs. SPWLA 48th Ann. Logg. Symp.*

model for deep water mudrock-rich turbidites formations. The vertical resolution of standard logging measurements can be much larger than the lithology change scale of the submarine channels, which can cause inaccurate formation evaluation results. For example, the thin-bed laminated turbidites formations, which are generally anisotropic, often exhibit low-resistivity readings measured by conventional resistivity logging tools. The electrical anisotropy greater than three in deepwater turbidite formations is generally caused by laminated water-bearing shale layers (Nelson et al., 2006). The measured apparent resistivity in laminated turbidites hydrocarbon-bearing formations is reduced because of the laminated shale thin-layers, which causes overestimated water saturation and even missed pay zones. Manescu et al. (2010) integrated the tensor resistivity from triaxial induction tool and NMR measurements to accurately estimate fluid saturation, sand porosity, and net sand thickness. The inverted horizontal resistivity R_h and vertical resistivity R_v from 1D inversion of triaxial induction resistivity measurements can be used to build laminated sand-shale models. The vertical resistivity is much more sensitive to the presence of hydrocarbon than horizontal resistivity which is biased by the shale/conductive laminations in laminated sand formations (Manescu et al., 2010). They used a workflow illustrated in Fig. 1.31 to calculate the true sandstone resistivity, volume fraction of sand, and the volume fraction of shale. The true resistivity of shale can be determined in thick shale layers. The true sandstone resistivity can then be used to calculate hydrocarbon saturation by using Archie equation in clean sand or Waxman-Smits equation in sand containing dispersed clay. For the analysis, sand

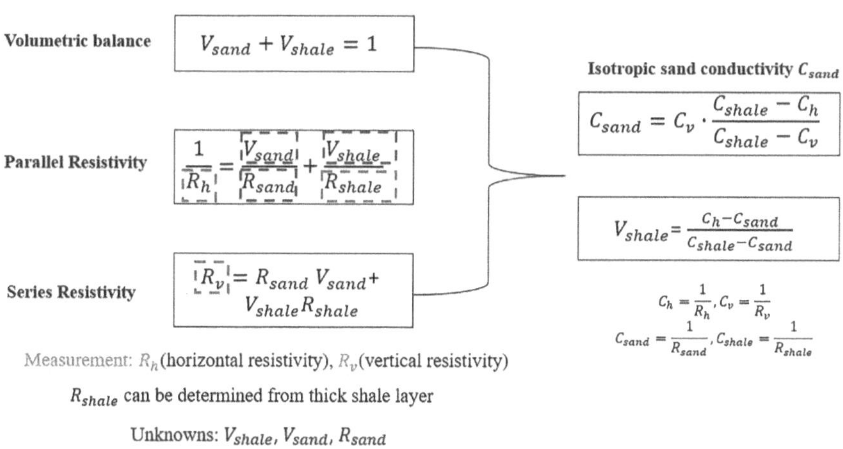

FIGURE 1.31 The workflow to determine the conductivity of isotropic sand and the shale lamination volume by processing horizontal/parallel and vertical/series resistivity. *According to Manescu A, Kho SF, Ha KT, Berhad SS, October 18–20, 2010. Case study of improved oil-in-place determination from petrophysical evaluation of resistivity anisotropy logging data in a Malaysian deepwater turbidite field. SPE Asia Pacific Oil Gas Confer. Exhibit.; Source: Authors.*

and shale are assumed to be isotropic. Such thin-bed shaly-sand analysis is only possible when triaxial induction tool is used and the anisotropy is transversely isotropic.

Clavaud et al. (2005) proposed a laminated sand-shale model for resistivity anisotropy which assumes shale is anisotropic. They also investigated the effects of shale anisotropy on water saturation calculation. Fig. 1.32 illustrates the schematic laminated isotropic sand and anisotropic shale turbidite formation model. The anisotropic induction resistivity logs in a thin-beds laminated sand-shale formation, borehole images, and the core images clearly show laminated shale layers are illustrated in Fig. 1.33. The horizontal and vertical resistivity of shale can be determined in pure and thick shale layers to solve the resistivity of sand and the volume fraction of shale, which are considered as the primary unknowns in the sand-shale model for thin-bed analysis. The resistivity of shale can be also inverted if the volume fraction of shale can be calculated by nuclear logs and NMR logs. The resistivities of shale and sand are calculated based on the tensor resistivity, volume fraction of shale, and shale anisotropy (Clavaud et al., 2005). They then calculated the water saturation in sand and shale component by any saturation-resistivity model, and the volumetric average of water saturation in the two components. Fig. 1.34 compares the estimated water saturation obtained from thin-bed analysis of triaxial induction resistivity logs, S_w(Anisotropy Model), and that from conventional log interpretation using dual-water model, S_w(Classic Model), in a laminated shale sand formation. Conventional log interpretation

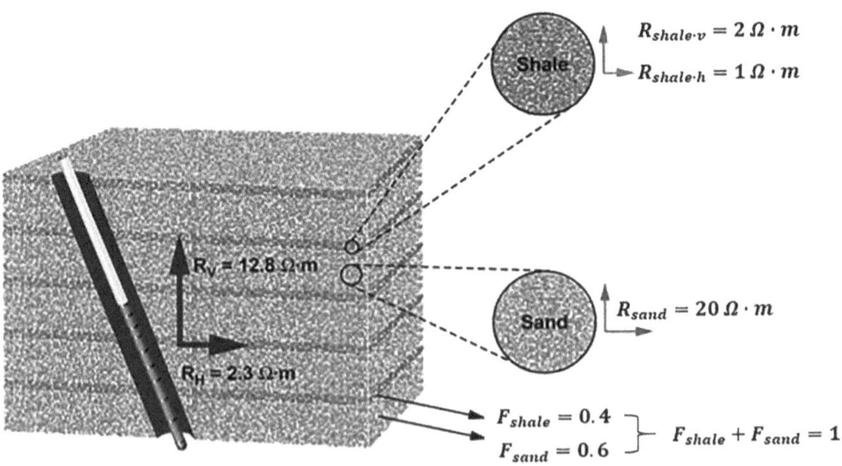

FIGURE 1.32 Schematic laminated isotropic sand and anisotropic shale turbidite formation model. Sand is isotropic and shale is anisotropic. *From Clavaud JB, Nelson R, Guru UK, Wang H, June 26–29, 2005. Field example of enhanced hydrocarbon estimation in thinly laminated formation with a triaxial array induction tool: a laminated sand-shale analysis with anisotropy shale. SPWLA 46th Ann. Logg. Symp.*

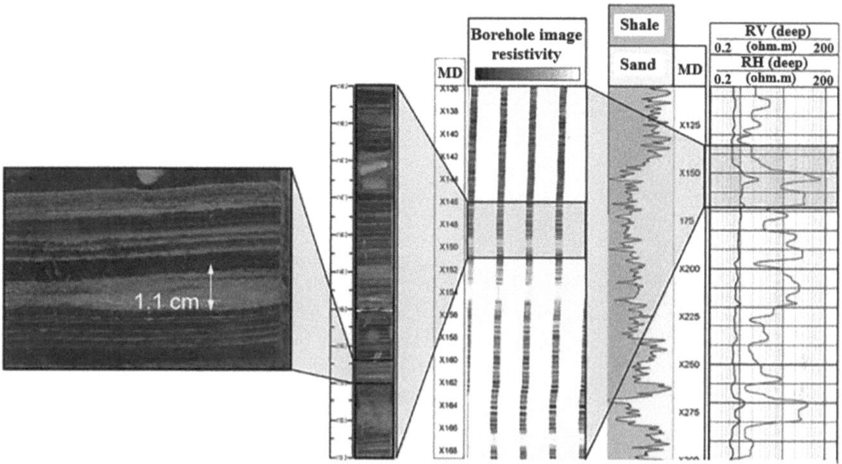

FIGURE 1.33 Log interpretation, borehole images, and core pictures in a thinly bedded laminated sand-shale formation. *Modified from Clavaud JB, Nelson R, Guru UK, Wang H, June 26–29, 2005. Field example of enhanced hydrocarbon estimation in thinly laminated formation with a triaxial array induction tool: a laminated sand-shale analysis with anisotropy shale. SPWLA 46th Ann. Logg. Symp.*

underestimates the hydrocarbon saturation, wherever the anisotropy ratio is greater than 1. EM logs interpretation methods in turbiditic formations, which have thin shale laminations, often require a two-component (sand-shale) rock physics model and triaxial multiarray induction resistivity measurements to accurately estimate fluid saturations.

5.3 Electromagnetic log-interpretation methodologies for shaly-sand formation

The hydrocarbon-bearing shaly-sand formations containing clay and pyrite minerals have low-resistivity readings, and are easily misinterpreted as water-bearing zones. There are various types of shaly-sand systems, which includes laminar-shale distributed sand, structural-shale distributed sand, and dispersed-shale distributed sand. The laminar-shale distributed sand has thin-beds or streaks of shale layers, and the effective porosity, fluid saturation, and horizontal permeability in such laminar-shale distributed sandstone reservoir are not affected by the presence of shale (Hamada et al., 2001). The shaly-sand with dispersed clays or shales has different properties from the laminar shale distributed sand because the dispersed clays coat the sand grains and partially fill the pore spaces, which generally increases the residual water saturation and reduces the permeability and porosity. The structural-shale distributed sand has shale grains forming the rock matrix along with sand grains, and the

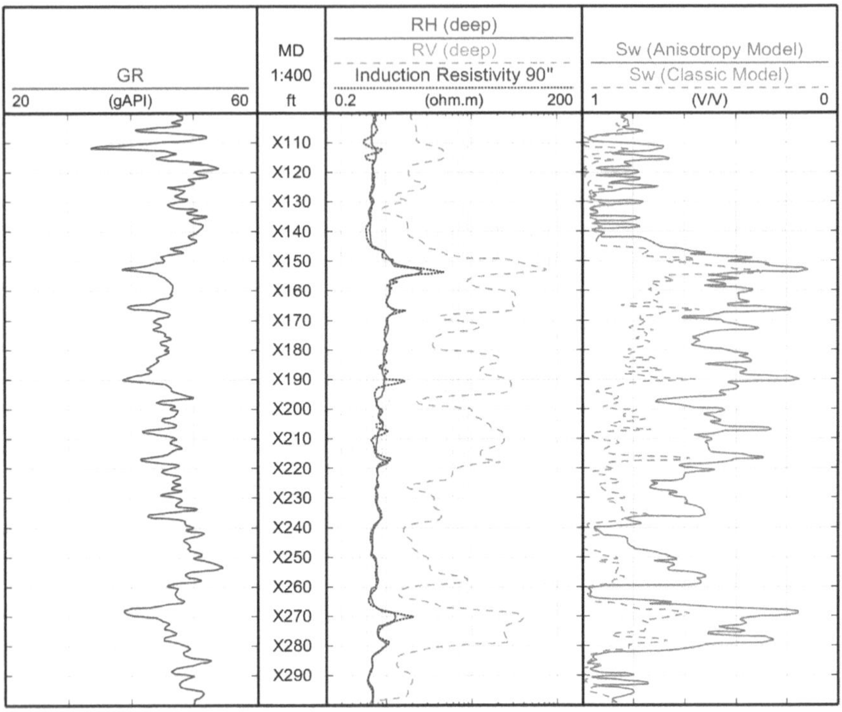

FIGURE 1.34 Comparison of water-saturation estimates obtained using conventional log interpretation versus thin-bed analysis of triaxial induction resistivity measurements in a laminated shale-sand formation. *According to Clavaud JB, Nelson R, Guru UK, Wang H, June 26–29, 2005. Field example of enhanced hydrocarbon estimation in thinly laminated formation with a triaxial array induction tool: a laminated sand-shale analysis with anisotropy shale. SPWLA 46th Ann. Logg. Symp.; Source: Authors.*

effects of clays or shales on permeability and resistivity of the rock are close to those of dispersed shale distributed sand (Hamada et al., 2001).

Bose et al. (2019) used a parallel-conductor model to quantify resistivity of laminated-clay system, as illustrated in Fig. 1.35. This model is only applicable in laminar shale/clay distributed sandstone formations. Waxman-Smits model can be used to interpret resistivity logs in dispersed clay distributed shaly-sand formations, and it is not applicable in laminated sand formations. The dual-water model incorporates free water and clay-bound water volumes in the pore space and can also be used in distributed shaly-sand formation. Simandoux model can be applied in most shaly-sand formations for resistivity logs interpretation regardless of the shale or clay distributions. Iheanacho (2014) derived shaly-sand models for different types of shale or clay distributions by varying the value of a parameter in these models. Kurniawan and Bassiouni (2007) derived petrophysical models to consider the electric current

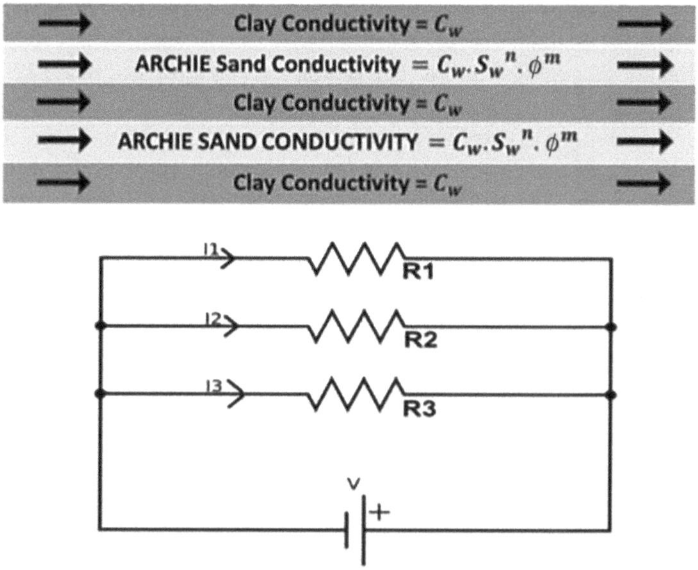

FIGURE 1.35 Parallel resistivity model for laminar clay/shale-distributed sand formation. *From Bose S, Myers MT, Chen P, Thakur G, June 17–19, 2019. Application of an integrated petrophysical modeling to improve log based reservoir characterization and oil in-place estimate of a fresh water shaly sand reservoir. SPWLA 60th Ann. Logg. Symp.*

through the free and clay-bound water, and they mentioned that the using of two cementation exponents for expressing the effects of free water and bound water is better than one cementation exponent. They also mentioned that the correlation between saturation exponent n and CEC exists based on experimental results. The dispersed shaly-sand model by Freed et al. (2016), which is illustrated in Fig. 1.26, includes dispersed spherical rock grains and dispersed charged spheroidal grains, is discussed in previous section and can also be used to interpret dielectric dispersion logs to estimate water-filled porosity, water salinity, CEC, and the aspect ratio of the dispersed clay mineral. The clay distribution affects the pore structure, fluid saturation and distribution, and conduction mechanisms; consequently, the resistivity. The EM log-interpretation workflow in shaly-sand formation depends on the type of clay-distributed sand or clay distribution.

5.4 Electromagnetic log-interpretation methodologies for carbonate formation

More than 60% of oil and 40% gas reserves of the world are in carbonate reservoirs. Archie-type resistivity-interpretation models always fail in carbonate formations because of the complex pore structure, microporosity, multifarious porosity types, and isolated pore space of carbonates. The formation evaluation in carbonate is challenging because of the variable sedimentary cycles and severe diagenesis geology nature. The biological and chemical processes during diagenesis leads to complex and highly heterogeneous spatial distribution of pore space, including isolated and connected pores. Fig. 1.36 shows a schematic heterogeneous carbonate rock which has complex pore system from micropores to vugs (Hulea et al., 2015). The porosity, tortuosity, and pore connectivity influence the transport properties of porous medium, such as electrical conductivity and permeability. The Archie's equation was developed and validated in sandstone formations that were typically water wet. However, the wettability in carbonate can be either water wet or oil wet. Tariq et al. (2018) proposed a dual pore-space network model which divides the pores to macropores and micropores to consider the two parallel electrical and hydrodynamical networks of carbonate. They defined the conductivity relation between the two pore systems, and derived water saturation model for dual porosity system.

Aboites et al. (2014) mentioned the application of dielectric dispersion logging in a highly fractured carbonate reservoir in the Gulf of Mexico. The depth of invasion of mud varies a lot in highly heterogeneous reservoirs, such as carbonate, which can even adversely affect the deepest resistivity log. In case of deep mud-filtrate invasion, the application of dielectric dispersion logs can provide accurate and independent remaining water-filled porosity estimates irrespective of the variations in the invaded-zone water salinity and

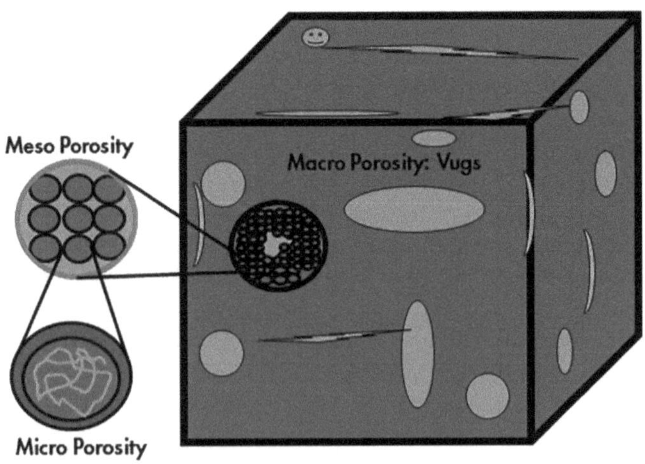

FIGURE 1.36 Schematic carbonate rock sample having complex pore systems including micropores, mesopores, and macropores such as vugs. *From Hulea IN, Frese D, Ramaswami S, July 18–22, 2015. Improved permeability prediction in heterogeneous carbonate reservoirs: a new approach using rock typing and wireline formation testing. SPWLA 56th Ann. Logg. Symp.*

cementation index of the formations. Compared to the conventional resistivity log interpretation, the use of dielectric dispersion removes the assumption of constant water salinity and cementation index along the well length. They used bimodal model, which considers the complex pore geometry of carbonate and the rock texture as a mixture of spheres and platy grains, to process the dielectric dispersion logs. They also mentioned that the rock matrix permittivity can be misinterpreted in highly heterogeneous carbonate rocks because of the complex rock texture, which gives rise to inaccurate water-filled porosity estimates. The water saturation estimates in carbonate formations derived from dielectric dispersion logs interpretation can be validated by water-cut measurement by wireline formation testing tools (Valiakhmetov et al., 2019).

Shahin et al. (2017) combined the pore-combination modeling technique, brine, and carbonate dielectric models for dual-porosity carbonate formations to jointly invert the resistivity and dielectric dispersion synthetic logs. They used very fast simulated annealing stochastic inversion algorithm to estimate water-filled porosity of vugs, water-filled porosity of intergranular pores, salinity, lithology exponents, water saturation, and matrix dielectric constant (Shahin et al., 2017). The pore-combination modeling is widely used to model the impact of pore geometry on resistivity and permeability, and can model multiple pore systems. They also applied the semi-empirical mixing law proposed by Myers (1996) to model the dielectric dispersion response of carbonate. The mixing law includes the Maxwell—Wagner terms or interfacial polarization terms related to intergranular porosity and vuggy porosity, water saturation and salinity. It also contains a high-frequency limit term which is modeled by the Hanai-Bruggeman equation (Shahin et al., 2017). The dielectric dispersion effects are more obvious for rocks with vuggy pore volumes. They applied the integrated resistivity and dielectric dispersion interpretation method on synthetic logs. Fig. 1.37 shows that the inverted water-filled intergranular porosity, vuggy porosity, total porosity, and water saturation match the true values well (Shahin et al., 2017). The complexity of pore system of carbonate formation was taken into account, and can be accurately quantified and described. The water saturation in different pore structures including intergranular pores and vugs can be separately and simultaneously estimated using the Shahin et al. (2017) model. In general, interpretation of EM logs in carbonate is challenging because of its heterogeneous nature; moreover, the presence of multiple pore types needs to be quantitively considered in the model. The pore-combination modeling developed by Shahin et al. (2017) can be used to quantify the influences of pore geometry and rock fabric on EM responses.

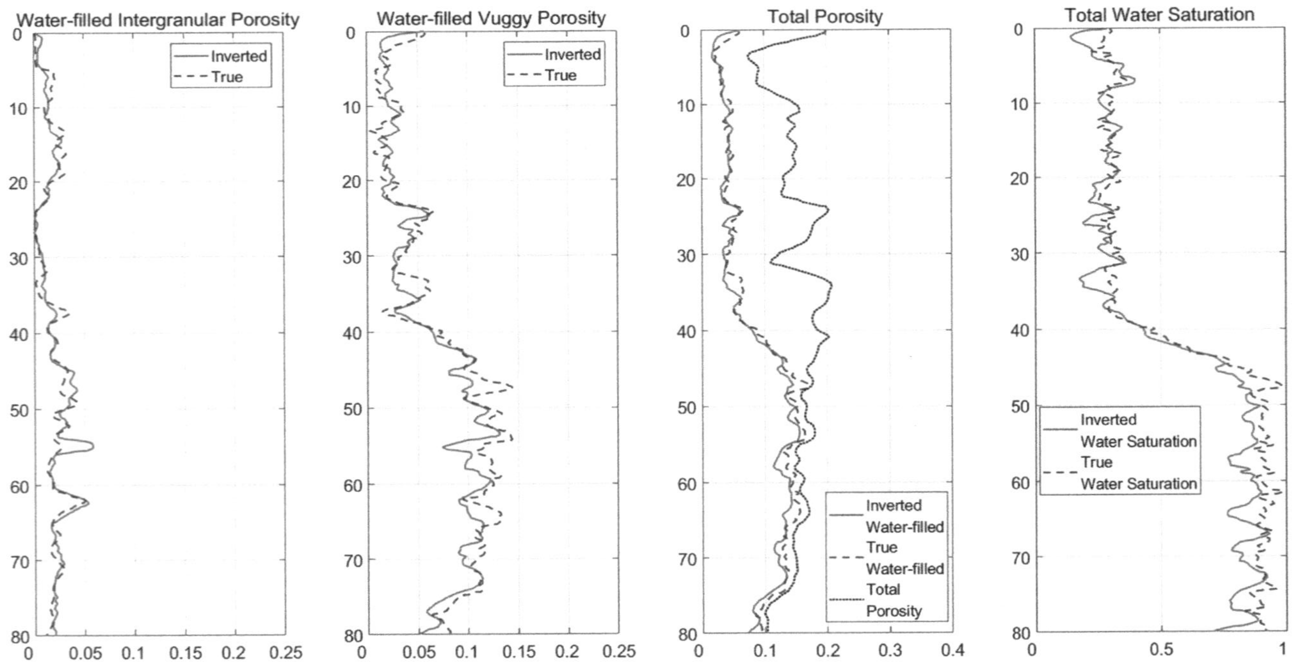

FIGURE 1.37 The comparison between the true and inverted water-filled intergranular porosity, vuggy porosity, total porosity, and water saturation. *According to Shahin A, Myers MT, Hathon LA, June 17–21, 2017. Carbonate formation evaluation using joint inversion of resistivity and dielectric logs. SPWLA 58th Ann. Logg. Symp.; Source: Authors.*

5.5 Electromagnetic log-interpretation methodologies for unconventional formation characterization

Unconventional reservoirs have become prolific producers of gas and oil because of the developments in horizontal drilling and hydraulic fracturing. Pores in shale gas reservoirs contain various types of gas, such as free gas, adsorbed gas, and solution gas. Holmes et al. (2017) defined four porosity components including total porosity, clay porosity, inorganic effective porosity, and organic effective porosity which is related to total organic carbon (TOC). Fig. 1.38 illustrates the typical components, including porosity components, fluids and solids, in an unconventional reservoir. The pore volume in organic matter is considered as the secondary pores related to TOC. These pores are generated during the thermal maturation process of organic matter (Holmes et al., 2017). Fig. 1.39 shows the pore system in organic rich rock, and the distributions of free gas in the kerogen pore, absorbed gas in kerogen pore wall, and water in inorganic matters.

Eubank (2008) used borehole image logs to detect gas zones in unconventional reservoirs, such as tight sands, by gas smear features on the image logs, which have white-out appearances. Moherek et al. (2016) used the high-resolution microresistivity logs, which have only 0.2-inch vertical resolution, to estimate the hydrocarbon saturation in unconventional Marcellus shale reservoir. They also combined gamma ray spectroscopy and microresistivity logs to reconstruct density and neutron, which were adversely affected by wash-out because of underbalanced drilling. Yu et al. (2018) used three models to estimate water saturation in shale gas reservoir; these models include modified Archie equation, modified Indonesia equation, and Waxman-Smits model. Fig. 1.40 shows the water saturation estimates by using these three models in unconventional reservoir (S_w_AR, S_w_INDO, S_w_WS, respectively).

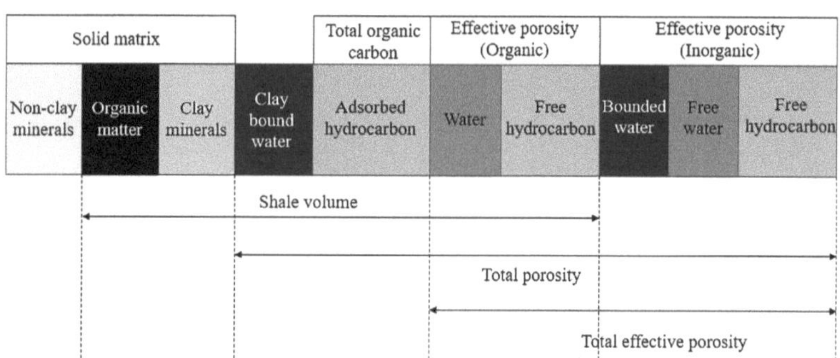

FIGURE 1.38 Petrophysical model of unconventional organic-rich shale reservoir and typical components, including nonshale matrix, organic matter, clay solids, and porosity components. *According to Holmes M, Holmes AM, Holmes DI, June 17–21, 2017. Mixed reservoir wetting in unconventional reservoirs and interpretation of porosity/resistivity cross plots, derived from triple-combo log data. SPWLA 58th Ann. Logg. Symp.; Source: Authors.*

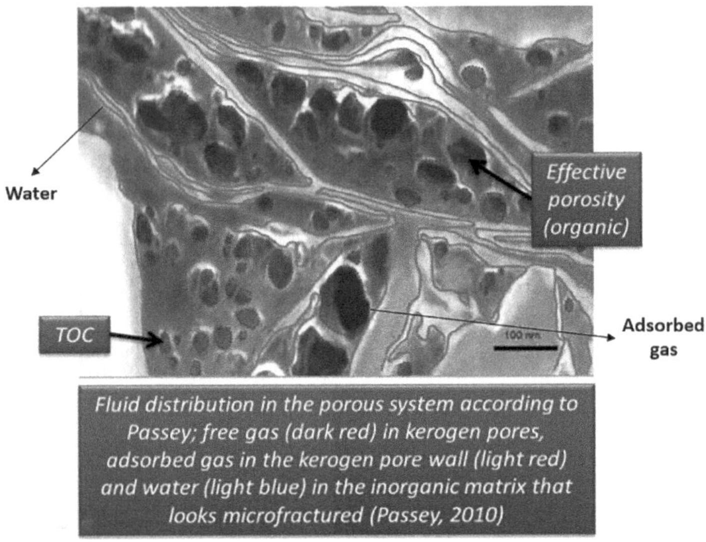

FIGURE 1.39 The effective porosity, total organic carbon, and fluid distributions in the pore space of one organic-rich shale sample. *Modified from Holmes M, Holmes AM, Holmes DI, June 17–21, 2017. Mixed reservoir wetting in unconventional reservoirs and interpretation of porosity/resistivity cross plots, derived from triple-combo log data. SPWLA 58th Ann. Logg. Symp.*

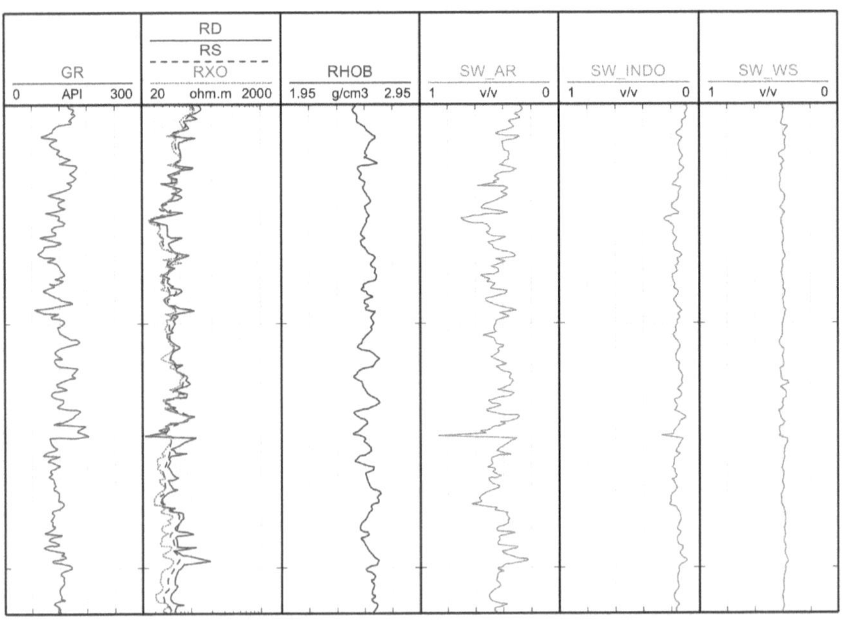

FIGURE 1.40 Logs of gamma ray, resistivity including deep resistivity (*RD*), shallow resistivity (*RS*), and invaded zone resistivity (*RXO*), and water saturation estimates by using modified Archie equation, modified Indonesia equation, and Waxman-Smits model in shale gas reservoirs. *According to Yu H, Wei X, Wang Z, Rezaee R, Zhang Y, Lebedev M, Iglauer S, October 23–25, 2018. Review of water saturation calculation methods in shale gas reservoir. SPE Asia Pacific Oil Gas Confer. Exhibit.; Source: Authors.*

They claimed that the modified Indonesia equation is most applicable to estimate water saturation (S_w_INDO), and Archie equation gives the most overestimated water saturation (S_w_AR). There are several uncertain parameters in Waxman-smit model such as equivalent conductivity of exchange cations, B, and the capacitance of exchange anions, Q_v. The water saturation estimated by Waxman-smit model (S_w_WS) can be misinterpreted because of those uncertain parameters.

Akbar et al. (2018) estimated TOC (TOC_avg in track 4 of Fig. 1.41) by using various methods such as $\Delta logR$, and then corrected the total porosity by including the effect of TOC. They then used Indonesia model with the TOC-corrected total porosity to estimate water saturation. The TOC-corrected porosity (PHI_TOC_corr in track three of Fig. 1.41) is expressed as,

$$\phi_T = \frac{(\rho_{ma} - \rho_b) + \rho_b\left(w_{TOC} - \rho_{ma}\dfrac{w_{TOC}}{\rho_{kerogen}}\right)}{\rho_{ma} - \rho_f} \tag{2.3}$$

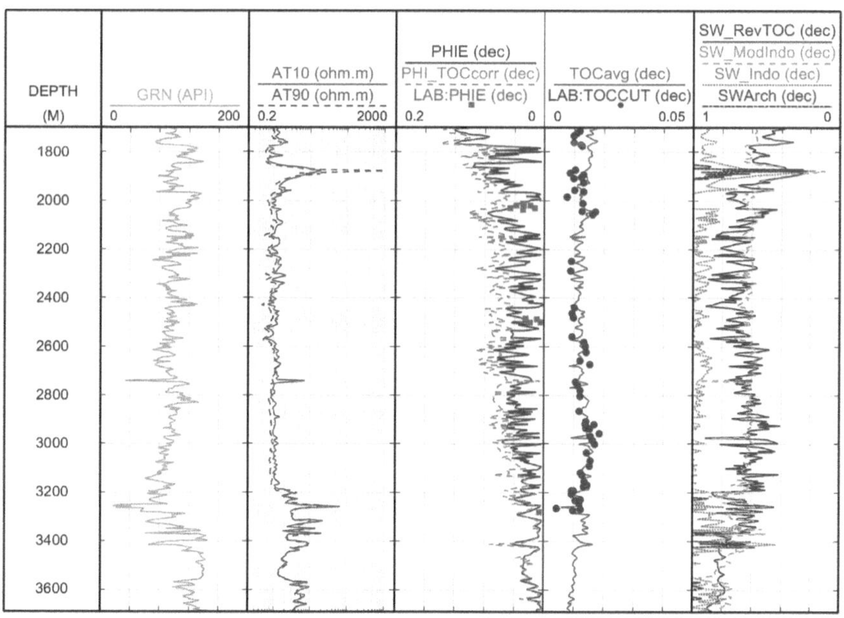

FIGURE 1.41 Petrophysical interpretation results including total organic carbon (TOC), total porosity, TOC-corrected total porosity, and water saturation estimates by various models in shale formations in North Sumatra Basin. *According to Akbar MNA, Musu JT, Milad B, July 23–25, 2018. Water saturation interpretation model for organic-rich shale reservoir: a case study of North Sumatra Basin. Unconvent. Res. Tech. Confer.; Source: Authors.*

where ϕ_T is total density porosity corrected for TOC, ρ_{ma} is matrix density, ρ_b is bulk density, w_{TOC} is weight fraction of TOC, $\rho_{kerogen}$ is the density of kerogen, and ρ_f is the density of fluid.

Akbar et al. (2018) mentioned that the TOC-modified Indonesia water saturation estimates (*Sw*_ModIndo) provided the most accurate water saturation estimates in organic-rich shale formations compared with simple Archie, Indonesia equation, and the revised TOC-water saturation model proposed by Xu et al. (2017). Fig. 1.41 shows petrophysical analysis results including water saturation estimates by the aforementioned models in organic-rich shale formations in North Sumatra Basin. The estimated water saturation by Archie's equation (*Sw*_Arch) and Indonesia methods (*Sw*_Indo) indicates almost fully water saturated formation, which is in contrast with prior knowledge of this organic-rich shale reservoir. The obtained water saturation by the revised TOC-water saturation model (*Sw*_RevTOC) has unreasonable large fluctuations. The revised TOC-water saturation model can be expressed as,

$$S_w = a \times S_{Wcon} \times \left(b - \frac{TOC}{TOC_{max}}\right) \tag{2.4}$$

where a and b are coefficients, TOC_{max} is the maximum TOC, S_w is water saturation estimated by revised TOC-water saturation model, and S_{Wcon} is water saturation estimated by Archie's equation.

Raina et al. (2017) used dielectric dispersion logs to estimate water saturation in source rock gas formation in Abu Dhabi, which is an organic-rich, argillaceous limestone formation. Estimated water saturation were compared with those obtained by resistivity log interpretation and NMR core measurements. Water saturation estimated using Bimodal model agrees well with NMR measured values, and the water saturation derived using CRIM was an underestimation. The EM measurements interpretation in unconventional formation such as shale is challenging because of the presence of clay, conductive minerals, and kerogen. Most of the previously proposed methods are empirical equations and are not based on rock physics modeling, or simply applying models developed in conventional formations.

The joint interpretation of induction resistivity or laterolog resistivity and dielectric dispersion logs was also used to estimate hydrocarbon saturation in unconventional reservoirs such as Bakken and Wolfcamp (Tathed et al., 2018a, 2018b). The recently developed clay and pyrite dispersion model, which accounted for the interfacial polarization effects of clay and pyrite, was applied for purposes of improved estimation of hydrocarbon saturation in organic-rich shale gas formation (Han and Misra, 2016, 2017; Han et al., 2019). More discussions and demonstrations on this topic are available in Chapters 8 and 9.

5.6 Electromagnetic interpretation methodologies for pore-scale digital rock

Micro-computed tomography (micro-CT) is a nondestructive imaging technique used to visualize the cross-section of internal structure and the pore space of porous materials. The rock samples can be scanned by micro-CT scanner and then the scanned gray-scale sequential images can be segmented to construct digital rock. Toumelin and Torres-Verdin (2009) summarized a pore-scale model to numerically simulate wideband EM dispersions of saturated rocks using the digital rock models incorporating pore and fluid distributions. The existing shaly-sand models simply assume that the clay/shaly material is laminated, dispersed, or structurally distributed in geological formation. Garcia et al. (2017) used the digital rock technology to quantify the clay network geometry or distribution and rock fabric parameters in rocks obtained from clay-rich and mudrock formations. They introduced a new resistivity model which is based on pore combination modeling and self-consistent effective-medium theory, and the model incorporates the directional connectivity of each conductive component, including clay and brine, to more accurately estimate water saturation. Fig. 1.42A shows the 2D segmented high-resolution image of rock sample obtained from a mudrock formation. The brown (light gray in printed version), gray, black, and white regions in the figure indicate clay minerals and nonconductive grains, such as quartz, kerogen, and nonconductive heavy minerals, respectively. Fig. 1.42B is the 2D segmented image of rock sample obtained from a clay-rich formation, and the

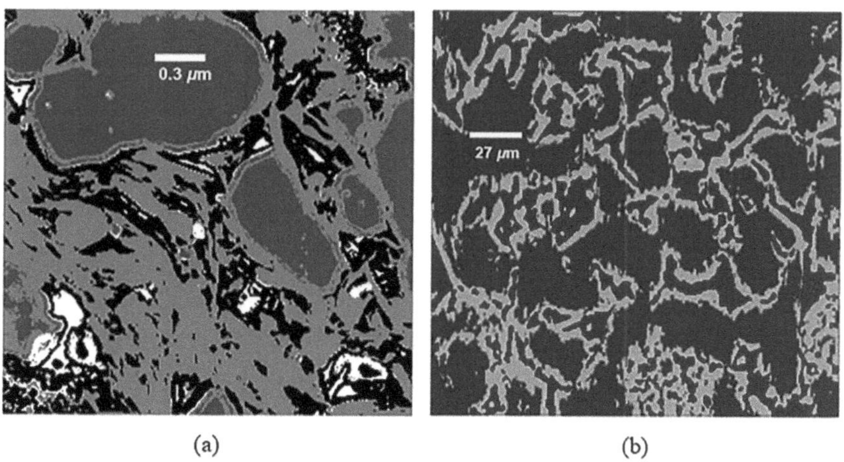

FIGURE 1.42 (A) Two-dimensional segmented high-resolution image of sample from a mudrock formation; (B) Two-dimensional segmented image of sample from a clay-rich formation. *From Garcia AP, Jagadisan A, Rostami A, Heidari Z, June 17–21, 2017. A new resistivity-based model for improved hydrocarbon saturation assessment in clay-rich formations using quantitative clay network geometry and rock fabric. SPWLA 58th Ann. Logg. Symp.*

gray, blue (dark gray in printed version), and green (light gray in printed version) areas indicate nonconductive grains, pores, and clay minerals, respectively. They compared the actual water saturation with the those obtained by their proposed new resistivity equation incorporating the spatial distribution and directional connectivity of clay, and Waxman-Smits model. The results indicate that in both clay-rich formation case (Fig. 1.43A) and mudrock formation case (Fig. 1.43B), the newly proposed resistivity model can estimate water saturation more accurately than the Waxman-Smits model.

Garcia and Heidari (2018) performed pore-scale numerical simulations to calculate the electrical conductivity of the constructed digital rock. They used transmissibility approach to calculate the spatial distributions of electric potential to ensure conservation of electrical charges. The continuity equation of electric charges for pore-scale simulation can be expressed as

$$\nabla \cdot J = \nabla \cdot (\sigma \nabla \phi) \tag{2.5}$$

where σ is electrical conductivity and $\nabla \phi$ is electric potential gradient. Garcia et al. (2018) also performed numerical simulations to calculate multifrequency electrical conductivity and dielectric permittivity on digital rocks. They used the electro-quasistatic approximation to solve the Maxwell equations and applied finite difference method to solve the continuity equation which can be express as,

$$\nabla \cdot J = \nabla \cdot (\sigma E + j\omega \varepsilon \varepsilon_0 E) = 0 \tag{2.6}$$

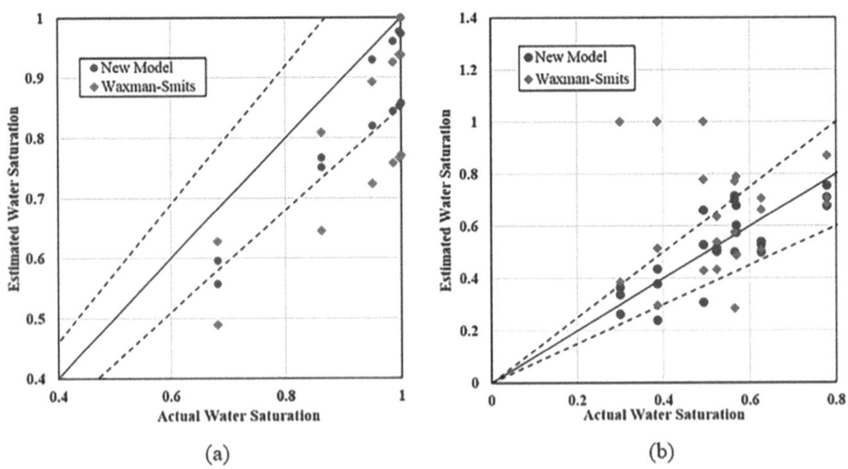

FIGURE 1.43 The comparison between the numerically simulated water saturation and those obtained by the proposed new model and Waxman-Smits model in (A) clay-rich formation and (B) mudrock formation. *From Garcia AP, Jagadisan A, Rostami A, Heidari Z, June 17–21, 2017. A new resistivity-based model for improved hydrocarbon saturation assessment in clay-rich formations using quantitative clay network geometry and rock fabric. SPWLA 58th Ann. Logg. Symp.*

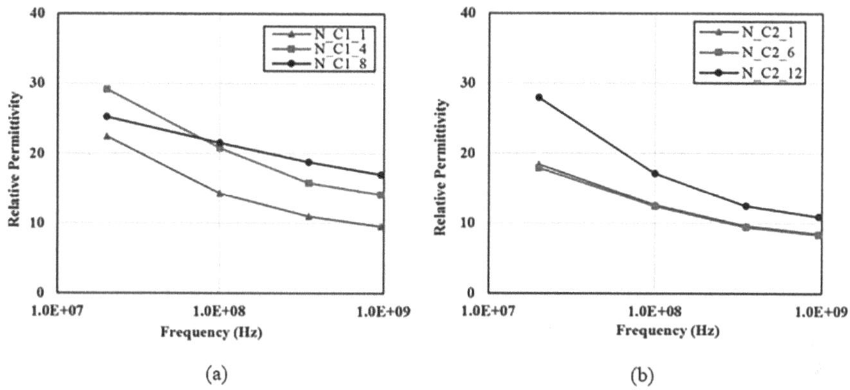

FIGURE 1.44 The numerically simulated multifrequency dielectric permittivity of digital carbonate rock samples having complex pore structure with bimodal pore-body size distribution and fully saturated with water. *From Garcia AP, Han Y, Heidari Z, October 2018. Integrated workflow to estimate permeability through quantification of rock fabric using joint interpretation of nuclear magnetic resonance and electric measurements. Petrophysics 59 (5), 672–693.*

where E is electric field, ω is angular frequency, ε is relative complex permittivity, ε_0 is permittivity of free space. Fig. 1.44 shows the numerically simulated relative dielectric permittivity at frequencies of the dielectric dispersion logs operation frequencies in six fully water-saturated digital carbonate rock samples. The trend that relative dielectric permittivity decreases with increasing frequency is physically correct, and consistent with core-scale laboratory and log-scale field measurements. The values of modeled relative dielectric permittivity are in the typical range of core-scale and log-scale measurements on carbonate. The proposed EM interpretation method for pore-scale digital rock can incorporate the complex pore structure, clay and fluid network and distribution, which brings extra values for formation evaluation in unconventional reservoirs.

6. Cases on the use of electromagnetic measurements with other types of measurement

6.1 Joint interpretation of electromagnetic measurements with nuclear magnetic resonance measurements

The integrated dielectric permittivity, electrical resistivity, and NMR measurements interpretation workflow proposed by Garcia et al. (2018) can estimate directional permeability, pore-throat size distribution, and tortuosity. Fig. 1.45 shows the proposed workflow to interpret multi-physics measurements for estimating rock-fabric parameters, which is purely based on rock physics and eliminates calibration parameters. The porosity and effective pore-body size interpreted by the NMR T_2 distribution, and constriction factor

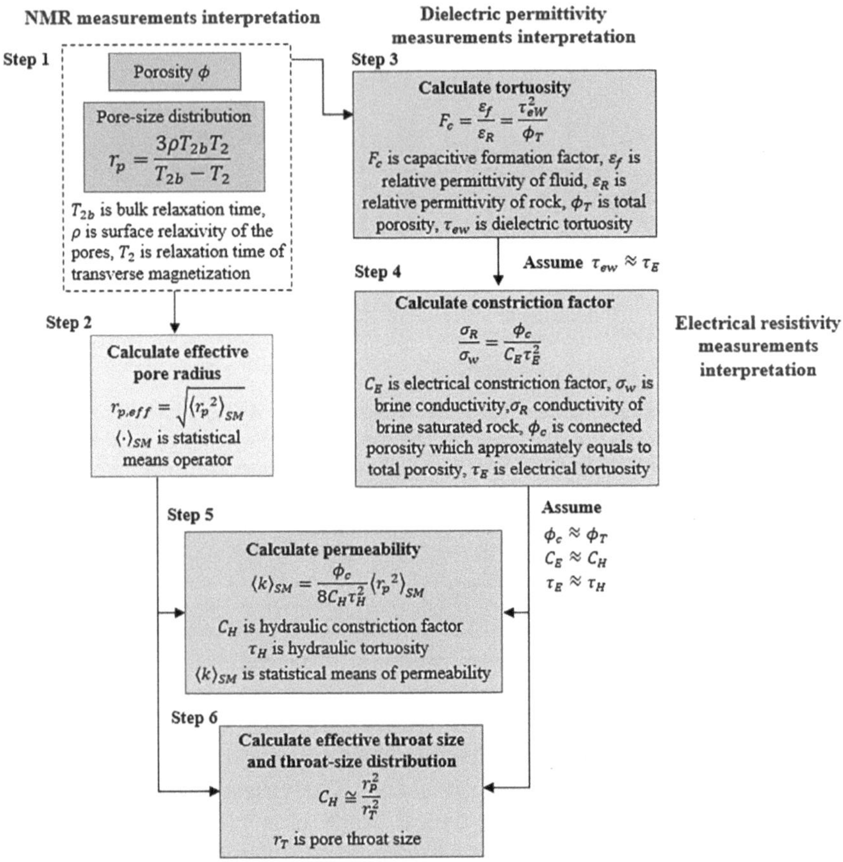

FIGURE 1.45 The proposed workflow for the joint interpretation of nuclear magnetic resonance, electrical resistivity, and dielectric permittivity to calculate rock fabric parameters, and estimate permeability and pore-throat-size distribution. *According to Garcia AP, Han Y, Heidari Z, October 2018. Integrated workflow to estimate permeability through quantification of rock fabric using joint interpretation of nuclear magnetic resonance and electric measurements. Petrophysics 59 (5), 672–693.*

obtained by electrical resistivity interpretation were used to calculate permeability. The tortuosity and pore-throat size distribution can be further estimated if dielectric permittivity measurement at frequency close to 1 GHz is available. They assumed that the measured rock properties are similar within the rock space under investigation because those multi-physics measurements have different depth of investigation. Fig. 1.46 shows that the estimated permeability agrees well with the laboratory measured results by core flooding in core samples mentioned in previous section having wide range of permeability from 0.1 to 100 mD. The peaks of estimated pore-throat size distribution are also comparable to measurements by mercury injection capillary pressure (MICP) (Garcia et al., 2018).

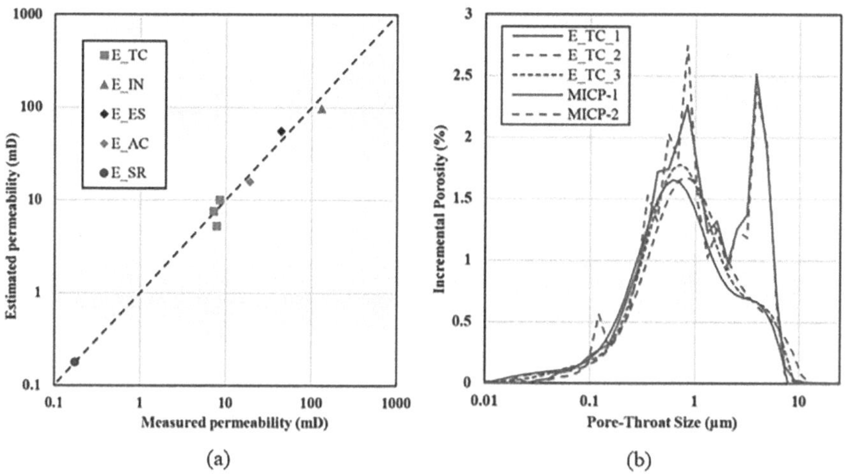

FIGURE 1.46 (A) The comparison between the estimated permeability by the proposed workflow and measured permeability by core-flooding experiments; (B) the comparison between the estimated pore-throat size distribution and these values obtained by mercury injection capillary pressure measurements. *From Garcia AP, Han Y, Heidari Z, October 2018. Integrated workflow to estimate permeability through quantification of rock fabric using joint interpretation of nuclear magnetic resonance and electric measurements. Petrophysics 59 (5), 672–693.*

Newgord et al. (2019) proposed a new workflow which is illustrated in Fig. 1.47 to jointly interpret electrical resistivity and NMR measurements, and minimize the difference between wettability index estimated by resistivity and NMR measurements to get consistent wettability index estimates. The water saturation can be simultaneously estimated with wettability index. The rock wettability can be quantified by the interpretation of NMR T_2 distribution or longitudinal T_1-T_2 measurements. Newgord et al. (2019) performed NMR measurements on water saturated water-wet rock, hydrocarbon saturated hydrocarbon-wet rock, and mixed-wet rock obtained from the same type of formation. They defined and calculated NMR-based wettability based on the harmonic mean of T_2 distribution of those rock samples, water saturation, and the bulk relaxation of water and hydrocarbon, which is illustrated in Fig. 1.47. They also calculated the resistivity-based wettability index, which is defined as the function of water saturation, resistivity index, the volume fraction of maximum isolated water, and depolarization factor of droplets of hydrocarbon (Newgord et al., 2019). Finally, they applied the Generalized Reduced Gradient method to find the water saturation value that minimizes the difference between the resistivity-based wettability and NMR-based wettability index. The estimated water saturation and wettability index were verified against laboratory measurements, including gravimetrically measured water saturation and contact angle. Fig. 1.48 shows the Amott cell setup for spontaneous imbibition of brine and diesel, and the contact angles of brine droplets

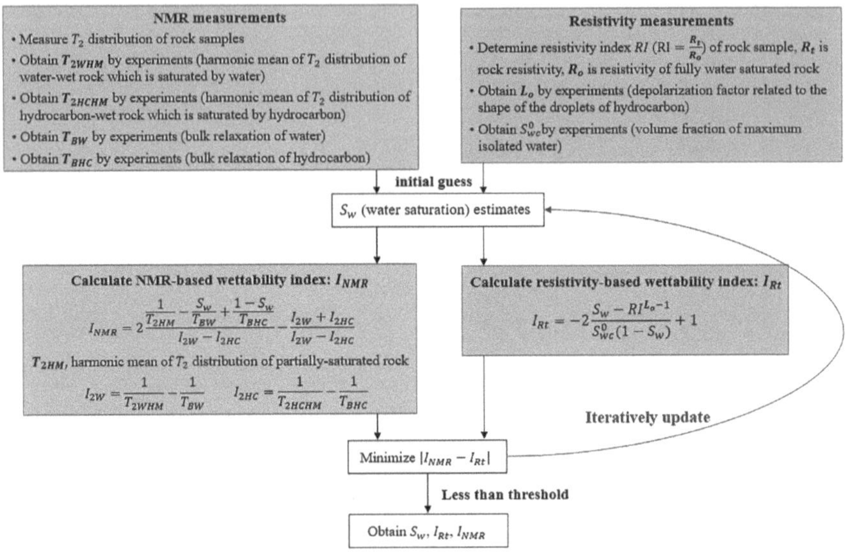

FIGURE 1.47 The integrated physics-based workflow for combining electrical resistivity and nuclear magnetic resonance measurements to simultaneously estimate wettability and water saturation. *According to Newgord C, Garcia AP, Heidari Z, June 17–19, 2019. A new workflow for joint interpretation of electrical resistivity and NMR measurements to simultaneously estimate wettability and water saturation. SPWLA 60th Ann. Logg. Symp.; Source: Authors.*

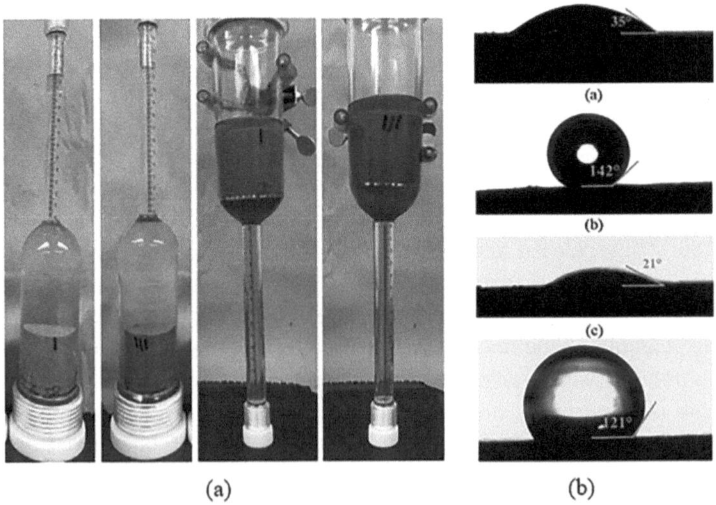

FIGURE 1.48 (A) Amott cell setup for spontaneous imbibition of brine and diesel for both water-wet and hydrocarbon rock; (B) contact angles of brine droplets measurements on the cores immersed in hydrocarbon. *From Newgord C, Garcia AP, Heidari Z, June 17–19, 2019. A new workflow for joint interpretation of electrical resistivity and NMR measurements to simultaneously estimate wettability and water saturation. SPWLA 60th Ann. Logg. Symp.*

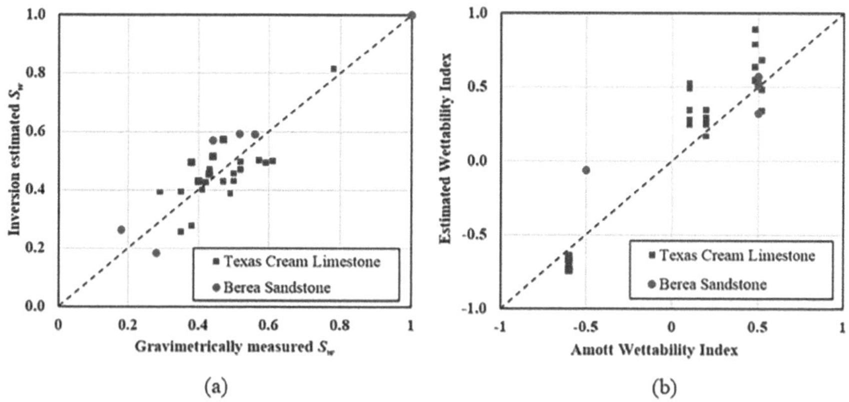

FIGURE 1.49 (A) The comparison between estimated and gravimetrically measured water saturation on Texas Cream limestone and Berea sandstone; (B) The comparison between estimated and Amott cell setup measured wettability index on the same core samples. *From Newgord C, Garcia AP, Heidari Z, June 17–19, 2019. A new workflow for joint interpretation of electrical resistivity and NMR measurements to simultaneously estimate wettability and water saturation. SPWLA 60th Ann. Logg. Symp.*

measurements on both water-wet and hydrocarbon-wet core samples. They used chemical treatments to alter the wettability of the core samples. Fig. 1.49 illustrates that the estimates of water saturation and wettability index based on the proposed workflow are comparable to those measured in laboratory on Texas Cream limestone and Berea sandstone core samples (Newgord et al., 2019).

Jacobsen et al. (2007) integrated resistivity, NMR, and porosity logs to predict the gas and water production percentages or water cut. The log interpretation can provide virgin zone water saturation, irreducible water saturation, and residual gas saturation, and the fluid viscosity can be estimated by the pressure, volume, and temperature correlation. They derived a log-based gas and water cut prediction technique by calculating relative mobility.

6.2 Joint interpretation of electromagnetic measurements with wireline formation tester measurements

Alpak et al. (2004) developed an inversion algorithm for joint inversion of induction resistivity and formation tester data including transient pressure, flowline water-cut, and salt production to estimate formation water saturation, vertical and horizontal permeability, relative permeability curves, and capillary pressure curves. The estimated petrophysical parameters by the joint inversion of multi-physics measurements are more accurate and generate less non-unique estimates. Angeles et al. (2008) integrated laterolog resistivity and formation-testing measurements and employed a multilayer formation model to estimate porosity, permeability, and water saturation. The proposed joint interpretation workflow assumes that the vertical wellbore trajectory, horizontal target formation with no lateral variations, and no gas phase that formation pressure is above bubble point during simulation or measurements. Fig. 1.50 shows their proposed simulation and interpretation workflow including required prior petrophysical parameters and measurements. The simulation process includes the mud filtrate invasion simulation which consequently affects the difference between deep and shallow resistivity.

Angeles et al. (2008) illustrated the application of the proposed joint interpretation workflow in a well drilled in a Jurassic formation in Norwegian Sea. This formation was deposited in marine inner shelf, and the lithology is fine grained sandstone with thin beds of siltstone. First, they used a numerical reservoir simulator coupled to a mud-filtrate invasion simulator to accurately model two-phase immiscible fluid flow in the multilayer formation. The 2D spatial distribution of simulated resistivity was then obtained based on the model predictions of water saturation and salt concentration distributions because of the two-phase flow. Fig. 1.51A illustrates the model prediction of

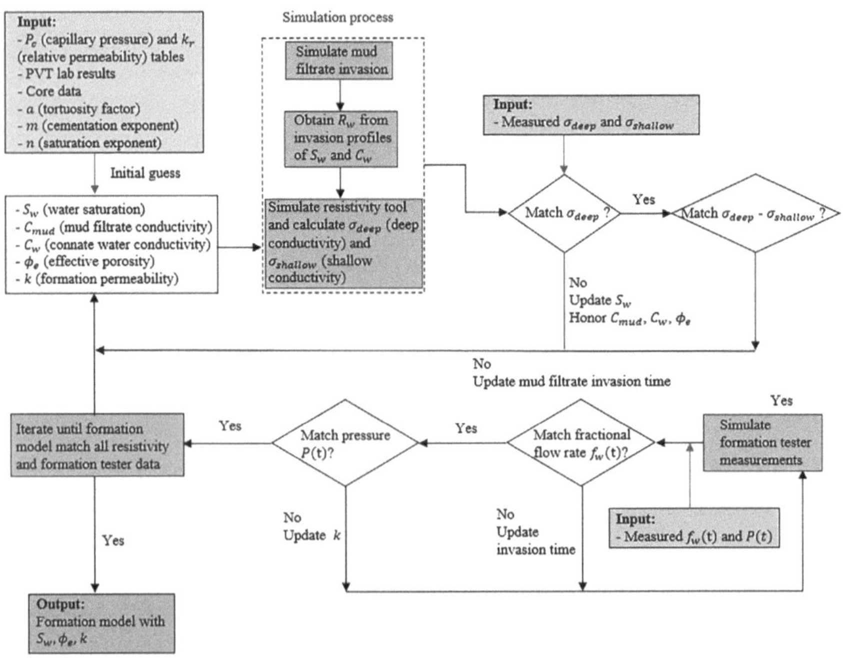

FIGURE 1.50 The joint interpretation workflow combining laterolog resistivity and formation-testing measurements with prior information to iteratively estimate porosity, permeability, and water saturation. *Modified from Angeles R, Skolnakorn J, Antonsen F, Chandler A, Torres-Verdin C, May 25–28, 2008. Advantages in joint-inversions of resistivity and formation-tester measurements: examples from a Norwegian field. SPWLA 49th Ann. Logg. Symp.; Source: Authors.*

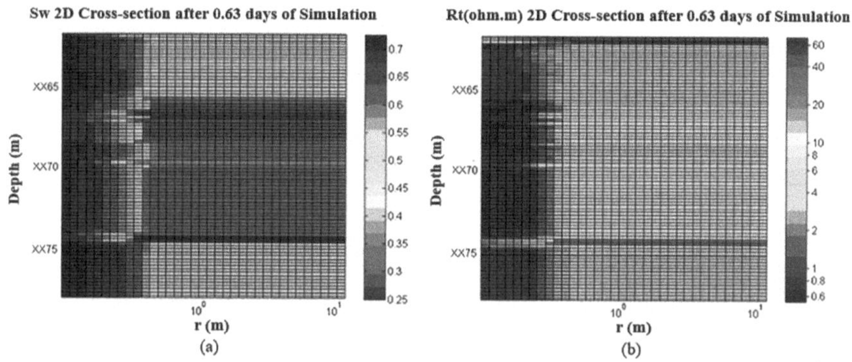

FIGURE 1.51 (A) The model predictions of spatial distribution of two-dimensional water saturation after the 0.63 days of two-phase flow; (B) the simulated laterolog tool resistivity spatial response after 0.63 days. *From Angeles R, Skolnakorn J, Antonsen F, Chandler A, Torres-Verdin C, May 25–28, 2008. Advantages in joint-inversions of resistivity and formation-tester measurements: examples from a Norwegian field. SPWLA 49th Ann. Logg. Symp.*

spatial distribution of 2D water saturation after the 0.63 days of two-phase flow, which demonstrates the heterogeneous nature of the formation and thin laminations. Fig. 1.51B indicates the laterolog tool resistivity response after 0.63 days. They initially used NMR log to determine 23 electrofacies (Angeles et al., 2008) in the interval under investigation. Then they processed electrofacies to obtain 18 numerical formation layers and six groups to reduce the number of parameters to be estimated. Finally, they simulated pressure and fractional flow curve to match the field data and estimate or update the permeability, which is illustrated in Fig. 1.52. Fig. 1.52A shows that the simulated pressure agrees well with pressure measurements except for the initial stage, which was affected by tool storage effects. The simulated fractional flow rate is overestimated compared with values obtained by downhole fluid analyzer LFA.

Fig. 1.53 illustrates that the proposed joint-interpretation workflow can be used to estimate porosity and water saturation, which agree well with values from core measurements and log interpretation. However, the estimated permeability in the sampling depth is lower than value obtained from log interpretation (Angeles et al., 2008). Liang et al. (2009) also developed a similar workflow which combines induction resistivity and pressure transient measurements to estimate petrophysical parameters and mud-filtrate invasion profile. Their proposed method can directly estimate the mud-filtrate invasion volume from the fluid-flow constrained inversion of induction resistivity without calculating or simulating the complicated and less reliable invasion rate.

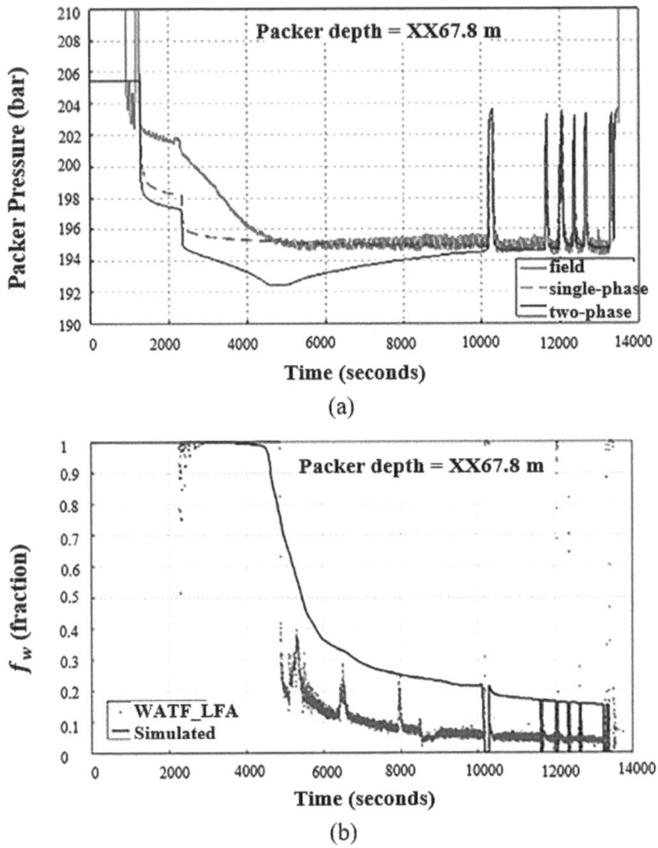

FIGURE 1.52 The comparisons between the simulated and measured (A) pressure and (B) fractional flow curve. *From Angeles R, Skolnakorn J, Antonsen F, Chandler A, Torres-Verdin C, May 25–28, 2008. Advantages in joint-inversions of resistivity and formation-tester measurements: examples from a Norwegian field. SPWLA 49th Ann. Logg. Symp.*

6.3 Joint interpretation of electromagnetic measurements with sonic or seismic measurements

Gao et al. (2010) introduced a joint interpretation of seismic and EM measurements to estimate porosity and fluid saturation distributions. The porosity and water saturation estimates were more accurate than the separate EM data inversion and full-wave seismic data inversion. Gao et al. (2012) mentioned that EM and sonic logging tools can be combined to accurately estimate porosity and fluid saturation distributions in the near-wellbore regions. The triaxial induction resistivity tool and sonic tool can make 3D measurements

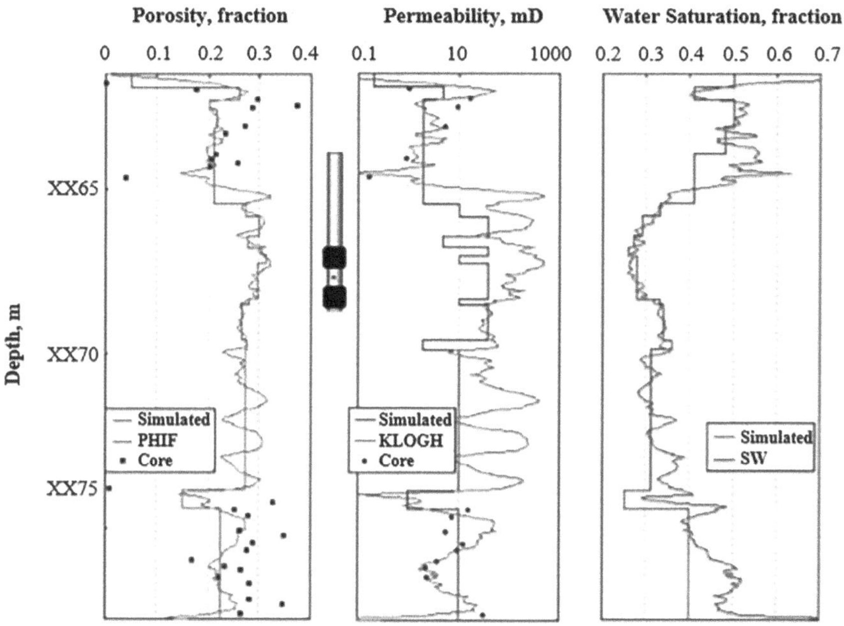

FIGURE 1.53 The comparisons among core-derived, log-derived, and the proposed joint interpretation–based estimations of porosity, permeability, and water saturation in the interval of the well under investigation. *From Angeles R, Skolnakorn J, Antonsen F, Chandler A, Torres-Verdin C, May 25–28, 2008. Advantages in joint-inversions of resistivity and formation-tester measurements: examples from a Norwegian field. SPWLA 49th Ann. Logg. Symp.*

including multifrequency EM measurements using triaxial arrays, monopole, and dipole sonic measurements. The triaxial induction resistivity tool has three tri-axial transmitters and receivers that can measure formation dip and azimuth. They mentioned that the porosity and water saturation can be obtained by the interpretation of electrical resistivity measurements using Archie equation or Waxman Smits model. The fluid substitution model can also be applied to calculate porosity and fluid saturation based on the seismic velocities including compressional and shear waves interpretation. They presented a numerical simulation example to prove the effectiveness of the integrated method. They used a 2.5D frequency-domain finite difference EM simulation at 15 and 30 kHz operation frequencies, and 2D elastic wave simulations at 0.5, 2.5, 5.0, and 10.0 kHz operating frequencies (Gao et al., 2012). The inversion results indicate that the porosity and water saturation distributions can be reliably estimated even in the mud-filtrate invaded zone, and the oil-water contacts and formation structures can also be detected through this approach.

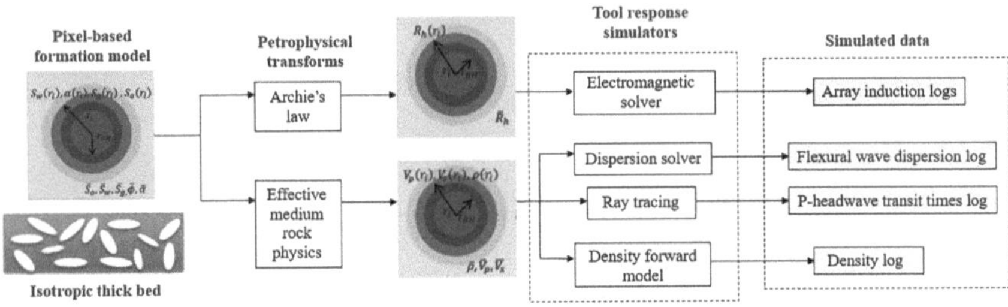

FIGURE 1.54 The forward modeling framework to simulate array induction, elastic waves, and density log responses. *According to Shetty S, Liang L, Simoes V, Canesin F, Boyd A, Zeroug S, Sinha B, Habashy T, Domingues AB, Amorim C, Abbots F, October 2017. Imaging near-wellbore petrophysical properties by joint inversion of sonic, resistivity, and density logging data. Petrophysics 58 (5), 501–516; Source: Authors.*

Shetty et al. (2017) integrated array induction, array sonic, and density log data to calculate porosity, radial distribution of fluid saturations, and pore shape which is the aspect ratio of oblate spheroidal grain. They used their sonic and EM forward-model solvers to generate sonic and EM data for the aforementioned tools. Fig. 1.54 demonstrates the forward-modeling framework including pixel-based formation modeling, effective-medium rock physics modeling, and petrophysical transformation to calculate multi-physics logs (Shetty et al., 2017). They took into account the different depths of investigation of different tools, mud-filtrate invasion, presence of gas phase, and formation mechanical damages, which are often not considered for most of the multi-physics interpretation methods. The porosity, radial distribution of fluid saturations, and pore shape were iteratively updated to match the field data. The proposed forward-modeling framework or interpretation method was applied to a gas-bearing offshore well drilled with oil-based mud in East Africa. Fig. 1.55 presents caliper log, mineral volume fractions, porosity estimates, and estimated water saturation of the virgin zone. The water-saturation estimates were obtained using both the proposed method, S_W (inversion), and conventional Archie equation, S_W_AR. The estimated porosity, PHIT (inversion), and virgin zone water saturation are in overall good agreement with results obtained by conventional log analysis. The radial distributions or 2D images of pore shape (Alpha) and fluid saturations are also illustrated in Fig. 1.55. The presence of gas at various depths is detected, and the radial distribution of water saturation indicates the mud invasion. The pore shape alterations from the borehole to the virgin zone indicate rock mechanical damages (Shetty et al., 2017).

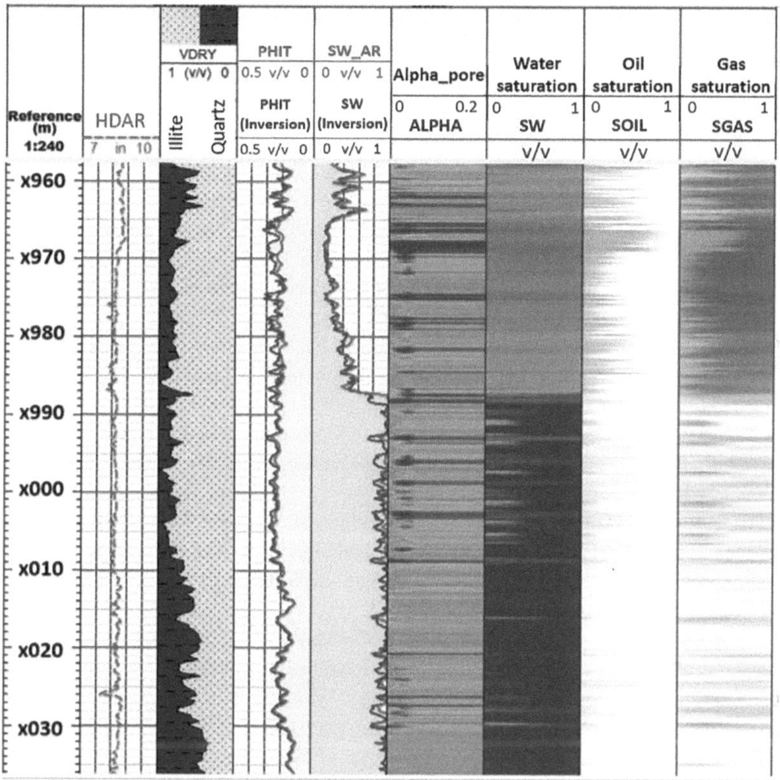

FIGURE 1.55 The comparison between the estimated porosity and virgin zone water saturation calculated using conventional log analysis and the newly proposed multi-physics method; and the interpretation results based on the proposed method includes pore shape and fluid saturations images, which indicate mud invasion. *Modified from Shetty S, Liang L, Simoes V, Canesin F, Boyd A, Zeroug S, Sinha B, Habashy T, Domingues AB, Amorim C, Abbots F, October 2017. Imaging near-wellbore petrophysical properties by joint inversion of sonic, resistivity, and density logging data. Petrophysics 58 (5), 501–516.*

7. Conclusions

The chapter discusses EM data acquisition methods in the laboratory and subsurface environments. The chapter also provides information about various interpretation models and workflows for geological characterization and formation evaluation. Following that, the chapter presents the use of various electromagnetic logging tools for subsurface multifrequency EM data acquisition. Several case studies of subsurface EM log acquisition and interpretation are elaborated and analyzed in the chapter. Different EM log-interpretation workflows are discussed for various formations, including conventional sandstone, shaly-sand, carbonate, turbidities, and unconventional formations. Finally, the chapter discusses the integration of EM measurements with other physical measurements, such as NMR and formation testing measurements, for purposes of reducing the uncertainties in estimation and to estimate new formation properties that can only be computed using joint-interpretation methods.

References

Aboites, D.L., del Ángel Morales, J.G., Hurtado, A.O., de Jesús Correa López, M., Buenrostro, C.A., Phillips, C.E., Anguiano-Rojas, P., May 18-22, 2014. Case studies of dielectric logging in a highly fractured carbonate reservoir in the Gulf of Mexico. SPWLA 55th Ann. Logg. Symp., SPWLA-2014-TTTT.

Adisoemarta, P.S., Anderson, G.A., Frailey, S.M., Asquith, G.B., March 21-23, 2000. Historical Use of M and a in Well Log Interpretation: Is Conventional Wisdom Backwards? SPE Perm. Basin Oil Gas Rec. Confer., SPE-59699-MS.

Akbar, M.N.A., Musu, J.T., Milad, B., July 23-25, 2018. Water saturation interpretation model for organic-rich shale reservoir: a case study of North Sumatra Basin. Unconvent. Res. Tech. Confer., URTEC-2879229-MS.

Alpak, F.O., Torres-Verdin, C., Habashy, T.M., Sepehrnoori, K., September 26-29, 2004. Simultaneous estimation of In-Situ multi-phase petrophysical properties of rock formations from wireline formation tester and induction logging measurements. SPE Ann. Tech. Confer. Exhibit., SPE-90960-MS.

Altman, R., Anderson, B., Rasmus, J., Luling, M.G., May 25–28, 2008. Dielectric effects and resistivity dispersion on induction and propagation-resistivity logs in complex volcanic lithologies: a case study. SPWLA 49th Ann. Logg. Symp., SPWLA-2008-JJJJ.

Anderson, B.I., January, 2001. Modeling and Inversion Methods for the Interpretation of Resistivity Logging Tool Response. Delft University of Technology.

Anderson, B.I., Barber, T.D., Luling, M.G., Rasmus, J., Sen, P.N., Tabanou, J.R., Haugland, M., June 3–6, 2007. Observations of large dielectric effects on LWD propagation-resistivity logs. SPWLA 48th Ann. Logg. Symp., SPWLA-2007-BB.

Angeles, R., Skolnakorn, J., Antonsen, F., Chandler, A., Torres-Verdin, C., May 25–28, 2008. Advantages in joint-inversions of resistivity and formation-tester measurements: examples from a Norwegian field. SPWLA 49th Ann. Logg. Symp., SPWLA-2008-SSS.

Barber, T., Anderson, B., Abubakar, A., Broussard, T., Chen, K.C., Davydycheva, S., Druskin, V., Habashy, T., Homan, D., Minerbo, G., Rosthal, R., Schlein, R., Wang, H., September 26-29, 2004. Determining formation resistivity anisotropy in the presence of invasion. SPE Ann. Tech. Confer. Exhibit., SPE-90526-MS.

Bassiouni, Z., 1994. Theory, Measurements and Interpretation of Well Logs, fourth ed. Society of Petroleum Engineers.

Basu, I., Machin, N., Tyagi, A., Saxena, K., Altman, R., Mathew, A., Kumar, R., June 3-6, 2007. Enhanced reservoir characterization in a deep water turbidite system using borehole images and spectroscopy logs. SPWLA 48th Ann. Logg. Symp., SPWLA-2007-RR.

Bean, C., Cole, S., Boyle, K., Kho, D., Neville, T.J., June 22-26, 2013. New wireline dielectric dispersion logging tool result in fluvio-deltaic sands drilled with oil-based mud. SPWLA 54th Ann. Logg. Symp., SPWLA-2013-ZZ.

Berg, C.F., 2012. Re-examining Archie's Law: conductance description by tortuosity and constriction. Phys. Rev. E 86 (4). https://doi.org/10.1103/PhysRevE.86.046314.

Bitterlich, V., Wobking, H., 1970. Eine method zur direkten bestimmung der sogenannten "Geometrischen Tortuosit" (A method for the direct determination of geometrical tortuosity). Zeitschift Für Geophysik. 36, 607–620.

Bona, N., Rossi, E., Capaccioli, S., March 2001. Electrical measurements in the 100 Hz to 10 GHz frequency range for efficient rock wettability determination. SPE J. 6 (1), 80–88.

Bose, S., Myers, M.T., Chen, P., Thakur, G., June 17-19, 2019. Application of an integrated petrophysical modeling to improve log based reservoir characterization and oil in-place estimate of a fresh water shaly sand reservoir. SPWLA 60th Ann. Logg. Symp., SPWLA-2019-DDDDD.

Clavaud, J.B., Nelson, R., Guru, U.K., Wang, H., June 26-29, 2005. Field example of enhanced hydrocarbon estimation in thinly laminated formation with a triaxial array induction tool: a laminated sand-shale analysis with anisotropy shale. SPWLA 46th Ann. Logg. Symp., SPWLA-2005-WW.

Claverie, M., 2012. Dielectric Scanner Multifrequency Dielectric Dispersion Service. Schlumberger, Publication.

Clennell, M.B., January, 1997. Tortuosity: a guide through the maze. Develop. Petrophys. 122, 299–344.

Donadille, J.M., ILyin, I., Luling, M.G., Meszaros, T., Reischman, R., Repnev, A., Yukhlin, V., Khan, N., Jones, M., July 24-26, 2017. Slim, high resolution laterolog array tool: first field experiences. Unconvent. Res. Tech. Confer., URTEC-2671192-MS.

Ellis, D.V., Singer, J.M., 2007. Well Logging for Earth Scientists. Springer.

Eubanks, D.L., June 25–28, 2008. Application of resistivity borehole image analysis in unconventional gas reservoir field development. SPWLA 49th Ann. Logg. Symp., SPWLA-2008-XX.

Freed, D., Seleznev, N., Hou, C.Y., Fellah, K., Little, J., Dumy, G., Sen, P., June 25-29, 2016. A physics-based model for the dielectric response of shaly sands. SPWLA 57th Ann. Logg. Symp., SPWLA-2016-L.

Garcia, A.P., Heidari, Z., October, 2018. Development of a resistivity model that incorporates quantitative directional connectivity and tortuosity for enhanced assessment of hydrocarbon reserves. SPE J. 23 (5), 1552–1565.

Gao, G., Abubakar, A., Habashy, T.M., November 4–9, 2012. Joint inversion of borehole electromagnetic and sonic measurements. SEG Int. Exposit. Ann. Meet., SEG-2012-0133.

Gao, G., Abubakar, A., Habashy, T.M., October 17-22, 2010. Simultaneous joint petrophysical inversion of electromagnetic and seismic measurements. SEG Int. Exposit. Ann. Meet., SEG-2010-2799.

Garcia, A.P., Han, Y., Heidari, Z., October, 2018. Integrated workflow to estimate permeability through quantification of rock fabric using joint interpretation of nuclear magnetic resonance and electric measurements. Petrophysics 59 (5), 672–693.

Garcia, A.P., Heidari, Z., Torres-Verdin, C., June 17–19, 2019. Multi-frequency interpretation of electric resistivity and dielectric permittivity measurements for simultaneous assessment of porosity, water saturation, and wettability. SPWLA 60th Ann. Logg. Symp., SPWLA-2019-III.

Garcia, A.P., Jagadisan, A., Rostami, A., Heidari, Z., June 17–21, 2017. A new resistivity-based model for improved hydrocarbon saturation assessment in clay-rich formations using quantitative clay network geometry and rock fabric. SPWLA 58th Ann. Logg. Symp., SPWLA-2017-G.

Griffiths, R., Barber, T., Faivre, O., June 4–7, 2000. Optimal evaluation of formation resistivities using array induction and array laterolog tools. SPWLA 41st Ann. Logg. Symp., SPWLA-2000-BBB.

Han, Y., Misra, S., May, 2018. Joint petrophysical inversion of multifrequency conductivity and permittivity logs derived from subsurface galvanic, induction, propagation, and dielectric dispersion measurements. Geophysics 83 (3), D97–D112.

Han, M., Cuadros, J., Suarez, C.A.P., Decoster, E., Faivre, O., Mosse, L., Seleznev, N., June 16–20, 2012. Continuous estimate of cation exchange capacity from log data: a new approach based on dielectric dispersion analysis. SPWLA 53rd Ann. Logg. Symp., SPWLA-2012-101.

Han, Y., Misra, S., 2018. Bakken petroleum system characterization using dielectric-dispersion logs. Petrophysics 59 (2), 201–217.

Hamada, G.M., Al-Awad, M.N.J., Almalik, M.S., 2001. Log evaluation of low - resistivity sandstone reservoirs. SPE Perm. Basin Oil Gas Rec. Confer., SPE-70040-MS.

Han, Y., Misra, S., September 24–29, 2017. Improved water-saturation estimates derived from inversion-based interpretation of broadband electromagnetic dispersion logs in organic-rich shale formations. SEG Int. Exposit. Ann. Meet., SEG-2017-17797784.

Han, Y., Misra, S., Wang, H., Toumelin, E., February 2019. Hydrocarbon saturation in a Lower-Paleozoic organic-rich shale gas formation based on Markov-chain Monte Carlo stochastic inversion of broadband electromagnetic dispersion log. Fuel 243, 645–658.

Hizem, M., Budan, H., Deville, B., Faivre, O., Mosse, L., Simon, M., September 21–24, 2008. Dielectric dispersion: a new wireline petrophysical measurement. SPE Ann. Tech. Confer. Exhibit., SPE-116130-MS.

Holmes, M., Holmes, A.M., Holmes, D.I., June 17-21, 2017. Mixed reservoir wetting in unconventional reservoirs and interpretation of porosity/resistivity cross plots, derived from triple-combo log data. SPWLA 58th Ann. Logg. Symp., SPWLA-2017-LLLL.

Hulea, I.N., Frese, D., Ramaswami, S., July 18-22, 2015. Improved permeability prediction in heterogeneous carbonate reservoirs: a new approach using rock typing and wireline formation testing. SPWLA 56th Ann. Logg. Symp., SPWLA-2015-GGG.

Iheanacho, P.C., May, 2014. Formation-resistivity theory: how Archie equations, shaly-reservoir models, conductive rock-matrix model, and dual-triple-porosity models are related. SPE Reservoir Eval. Eng. 17 (2), 141–151.

Jacobsen, S., May, D., Little, J., December, 2007. Producibility prediction in gas sands through effective integration of NMR, resistivity, and porosity log data. Petrophysics 48 (6), 438–453.

Jin, Y., Misra, S., Homan, D., Rasmus, J., 2020. Effects of wettability of conductive particles on the multifrequency conductivity and permittivity of fluid-filled porous material. Fuel 268, 117411.

Juhasz, I., June 23-26, 1981. Normalised Qv - the key to shaly sand evaluation using the Waxman-Smits equation in the absence of core data. SPWLA 22nd Ann. Logg. Symp., SPWLA-1981-Z.

Kurniawan, B., Bassiouni, Z., June 3-6, 2007. Use of CEC-dependent cementation and saturation exponents in shaly sand resistivity models. SPWLA 48th Ann. Logg. Symp., SPWLA-2007-FFF.

Leroy, P., Revil, A., Kemna, A., Cosenza, P., Ghorbani, A., 2008. Complex conductivity of water-saturated packs of glass beads. J. Colloid Interface Sci. 321, 103–117.

Li, Q., Liu, C., Maeso, C., Wu, P., Smits, J., Prabawa, H., Bradfield, J., 2004. Automated interpretation for LWD propagation resistivity tools through integrated model selection. Petrophysics 45 (1), 14–26.

Liang, L., Abubakar, A., Habashy, T., October 4-7, 2009. Determination of petrophysical parameters and mud filtrate invasion profile using joint inversion of induction logging and pressure transient measurements. SPE Ann. Tech. Confer. Exhibit., SPE-124049-MS.

Misra, S., Han, Y., August 1-3, 2016. Petrophysical interpretation of multi-frequency electromagnetic measurements in clay- and conductive-mineral-rich mudrocks. Unconvent. Res. Tech. Confer., URTEC-2433541-MS.

Misra, S., Han, Y., September 25–27, 2016. Improved oil-in-place estimates in clay-and pyrite-bearing shales based on inversion of multi-frequency electromagnetic measurements. AAPG East. Section Meet., 90258.

Misra, S., Torres-Verdín, C., Homan, D., Minerbo, G., Rasmus, J., August, 2015. Laboratory investigation of petrophysical applications of multi-frequency inductive-complex conductivity tensor measurements. In: SPWLA 56th Annual Logging Symposium. Society of Petrophysicists and Well-Log Analysts.

Misra, S., Torres-Verdín, C., Homan, D., Rasmus, J., July, 2015. Complex electrical conductivity of mudrocks and source-rock formations containing disseminated pyrite. In: Unconventional Resources Technology Conference, San Antonio, Texas, 20-22 July 2015. Society of Exploration Geophysicists, American Association of Petroleum Geologists, Society of Petroleum Engineers, pp. 2478–2490.

Manescu, A., Kho, S.F., Ha, K.T., Berhad, S.S., October 18-20, 2010. Case study of improved oil-in-place determination from petrophysical evaluation of resistivity anisotropy logging data in a Malaysian deepwater turbidite field. SPE Asia Pacific Oil Gas Confer. Exhibit., SPE-133106-MS.

Misra, S., 2015. Wideband Directional Complex Electrical Conductivity of Geomaterials: A Mechanistic Description (Doctoral Dissertation).

Moherek, A., Mukherjee, S., Garrison, N., Caraway, A., Medina, R., Sarmah, B.B., August 1-3, 2016. Application of high-resolution blended resistivity measurement in characterization of unconventional plays. Unconvent. Res. Tech. Confer., URTEC-2461234-MS.

Mosse, L., Carmona, R., Decoster, E., Faivre, O., Hizem, M., June 21–24, 2009. Dielectric dispersion logging in heavy oil: a case study from the Orinoco Belt. SPWLA 50th Ann. Logg. Symp., SPWLA-2009-52901.

Myers, M.T., 1996. Pore geometry dependent dispersion model for the dielectric constant. Confer. Soci. Core Analyst, 9626.

Nelson, R., Clavaud, J.B., Guru, U.K., Wang, H., July, 2006. Enhancing hydrocarbon estimates in a deepwater turbidite sequence. J. Petrol. Technol. 58 (7), 44–47.

Newgord, C., Garcia, A.P., Heidari, Z., June 17–19, 2019. A new workflow for joint interpretation of electrical resistivity and NMR measurements to simultaneously estimate wettability and water saturation. SPWLA 60th Ann. Logg. Symp., SPWLA-2019-GG.

North, R.E., 2017. The Effects of Wideband Complex Electromagnetic Properties of Soils on Geophysical Sensors Performance. Colorado School of Mines.

Ortenzi, L., Dubourg, I., Os, R.V., Han, S.Y., Koepsell, R., Chow, S., Ha, Y., May 14–18, 2011. New azimuthal resistivity and high-resolution imager facilitates formation evaluation and well placement of horizontal slim boreholes. SPWLA 52nd Ann. Logg. Symp., SPWLA-2011-LLL.

Raina, I., Dasgupta, S., Mosse, L., Baig, M.Z., Laer, P.V., November 13-16, 2017. Quantifying water saturation in a source rock gas play using multi-frequency dielectric measurements. Abu Dhabi Int. Petroleum Exhibit. Confer., SPE-188511-MS.

Revil, A., February, 2012. Spectral induced polarization of shaly sands: influence of the electrical double layer. Water Resour. Res. 48 (2), W02517.

Revil, A., January, 2013. Effective conductivity and permittivity of unsaturated porous materials in the frequency range 1 mHz–1GHz. Water Resour. Res. 49 (1), 306–327.

Sabouroux, P., Ba, D.S., March, 2011. EpsiMu, a tool for dielectric properties measurement of porous media: application in wet granular materials characterization. Prog. Electromag. Res. B 29, 191–207.

Sabouroux, P., Boschi, P., 2005. EpsiMu: a new microwave materials measurements kit. Rev. Electr. Electron. 10, 58–62.

Schlumberger, 2007. arcVISION312 Slimhole Formation Evaluation while Drilling. Schlumberger, Publication.

Seleznev, N., Hou, C.Y., Freed, D., Habashy, T., Feng, L., Fellah, K., June 17–21, 2017. Coherent interpretation of wideband electromagnetic measurements in the millihertz to gigahertz frequency range. SPWLA 58th Ann. Logg. Symp., SPWLA-2017-MMMM.

Shahin, A., Myers, M.T., Hathon, L.A., June 17-21, 2017. Carbonate formation evaluation using joint inversion of resistivity and dielectric logs. SPWLA 58th Ann. Logg. Symp., SPWLA-2017-QQ.

Shetty, S., Liang, L., Simoes, V., Canesin, F., Boyd, A., Zeroug, S., Sinha, B., Habashy, T., Domingues, A.B., Amorim, C., Abbots, F., October, 2017. Imaging near-wellbore petrophysical properties by joint inversion of sonic, resistivity, and density logging data. Petrophysics 58 (5), 501–516.

Tathed, P., Han, Y., Misra, S., May, 2018. Hydrocarbon saturation in upper Wolfcamp shale formation. Fuel 219, 375–388.

Tariq, Z., Mahmoud, M., Hassan, A., Khan, M.R., Sadeed, A., April 23-26, 2018. Carbonate rock resistivity index measurement using equivalent rock model. SPE Ann. Tech. Symp. Exhibit., SPE-192353-MS.

Tathed, P., Han, Y., Misra, S., December, 2018. Hydrocarbon saturation in Bakken Petroleum System based on joint inversion of resistivity and dielectric dispersion logs. Fuel 233, 45–55.

Tiab, D., Donaldson, E.C., 2004. Petrophysics-theory and Practice of Measuring Reservoir Rock and Fluid Transport Properties, 2rd edition. Gulf Professional.

Toumelin, E., Torres-Verdin, C., June 21–24, 2009. Pore-scale simulation of kHz-GHz electromagnetic dispersion of rocks: effects of rock morphology, pore connectivity, and electrical double layers. SPWLA 50th Ann. Logg. Symp., SPWLA-2009-39488.

Valiakhmetov, R., Mendybaev, N., Ramatullayev, S., Charupa, M., Seilov, Y., Abishev, A., Rziyeva, Z., Salikova, N., October 22-24, 2019. Integration of a multi-frequency dielectric measurement and advanced wireline logs for the enhanced characterization of carbonate reservoirs of the Western Kazakhstan. SPE Russian Petroleum Tech. Confer., SPE-196954-MS.

Wang, H., Poppitt, A., June 22-26, 2013. The broadband electromagnetic dispersion logging data in a gas shale formation: a case study. SPWLA 54th Ann. Logg. Symp., SPWLA-2013-AAA.

Wu, P., Lovell, J., Clark, B., Bonner, S., Tabanou, J., October 3–6, 1999. Dielectric-independent 2-MHz propagation resistivities. SPE Ann. Tech. Confer. Exhibit., SPE-56448-MS.

Xiao, J., Buchanan, J., Bittar, M., Davis, E., Sanmartin, L., Hu, G., Zannoni, S., Morys, M., Liu, W., September 24–27, 2006. A new asymmetrical array induction logging tool. SPE Ann. Tech. Confer. Exhibit., SPE-101930-MS.

Xu, J., Lei, X., Yuxing, Q., March 2017. Two effective methods for calculating water saturations in shale-gas reservoirs. Geophysics 82 (3), D187–D197.

Yu, H., Wei, X., Wang, Z., Rezaee, R., Zhang, Y., Lebedev, M., Iglauer, S., October 23-25, 2018. Review of water saturation calculation methods in shale gas reservoir. SPE Asia Pacific Oil Gas Confer. Exhibit., SPE-192115-MS.

Zurich Instruments, 2019. MFIA User Manual. https://www.zhinst.com/sites/default/files/ziMFIA_UserManual_64000.pdf.

Chapter 2

Petrophysical models for the interpretation of electromagnetic logs: A brief review

Pratiksha Tathed[1], Siddharth Misra[2,3]
[1]*Petrophysicist, BP America Inc., Houston, TX, United States;* [2]*Harold Vance Department of Petroleum Engineering, College of Engineering, Texas A&M University, TX, United States;* [3]*Department of Geology and Geophysics, Texas A&M University, TX, United States*

1. Archie, Simandoux, Waxman–Smits, and Dual water model

Electrical and electromagnetic logging tools are widely used to determine the hydrocarbon saturation. Archie equation is an empirical petrophysical model to interpret subsurface resistivity logs acquired in clean sandstone formations at low electromagnetic (EM) frequencies. Archie equation (Archie, 1942) was developed for clay-free sandstone reservoirs by relating the true bulk resistivity (R_t) of the rock to porosity (ϕ), formation water resistivity (R_w), and water saturation (S_w), formulated as

$$R_t = \frac{aR_w}{\phi^m S_w^n} \tag{2.1}$$

where a is tortuosity factor, m is cementation exponent, and n is saturation exponent. Archie equation does not account for the clay and conductive minerals and the interfacial polarization (IP) effects around the clays and conductive minerals. Surface charge–bearing clay minerals in shaly-sand formations or conductive minerals, such as pyrite, can significantly change the bulk electrical resistivity; therefore, the Archie-derived water saturation in shaly-sand formations is unreliable. Simandoux (1963) proposed a shaly-sand model that considers both the brine conductivity and the conductivity of clay minerals. Simandoux equation was developed based on laboratory

Multifrequency Electromagnetic Data Interpretation for Subsurface Characterization
https://doi.org/10.1016/B978-0-12-821439-8.00010-0

investigations on synthetic homogeneous mixtures of sand, conductive clay minerals, and water. Simandoux equation is expressed as

$$\frac{1}{R_t} = \frac{\phi_T^m}{aR_w} S_w^n + \frac{V_{sh}}{R_{sh}} S_w^{n-1} \tag{2.2}$$

where R_t is the formation resistivity (in Ωm), ϕ_T is total formation porosity, S_w is water saturation, R_w is connate water resistivity (in Ωm), V_{sh} is volumetric fraction of shale, R_{sh} is the resistivity of shale (in Ωm), m is cementation exponent, n is saturation exponent, and a is tortuosity factor. For $V_{sh} = 0$, Simandoux equation is equal to the Archie equation.

Waxman and Smits (1968) developed petrophysical model for water saturation estimation in clay-rich formation by accounting for additional conductivity because of the presence of clay with high cation exchange capacity (CEC). Unlike Simandoux equation that uses V_{sh}, Waxman–Smits (WS) model uses CEC and is expressed as

$$\frac{1}{R_t} = \phi_T^m S_{wt}^n \left(C_w + \frac{BQ_v}{S_{wt}} \right) \tag{2.3}$$

where R_t represents the true resistivity of the formation, ϕ_T is the total porosity, S_{wt} is water saturation of the total porosity, C_w is the brine conductivity in S/m (which is reciprocal of formation water resistivity, R_w), Q_v is the CEC per unit pore volume in meq/cc or meq/mL, B is the equivalent conductance of sodium clay-exchange cations (also, referred as specific cation conductance) of the total pore volume in mS/meq, which depends on the salinity of brine and temperature of formation, m is the cementation exponent of the total pore volume, and n is the saturation exponent of the total pore volume. Peeters and Holmes (2014) showed that WS method is not very sensitive to variation of most of the parameters when B is fixed and reliable Q_v values are used. WS model is developed for dispersed shaly-sand formations. WS model does not account for the effects of shale laminations and thin beds.

The Dual-water model (Clavier et al., 1977) accounts for the conductivity near and within the double layer of surface-charge bearing clay minerals in addition to bulk conductivity of the pore space. Unlike WS equation, Dual-water model uses clay-bound water conductivity R_{cw} instead of the BQ_v factor. The dual-water model is formulated as

$$\frac{1}{R_t} = \phi_T^m S_{wt}^n \left[\frac{1}{R_w} + \frac{S_{wb}}{S_{wt}} \left(\frac{1}{R_{cw}} - \frac{1}{R_w} \right) \right] \tag{2.4}$$

where ϕ_T is the total porosity, R_{cw} is resistivity of the clay-bound water, R_w is the resistivity of the free water, S_{wt} is water saturation of the total porosity, and S_{wb} is the saturation of the clay-bound water, which serves as an alternative to shale volume. The m and n exponents of dual-water model should be smaller than those of the WS model. Effective porosity of such a dispersed shale-sand system is expressed as $\phi_T(1-S_{wb})$.

2. Complex refractive index and Lichteneker-Rother model

Complex refractive index (CRI) model is widely used to process complex permittivity logs at around 1 GHz, and it assumes that the IP effects at grain surface are negligible around 1 GHz. CRI model only takes into account the effects of orientation/dipole polarization. CRI model is a specific case of the Lichteneker-Rother (LR) model (Lichtenecker and Rother, 1931), which is expressed as

$$\left(\varepsilon_r^*\right)^\alpha = (1-\phi_T)(\varepsilon_{rm})^\alpha + S_w\phi_T\left(\varepsilon_{rw}^*\right)^\alpha + (1-S_w)\phi_T\left(\varepsilon_{rhy}\right)^\alpha \tag{2.5}$$

where ε_r^* is complex relative permittivity of the formation, ε_{rm} is relative permittivity of the matrix, ε_{rw}^* is complex relative permittivity of water, ε_{rhy} is relative permittivity of hydrocarbon, ϕ_T is total porosity, S_w is water saturation, and α is a geometrical fitting parameter. With $\alpha = 0.5$, LR model reduces to CRI model. LR and CRI models do not account for the tortuosity, rock fabric, pore network topology, and spatial distribution of minerals (Brovelli and Cassiani, 2008). Complex relative permittivity ε_{rw}^* is expressed as

$$\varepsilon_w^* = \varepsilon_{rw} + i\frac{C_w}{\varepsilon_0\omega} \tag{2.6}$$

where ε_0 is vacuum permittivity ($\varepsilon_0 = 8.85 \times 10^{-12}$ F/m), ω is angular frequency ($\omega = 2\pi f$, f is frequency), ε_{rw} is the real part of the complex relative permittivity of water, and C_w is water conductivity (in S/m, $C_w = 1/R_w$). The relative permittivity of water ε_{rw} decreases with increasing temperature because the thermal agitation reduces the overall alignment of water molecules along the direction of EM field. ε_{rw} is expressed as a function of temperature and salinity (Sabouroux and Ba, 2011) as

$$\varepsilon_{rw}(0,T) = 94.88 - 0.2317T + 0.000217T^2 \tag{2.7}$$

$$\varepsilon_{rw}(S,T) = \left[\frac{1}{\varepsilon_{rw}(0,T)} + \frac{2.4372 \times S}{58.443(1000-S)}\right]^{-1} \tag{2.8}$$

where S is salinity (in ppt) and T is temperature (in °F). The water conductivity C_w, in the imaginary part of Eq. (2.6), can also be formulated as a function of temperature and salinity (Sabouroux and Ba, 2011) expressed as

$$C_w(S,T) = \frac{T+7}{82}\left[0.0123 + \frac{3647.5}{(1000S)^{0.955}}\right] \tag{2.9}$$

3. Effective medium models (mixing formulas)

Effective medium model is analytical or theoretical model that describes the macroscopic properties of composite materials based on the microscopic properties of the representative volume (i.e., material constituent). Effective medium model is developed by averaging the properties of the constituents of the composite material. The effective medium model generally assumes that the macroscopic system is homogeneous. This modeling approach fails to predict the properties of a multiphase system, especially close to the percolation threshold.

3.1 Stroud-Milton-De model

Stroud et al. (1986) introduced an analytical mixing formula to calculate complex dielectric permittivity and frequency dependence (dispersion) of brine-saturated rocks as a function of total porosity, water saturation, brine conductivity, dielectric permittivity of rock matrix, brine, and oil, and rock textural parameter (m') that describes cementation/tortuosity (Han et al., 2012). Stroud-Milton-De (SMD) model is a commonly used model to process the dielectric dispersion measurements/logs acquired in the frequency range of 5–500 MHz. In SMD model, the complex relative permittivity of the porous material filled with conductive fluid (brine/water) and nonconductive fluid (hydrocarbon) is expressed as

$$\varepsilon_r^* = (\phi_T S_w)^{m'} \varepsilon_w^*(S,T) + [1-(\phi_T S_w)^{m'}]\varepsilon_{rn} - \varepsilon_{rn}\Gamma\left(\phi_T S_w, m', \frac{\varepsilon_w^*(S,T)}{\varepsilon_{rn}}\right) \tag{2.10}$$

where Γ is gamma function and ε_{rn} is relative permittivity of nonconductive phase, which includes the rock matrix and the hydrocarbon. ε_{rn} is expressed as

$$(\varepsilon_{rn})^{1/2} = \frac{(1-\phi_T)(\varepsilon_{rm})^{1/2} + (1-S_w)\phi_T(\varepsilon_{rhy})^{1/2}}{1-\phi_T S_w} \tag{2.11}$$

In strict terms, SMD model was not developed for clay- and pyrite-rich formations. SMD model is popularly used in the MHz frequency range. Implementations of SMD model is discussed in Tathed et al. (2018) and Han et al. (2019).

3.2 Bimodal model

The bimodal model was developed by Mendelson and Cohen (1982) and Sen (1984) that describes the geometrical relationship of rock grains and fluids. The model represents water as a continuous phase, hosting dispersed inclusions consisting of rock grains and oil. It is called bimodal because each effective inclusion can take two shapes, either sphere or ellipsoid with a fixed aspect ratio (Gkortsas et al., 2018). In the bimodal model, the water-filled porosity can be expressed as

$$\phi_w = \left(\frac{\varepsilon_w}{\varepsilon_r^*}\right)^{p_1} \left(\frac{\varepsilon_r^* - \varepsilon_m}{\varepsilon_w - \varepsilon_m}\right)^{p_2} \left(\frac{\varepsilon_w + A\varepsilon_m}{\varepsilon_r^* + A\varepsilon_m}\right)^{p_3} \left(\frac{\varepsilon_w + B\varepsilon_m}{\varepsilon_r^* + B\varepsilon_m}\right)^{p_4} \tag{2.12}$$

where ε_r^* is the complex dielectric permittivity of the rock, p_1, p_2, p_3, p_4, A, and B are functions of p and δ, wherein p represents the effective percentage of ellipsoidal inclusions, and δ is related to the depolarization factor (polarizability of the inclusion). Water saturation can be calculated from the water-filled porosity when the total porosity is known. For $\omega \rightarrow 0$,

$$\sigma = \sigma_w (\phi S_w)^m \tag{2.13}$$

where σ is the DC medium conductivity, σ_w is the DC water medium conductivity, and m is the cementation exponent computed as

$$m = \frac{3}{2}(1-p) + \frac{p}{3\delta}\frac{2+3\delta}{2-\delta} \tag{2.14}$$

3.3 Maxwell–Garnet model

Maxwell–Garnett (MG) is valid for dielectric composites with dilute conductive phases below the percolation threshold (Maxwell, 1904). It is based on the alteration of distribution of electric field in a composite material by the introduction of ellipsoidal inclusions into a host material (background) (Seleznev et al., 2004). The effective permittivity of composite/mixture using MG model is calculated as

$$\varepsilon_{ef} = \varepsilon_b + \frac{\sum_{i=1}^{n} f_i(\varepsilon_i - \varepsilon_b)\sum_{k=1}^{3}\frac{\varepsilon_b}{\varepsilon_b + N_{ik}(\varepsilon_i - \varepsilon_b)}}{1 - \frac{1}{3}\sum_{i=1}^{n} f_i(\varepsilon_i - \varepsilon_b)\sum_{k=1}^{3}\frac{N_{ik}}{\varepsilon_b + N_{ik}(\varepsilon_i - \varepsilon_b)}} \tag{2.15}$$

where ε_b is the permittivity of base dielectric (background), ε_i is the permittivity of i-th inclusion, f_i is the volume fraction of i-th inclusion, and N_{ik} is the depolarization factor of i-th inclusion in k-th direction, which depends on the shape of the ellipsoidal inclusion, index i represents each type of inclusion, and index $k = 1$, 2, or 3 corresponds to x, y, or z coordinates, respectively.

It is assumed that the volume fraction of the inclusions is below 20%, and the inclusions are assumed to be homogenously distributed with random orientations. The MG model assumes linearity of all parameters that does not depend on the intensity of EM field. MG model assumes the mixture to be electrodynamically isotropic such that ellipsoids are randomly oriented and separated by distances greater than their characteristic sizes (Koledintseva et al., 2006). Various applications of MG model for petrophysical interpretation of electromagnetic data are presented in Jin et al. (2020a,b).

3.4 Hanai–Bruggeman formula

The Hanai–Bruggeman (HB) equation provides estimates of effective medium properties when small amount of an inclusion is added to a host. HB formula was derived using differential analysis. HB formula provides an alternative to MG formula, especially for the mixed-wet cases, where the fractions of water-

and hydrocarbon-wet grains are similar to each other. The effective permittivity of composite/mixture (ε_{ef}) using HB formula (Garcia and Heidari, 2018) is expressed as

$$\frac{\varepsilon_{nw} - \varepsilon_{ef}}{\varepsilon_{nw} - \varepsilon_{w}} = X_w \left(\frac{\varepsilon_{ef}}{\varepsilon_w} \right)^{N_g} \tag{2.16}$$

where ε_{nw} is the permittivity of hydrocarbon-wet mineral grains that are considered as the inclusions, ε_w is the permittivity of the water-wet mineral grains coated with the electric double layer that is considered as the background, X_w is the volume fraction of the water-wet grains, and N_g is the depolarization factor corresponding to the hydrocarbon-wet grains, which is 1/3 for spherical grains. The HB equation is more reliable than the MG model for the intermediate values of X_w.

3.5 Clay-particle model for shaly-sand formations

Freed et al. (2018) developed physics-based model for the dielectric dispersion in low-salinity shaly sands, in the frequency range of 20 MHz to 1 GHz, which accounts for water-filled porosity, salinity, CEC, and a water-phase tortuosity exponent. Clay-particle model considers differential effective medium approach that incorporates a first-principle solution for the polarization of a charged, platy grain in the rock matrix. The water-filled porosity is expressed as

$$\phi_w = \Pi_{i=0}^{3} \left(\frac{\varepsilon_w^* - r_i}{\varepsilon_r^* - r_i} \right)^{p_i} \tag{2.17}$$

where r_i are the poles, and p_i are the residues of the function $1/(3\varepsilon F(\varepsilon))$. $F(\varepsilon)$ is the sum of the polarization coefficients of the spheres and spheroids expressed as

$$F(\varepsilon) = (1 - p)P_s(\varepsilon) + pf_aP_p(\varepsilon) + p(1 - f_a)P_n(\varepsilon) \tag{2.18}$$

where $P_s(\varepsilon)$ is the polarization coefficient of a spherical grain expressed as

$$P_s(\varepsilon) = \left(\frac{\varepsilon_m - \varepsilon}{\varepsilon_m + 2\varepsilon} \right) \tag{2.19}$$

and $P_p(\varepsilon)$ and $P_n(\varepsilon)$ are the polarization coefficients of ellipsoids along the axis of symmetry and perpendicular to the axis of symmetry, respectively, which are expressed as

$$P_{p,n} = \frac{1}{3} \left(\frac{\varepsilon_{p,n}^* - \varepsilon_w^*}{3L_{p,n}\varepsilon_{p,n}^* + (1 - L_{p,n})\varepsilon_w^*} \right) \tag{2.20}$$

where L_p and L_n are the depolarization factors along the axis of symmetry and perpendicular to the axis of symmetry, respectively, ε_w^* is the complex permittivity of brine, and ε_p^* and ε_n^* are the effective complex permittivities of clay particle parallel and perpendicular to the short axis.

For an anisotropic formation, a parameter related to the alignment of the grains may need to be adjusted. The overall scale of the CEC may also depend on the type of clay and cations, for example, in cleaner or higher salinity formations, clay-particle model reduces to bimodal for CEC value equal to zero.

3.6 Wideband electromagnetic dispersion model

Seleznev et al. (2018) developed a wideband rock model that accounts for two key polarization mechanism present in the spectral induced polarization frequency range and dielectric dispersion frequency range. This model is based on a differential effective-medium approach that accounts for both the Maxwell–Wagner IP related to the rock texture and the electric double-layer polarization because of the presence of charged grains. The wideband model applies to well-sorted clean sandstone formation and enables the estimation of water-filled porosity, water salinity, CEC, dominant grain size, and cementation exponent. Formation is modeled as a collection of charged spherical grains and noncharged spheroidal grains dispersed in conductive brine as shown in Fig. 2.1.

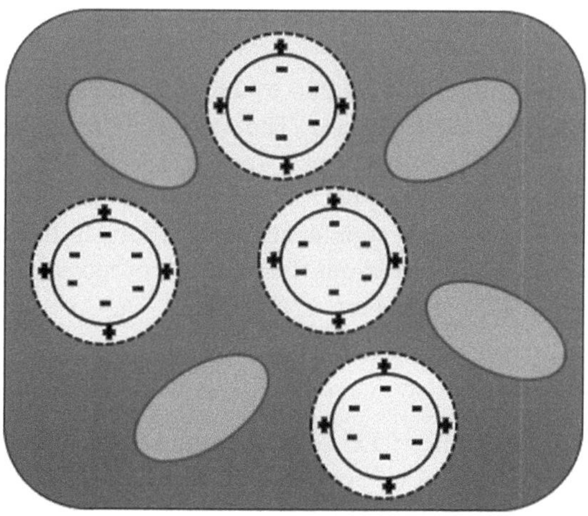

FIGURE 2.1 Representation of wideband model with charged spherical grains and noncharged spheroidal inclusions. *Modified from Seleznev, N., Hou, C.-Y., Freed, D., Habashy, T., Feng, L., Fellah, K., Nadeev, A., 2018. Coherent Interpretation of Wideband Electromagnetic Measurements in the Millihertz to Gigahertz Frequency Range. Society of Petrophysicists and Well-Log Analysts.*

The water-filled porosity ϕ_w is derived as

$$\phi_w = \Pi_{i=0}^{3}\left(\frac{\varepsilon_w^* - p_i}{\varepsilon_r^* - p_j}\right)^{r_j} \tag{2.21}$$

where r_i are the poles, and p_i are the residues of the rational function

$$f(\varepsilon) = \frac{D(\varepsilon)}{\varepsilon N(\varepsilon)} \tag{2.22}$$

$$\begin{aligned} N(\varepsilon) =& pf_a(\varepsilon_q - \varepsilon)\left[d_L\varepsilon_q + (2-d_L)\varepsilon\right]\left[\left(\varepsilon_q^* + \frac{2i\sigma_w^{ex}}{\omega\varepsilon_0}\right) + 2\varepsilon\right] + 2p(1-f_a)(\varepsilon_q - \varepsilon) \\ & \left[d_L\varepsilon + (1-d_L)\varepsilon_q\right]\left[\left(\varepsilon_q^* + \frac{2i\sigma_w^{ex}}{\omega\varepsilon_0}\right) + 2\varepsilon\right] + 3(1-p)\left[\left(\varepsilon_q^* - \frac{i\sigma_w^{ex}}{\omega\varepsilon_0}\right) - \varepsilon\right] \\ & + \left[\varepsilon_q(1-d_L) + d_L\varepsilon\right]\left[d_L\varepsilon_q + (2-d_L)\varepsilon\right] \end{aligned} \tag{2.23}$$

$$D(\varepsilon) = \left[\left[(1-d_L)\varepsilon_q + d_L\varepsilon\right]\left[d_L\varepsilon_q + (2-d_L)\varepsilon\right]\right]\left[\left(\varepsilon_q^* + \frac{2i\sigma_w^{ex}}{\omega\varepsilon_0}\right) + 2\varepsilon\right] \tag{2.24}$$

where f_a is an alignment factor defined along the z-axis, p is the volume fraction of spheroids, ε_q^* is the complex permittivity of charged grains, ε_q is the real part of dielectric constant of grain matrix, and L is the depolarization factor related to the aspect ratio of spheroid. Cementation exponent is calculated as

$$m = \frac{3}{2}(1-p) + p\frac{2f_a + d_L(2-3f_a)}{d_L(2-d_L)} \tag{2.25}$$

The wideband rock model is limited to well-sorted clean rocks, and hence requires rock with dominant grain size. The model captures physics governing the low-frequency and high-frequency response and hence the predicted response in mid-frequency may not be consistent with true response from formation. Effective medium approaches to model similar phenomena in geological materials and formations have been proposed by Misra et al. (2016), Han et al. (2017) and Misra et al. (2021).

3.7 Shaly-sand model

Shaly-sand model determines CEC-related parameters by including a dispersive phase in the SMD model to incorporate clay water (Han et al., 2012). The dispersive phase has both a Debye term to describe the extra dispersion from the clay and a conductivity term, similar to the extra conductivity in the dual-water model contributed by the clay as described above. Fig. 2.2 represents the

Model	SMD Model	New Approach
Conductive Phase	Conductive Water	Conductive Water, High Dispersive Phase
Isolating Phase	Dry Matrix	Dry Matrix
Rock geometry	By SMD Γ function and μ	By SMD Γ function and μ^*

FIGURE 2.2 Comparison of Stroud-Milton-De model and Shaly-Sand model. *Adapted from Han, M., Cuadros, J., Suarez, C. A. P., Decoster, E., Faivre, O., Mosse, L., Seleznev, N., 2012. Continous Estimate of Cation Exchange Capacity from Log Data: A New Approach Based on Dielectric Dispersion Analysis. Society of Petrophysicists and Well-Log Analysts.*

shaly-sand model approach as compared to SMD model. In this model, the fluid bearing rock permittivity is expressed as

$$\varepsilon^* = (\phi_w + p_d)^{\mu^*} \varepsilon_C^* + [1 - (\phi_w + p_d)^{\mu^*}]\varepsilon_m - \varepsilon_m \Gamma\left((\phi_w + p_d), \mu^*, \frac{\varepsilon_C^*}{\varepsilon_m}\right) \tag{2.26}$$

The petrophysical parameters obtained by the inversion of this model include not only ϕ_w, the water salinity, and textural parameter μ^* but also the newly introduced dispersive phase fraction p_d. It is expected that the inverted dispersive phase fraction p_d will increase with the CEC. The higher the CEC of a rock, the higher the clay effects are, and this is characterized by the dispersive phase.

3.8 Textural model

Textural model is based on a MG approximation for calculating electrical properties of the compound media (Seleznev et al., 2006). This model is applied to ellipsoidal grains and pores dispersed in the background media, described by a CRI mixing rule.

$$\varepsilon_{eff} = \varepsilon_{CRI} + \frac{\frac{1}{3}\sum_{i=1}^{n} f_i(\varepsilon_i - \varepsilon_{CRI})\sum_{j=1}^{3}\frac{\varepsilon_{CRI}}{\varepsilon_{CRI} + N_j^i(\varepsilon_i - \varepsilon_{CRI})}}{1 - \frac{1}{3}\sum_{i=1}^{n} f_i(\varepsilon_i - \varepsilon_{CRI})\sum_{j=1}^{3}\frac{N_i^i}{\varepsilon_{CRI} + N_j^i(\varepsilon_i - \varepsilon_{CRI})}} \tag{2.27}$$

where ε_i is the complex permittivity of the i-th ellipsoidal inclusion phase with the depolarization factors N_j^i ($i = 1, 2, 3$) and volumetric fraction f_i. N_j^i is the depolarization factor of i-th inclusion in j-th direction, which depends on the shape of the ellipsoidal inclusion, index i represents each type of inclusion, and index $j = 1, 2,$ or 3 corresponds to x, y, or z coordinates, respectively. Inclusion fractions are assigned as follows: $f_m = (1-\phi)$, $f_w = \phi S_w$, $f_{oil} =$

$\phi(1 - S_w)$, subscript m and w represent the matrix and water, respectively. This model calculates the rock permittivity and conductivity for a mixture of oblate spheroidal pores, grains, and oil droplets dispersed in a background and provides the water saturation, water resistivity, and Archie's cementation exponent.

4. Use of petrophysical models for interpreting electromagnetic measurements on core samples

4.1 Case study 1

Salah Al-Ofi et al. (2017) investigated the influence of wettability alteration on dielectric dispersion data by laboratory experiments using bimodal, SMD, and textural models in frequency range of 10 MHz to 1 GHz (Al-Ofi et al., 2017). A reflection open-ended coaxial probe to measure dielectric dispersion of a 1.5-in. core plug from a dolomitic outcrop sample (D4), which was selected because of its moderate range of porosity and permeability. The sample was initially fully saturated with a 50-kppm NaCl brine and placed in a core-flooding system. The outcrop sample was coreflooded with solutions of different concentrations of surface-active stearic acid in low-polarity oil. NMR measurements, contact angle measurements, and production data indicated that wettability transitioned from water wet (WW) to a more oil wet (OW) state. The original wettability state of a rock was altered by using different concentrations of stearic acid. This aging technique enabled tracking the influence of different wettability states on the geometrical parameters interpreted from dielectric dispersion models, particularly MN and the depolarization factor. In this case study, SMD gives better fitting than the bimodal model, although the outputs of both models are close for S_w and the apparent cementation factor or water-phase tortuosity parameter (MN). Fig. 2.3 depicts SMD model fitting with experimental measurements at 100% water saturated state. During flooding of the sample with different acid concentrations, not only is the fluid volume altered but also the geometry of the fluids inside the rock changes. For bimodal inversion, both the inverted MN and platy grain fraction p increase with the decrease of water saturation.

Tortuosity MN-parameter from all three interpretation models was observed to increase with increasing wettability toward the OW state. Interpretation of wettability from MN alone tends to be uncertain as increase of MN can be attributed to different water saturation levels or to the different Archie's cementation exponent m for different samples. With the information about the fluid saturation and Archie cementation exponent m, Archie saturation exponent n was calculated from MN. Saturation exponent n and bimodal platy grains concentration p increase with the change of wettability toward the OW state from a WW state. The inverted water saturation is not significantly impacted whereas the MN parameter increases when the sample changes from one state to another (from Soltrol drainage to 1 wt% acid drainage) as shown in Table 2.1.

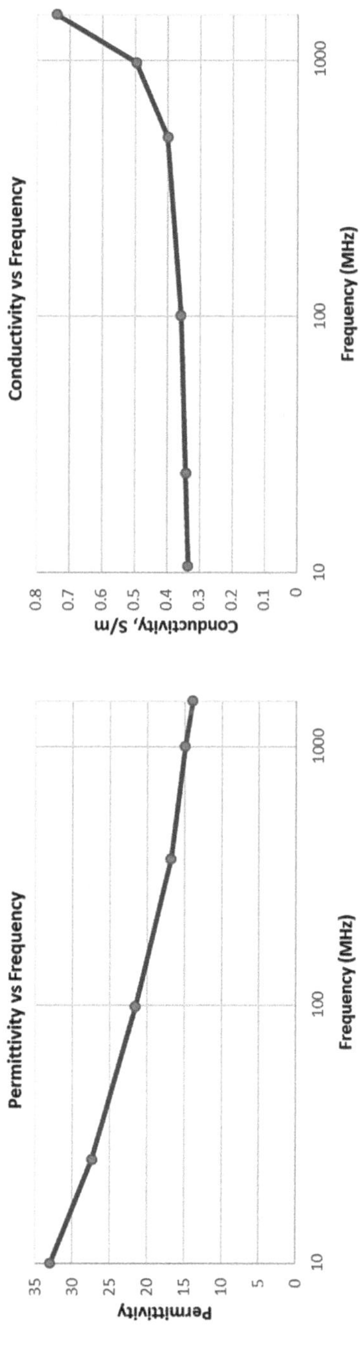

FIGURE 2.3 Stroud-Milton-De Model fitting at 100% water saturated state. *Adapted from Al-Ofi, S., Sauerer, B., Dyshlyuk, E., Valori, A., Abdallah, W., 2017. Influence of Rock Wettability Alteration on Dielectric Dispersion Data. Society of Petroleum Engineers. https://doi.org/10.2118/188028-MS.*

TABLE 2.1 Petrophysical parameters estimated by using Bimodal, Stroud-Milton-De, and Textural models to fit the dielectric dispersion measurements on a 1.5-in. core plug from a dolomitic outcrop sample (D4).

Model	Bimodal				Stroud-Milton-De				Textural model				
State	S_w	MN	n	p^*	S_w	MN	n	μ	S_w	MN	n	$DFWA$	$DFMO$
Fully brine saturated	100	1.85	1.85	0.14	100	2.06	2.06	0.08	100	1.66	1.66	1.43	0.29
Soltrol drainage	28	2.1	2.33	0.25	28	2.22	2.37	0.21	25	1.96	2.24	1.38	0.31
1% acid drainage	11	3.04	4.55	0.64	13	3.34	4.9	0	18	2.93	4.23	0.36	0.82

The connectivity parameter (μ), inverted from the SMD model decreased with water saturation decrease and with the rock wettability changing to a more OW state. The geometrical parameters can be used to distinguish wettability variation if the fluid saturation is constant. For the textural model inversion, water saturation is slightly underestimated ($\sim$3%) in comparison to the results of the other two models as shown in Table 2.1. The water saturation value after 1 wt% acid drainage is significantly overestimated because of poor fitting of the modeled data to experimental data. The *MN*-parameter is also smaller than the Bimodal and SMD inversion outputs by roughly 0.2 units, although the same increasing trend with saturation and wettability change is observed. The two additional textural parameters, water phase depolarization factor (DFWA) and matrix-oil depolarization factor (DFMO), change with water desaturation and wettability, shifting to the OW state. The DFWA factor is a unique parameter among all the used interpretation models because it is the only parameter directly related to water phase average shape in the pore space. The DFWA was observed to take values of 1.4 for a WW sample, and drops down to a value of 0.36 for more OW sample, which is very close to the depolarization factor of a sphere (0.33). The opposite trend is observed for the DFMO as it increases from 0.3 to 0.82 as they are apparently becoming more oblate in shape. DFWA also shows the strongest dependence on wettability among all the inverted parameters. Water inclusions tend to form disconnected and, possibly, spherical droplets inside the pore system in the OW formation.

Based on this case study, the water depolarization factor (DFWA) and the oil-matrix depolarization factor (DFMO), inverted from the textural model, seem to be the most promising parameters for wettability characterization. Both factors demonstrate very strong variation with wettability alteration and seem to be practically unaffected by saturation variation. The DFMO parameter should correlate with the platy grain fraction p from the bimodal model, according to the underlying physics principles. Like the platy grain fraction, it also changes with wettability, but in contrast to the platy grain fraction, it seems not to be affected by saturation variation.

4.2 Case study 2

Gkortsas et al. (2018) performed an experimental study to benchmark water saturation from dielectric measurements on eight carbonate cores and estimated their cementation exponent m and saturation exponent n in Archie equation from dielectric data [24]. The eight cores consisted of seven limestones and a dolomite (Table 2.2). All cores went through a series of drainage/imbibition steps, creating varying saturations of brine/fluorocarbon. The dielectric and NMR measurements were performed on axial cores, using network analyzer. The dielectric measurements were performed in a frequency range of 1 MHz–4 GHz, and complex refractive index model (CRIM), bimodal model, and SMD model were used to get inverted parameters.

TABLE 2.2 Petrophysical properties of the eight cores used in the experiments as presented by Gkortsas et al. (2018).

Core number	Description	$\phi(p.u.)$	m	$k(mD)$
1	Limestone 1	26.82	2.09	20
2	Dolomite	12.13	2.31	25
3	Limestone 2	26.89	1.61	21.5
4	Limestone 3	21.67	1.83	3.5
5	Limestone 4	26.92	2.64	74.6
6	Limestone 5	27.06	2.65	75
7	Limestone 6	26.79	1.86	13.3
8	Limestone 7	24.89	1.96	6.7

Fig. 2.4 shows the permittivity predictions of the bimodal (the red [black in print version] trace) and SMD (the green [light gray in print version] trace) models against the permittivity measurements (the blue [dark gray in print version] trace) for 2 saturation stages (100% water saturation and forced imbibition 20 psi) of cores 1 and 4. Unlike Figs. 2.4 and 2.5 compares the predictions and measurements of conductivity. Except for permittivity at low frequencies and conductivity at high frequencies, both models fit the experimental data well. For the bimodal model, key parameters are water saturation S_w, apparent cementation factor MN, and percent of ellipsoidal inclusions p. For the SMD model, the key parameters are water saturation S_w, apparent cementation factor MN, and connectivity factor α. The value of matrix permittivity used in these comparisons is ε_m is 8.5. Both models predict similar values of apparent cementation factor mn and the water saturation S_w.

This case study compares water-filled porosities from NMR and CRIM for each saturation stage for each of the eight cores. First, the average difference of water-filled porosity obtained from NMR and that from CRIM-based dielectric interpretation is approximately 1.5 p.u. Most of the CRIM-based water-filled porosities of each core fall within the 2 p.u. variation from the NMR water-filled porosity. The average difference in water-filled porosities obtained from NMR and that from SMD model based dielectric interpretation for all saturation stages of each core ranges from 1 to 3.5 p.u.

CRIM works well for the estimation of water-filled porosity because it works well at 1 GHz. At frequency 1 GHz, the IP can be ignored, and permittivity depends mainly on S_w. For frequencies of less than 1 GHz, CRIM does not work well. This is expected because, at frequencies lower than 1 GHz, the IP is significant and cannot be explained by the volumetric fractions of the different constituents of the fluid-filled porous material. Although

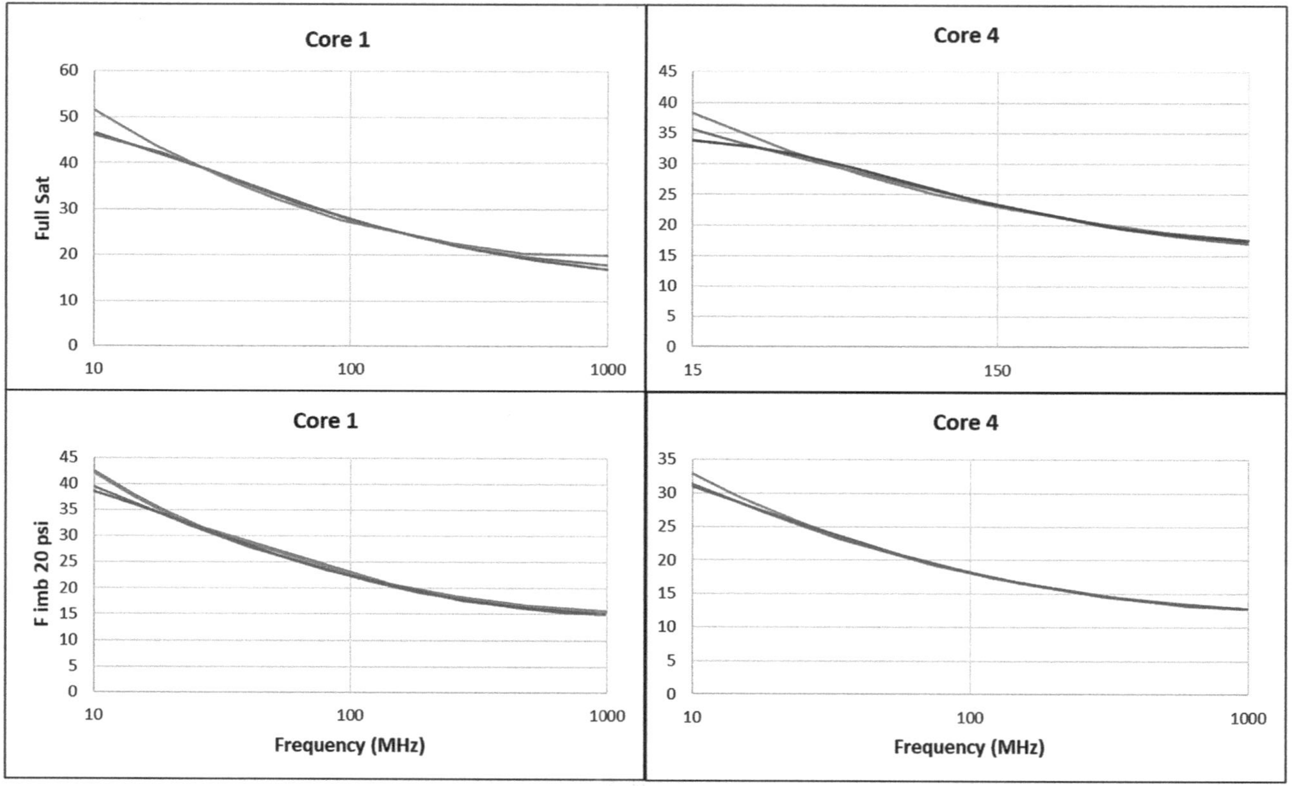

FIGURE 2.4 Comparison of permittivity predictions of the bimodal (the red [black in print version] trace) and Stroud-Milton-De (SMD) (the green [light gray in print version] trace) models against the permittivity measurements (the blue [dark gray in print version] trace) for 2 saturation stages (100% water saturation and forced imbibition 20 psi) of cores 1 and 4. *Modified from Gkortsas, V.-M., Vekataramanan, L., Fellah, K., Ramsdell, D., Hou, C.-Y., Seleznev, N., 2018. Comparison of different dielectric models to calculate water saturation and estimate textural parameters in partially saturated cores. Geophysics 83, E303–E318.*

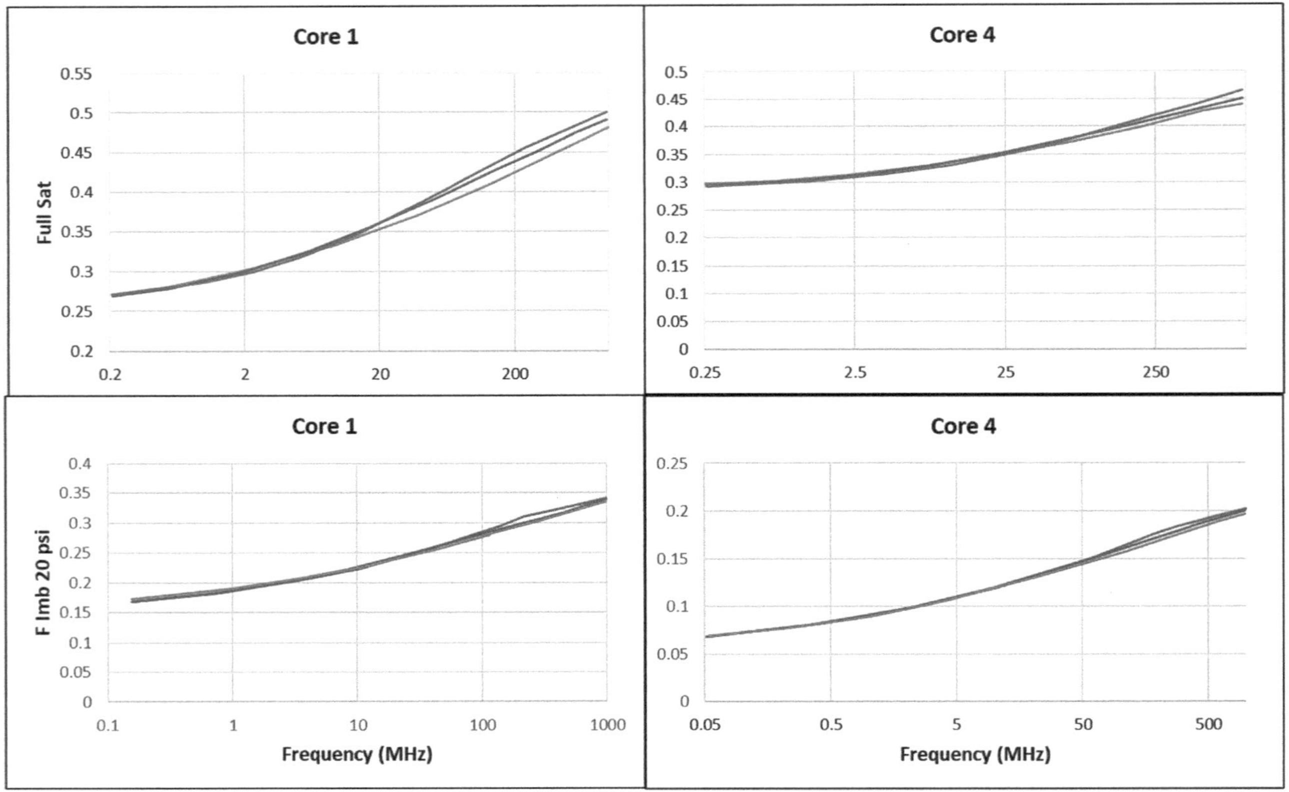

FIGURE 2.5 Comparison of conductivity predictions of the bimodal (the red [black in print version] trace) and Stroud-Milton-De (SMD) (the green [light gray in print version] trace) models against the conductivity measurements (the blue [dark gray in print version] trace) for 2 saturation stages (100% water saturation and forced imbibition 20 psi) of cores 1 and 4. *Modified from Gkortsas, V.-M., Vekataramanan, L., Fellah, K., Ramsdell, D., Hou, C.-Y., Seleznev, N., 2018. Comparison of different dielectric models to calculate water saturation and estimate textural parameters in partially saturated cores. Geophysics 83, E303–E318.*

bimodal and SMD models are needed to capture the textural effects of the rock, it implies the rock as having an apparent cementation factor of $MN = 2$ from CRIM model. The salinity measured from the experiments was 32 ppk. With the inversion of the dielectric data, salinity for CRIM, bimodal, and SMD were 28 ± 6, 34 ± 9, and 39 ± 8 ppk, respectively. The average value of inverted salinity is very close to the true value of 32 ppk with a small variance.

Further using the MN and S_w estimated using the bimodal and SMD models, the m and n exponents of each of the cores were estimated. The values of m and n of each core are different for each model because the values of MN and S_w are slightly different for bimodal and SMD models. The difference in the values of m and n between the two models is within 10%, except core 2. The estimated values of m and n for all the cores are shown in Table 2.3.

TABLE 2.3 Estimates of cementation exponent *m* and saturation exponent *n* using (A) bimodal model and (B) Stroud-Milton-De model.

A. Bimodal model			
Core number	$\phi(p.u.)$	m	n
1	26.82	2.23	1.95
2	12.13	2.54	2.12
3	26.89	2.03	2.37
4	21.67	2.03	2.47
5	26.92	2.67	1.16
6	27.06	2.83	1.29
7	26.79	2.06	2.37
8	24.89	2.05	2.16
B. Stroud-Milton-De (SMD) model			
1	26.82	2.04	1.9
2	12.13	2.51	3.3
3	26.89	1.87	2.35
4	21.67	1.9	2.36
5	26.92	2.66	1.1
6	27.06	2.86	1.23
7	26.79	1.91	2.23
8	24.89	1.94	2

Symbols and abbreviations

B Equivalent conductance of sodium clay exchange cations (mS/meq)
CEC Cation exchange capacity
CRI Complex Refractive Index
C_w Brine Conductivity(S/m)
EM Electromagnetic
GHz GigaHertz
HB Hanai–Bruggeman
IP Interfacial polarization
kHz KiloHertz
LR Lichteneker–Rother model
MG Maxwell–Garnett
MHz MegaHertz
MW Maxwell–Wagner
OW Oil Wet
RLA5 Resistivity Laterolog measured using galvanic resistivity tool at 1 kHz
S salinity (in ppt)
SIP Spectral induced polarization
SMD Stroud-Milton-De
T Temperature (in $^{\circ}F$)
WS Waxman–Smits
WW Water Wet
R_t True resistivity of the formation measured using galvanic resistivity tool at 1 kHz
R_w Formation Water Resistivity
ϕ_T Total porosity of the formation
a Tortuosity factor
S_w Water Saturation
m Cementation exponent
n Saturation exponent
V_{sh} Volumetric fraction of shale
R_{sh} Resistivity of shale
Q_v Cation exchange capacity per unit pore volume (meq/cc)
V_Q Amount of clay water associated with 1 unit (meq) of clay counterions
F_o Formation factor,
R_{cw} is resistivity of the bound water
ε^* Bulk complex permittivity of the formation
ε_{rm} Matrix relative permittivity
ε_{rhy} Relative permittivity of hydrocarbon,
ε_m Matrix permittivity
ω Angular frequency ($\omega = 2\pi f, f$ is frequency)
ε_{rw} Relative permittivity of water
ε_w^* Complex permittivity of water
ε_o Hydrocarbon permittivity
ε_b Real value of water permittivity assumed as 80
ε_o Vacuum permittivity with universal value of 8.85×10^{-12}
m' Rock textural parameter
Γ Gamma function
ε_{rn} Relative permittivity of nonconductive phase
p Effective percentage of ellipsoidal inclusions

δ Depolarization factor
σ DC medium conductivity
σ_w DC water medium conductivity
ε_i Relative permittivity for i-th inclusions,
ε_b Permittivity of background,
f Volume fraction of inclusions
N Depolarization factor that depends on the shape of the inclusions
ε_g Relative permittivity of mineral constituent of grains
ε_{solid} Permittivity of solid grains
ε_{SC} relative dielectric permittivity of the grains coated with the electric double layer
X_w water-wet fraction of the grains
N_g Depolarization factor corresponding to the hydrocarbon-wet grains.
p_i Residues of the function $\frac{1}{(3\varepsilon F(\varepsilon))}$
$F(\varepsilon)$ Sum of the polarization coefficients of the spheres and spheroids
$P_s(\varepsilon)$ Polarization coefficient of a spherical rock
$P_p(\varepsilon)P_n(\varepsilon)$ Polarization coefficients of spheroids
L_p Depolarization factors along the axis of symmetry
L_n Depolarization factors perpendicular to the axis of symmetry respectively
ε_w^* Complex permittivity of brine
ε_p^* Effective complex permittivities of clay particle parallel to the short axis
ε_n^* Effective complex permittivities of clay particle perpendicular to the short axis
MN Water-phase tortuosity parameter
α Geometrical fitting parameter assumed as 0.5

References

Al-Ofi, S., Sauerer, B., Dyshlyuk, E., Valori, A., Abdallah, W., June 1, 2017. Influence of Rock Wettability Alteration on Dielectric Dispersion Data. Society of Petroleum Engineers. https://doi.org/10.2118/188028-MS.

Archie, G.E., 1942. The electrical resistivity log as an aid in determining some reservoir characteristics. Trans., AIME 146, 54–67.

Brovelli, A., Cassiani, G., 2008. Effective permittivity of porous media: a critical analysis of the complex refractive index model. Geophys. Prospect. 56 (5), 715–727.

Clavier, C., Coates, G., Dumanoir, J., 1977. The theoretical and experimental basis for the Dual-Water model for the interpretation of shaly sands. In: SPE-6859: Society of Petroleum Engineers, 52nd Annual Technical Meeting and Exhibition, preprint, p. 16.

Freed, D.E., Seleznev, N., Hou, C.-Y., Fellah, K., Little, J., Dumy, G., Sen, P., June 1, 2018. A Physics-Based Model for the Dielectric Response of Shaly Sands and Continuous CEC Logging. Society of Petrophysicists and Well-Log Analysts.

Garcia, A.P., Heidari, Z., 2018. Integrated characterization of multi-frequency dielectric dispersion measurements in mixed-wet rocks. In: Presented at the SPWLA 59th Annual Logging Symposium, London, UK, 2-6 June. SPWLA-2018-BBB.

Gkortsas, V.-M., Vekataramanan, L., Fellah, K., Ramsdell, D., Hou, C.-Y., Seleznev, N., 2018. Comparison of different dielectric models to calculate water saturation and estimate textural parameters in partially saturated cores. Geophysics 83, E303–E318.

Han, M., Cuadros, J., Suarez, C.A.P., Decoster, E., Faivre, O., Mosse, L., Seleznev, N., June 16, 2012. Continous Estimate of Cation Exchange Capacity from Log Data: A New Approach Based on Dielectric Dispersion Analysis. Society of Petrophysicists and Well-Log Analysts.

Han, Y., Misra, S., Simpson, G., 2017. Dielectric dispersion log interpretation in Bakken petroleum system. In: *SPWLA 58th Annual Logging Symposium*. Society of Petrophysicists and Well-Log Analysts.

Han, Y., Misra, S., Wang, H., Toumelin, E., 2019. Hydrocarbon saturation in a Lower-Paleozoic organic-rich shale gas formation based on Markov-chain Monte Carlo stochastic inversion of broadband electromagnetic dispersion logs. Fuel 243, 645–658.

Jin, Y., Misra, S., Homan, D., 2020b. Estimation of in-situ wettability using multi-frequency electromagnetic measurements. In: Unconventional Resources Technology Conference, 20–22 July 2020. Unconventional Resources Technology Conference (URTeC), pp. 4181–4199.

Jin, Y., Misra, S., Homan, D., Rasmus, J., 2020. Effects of wettability of conductive particles on the multifrequency conductivity and permittivity of fluid-filled porous material. Fuel 268, 117411.

Koledintseva, M.Y., DuBroff, R.E., Schwartz, R.W., 2006. A Maxwell Garnett model for dielectric mixtures containing conducting particles at optical frequencies. Prog. Electromag. Res. PIER 63, 223–242.

Lichtenecker, K., Rother, K., 1931. Die Herleitung des logarithmischen Mischungsgesetzes aus allgemeinen Prinzipien der Stationiiren Stromung. Phys. Z. 32, 255–260.

Maxwell, G., 1904. Colors in metal glasses and in metal films. Trans. Royal Soc. CCIII, 385–420.

Mendelson, K., Cohen, M., 1982. The effects of grain anisotropy on the electrical properties of sedimentary rocks. Geophysics 47, 257–263. https://doi.org/10.1190/1.1441332.

Misra, S., Homan, D., Jin, Y., Rasmus, J., 2021. Systems and Methods for Determining Properties of Porous, Fluid-Filled Geological Formations Based on Multi-Frequency Measurements. U.S. Patent Application 16/940,492, filed February 4, 2021.

Misra, S., Torres-Verdín, C., Revil, A., Rasmus, J., Homan, D., 2016. Interfacial polarization of disseminated conductive minerals in absence of redox-active species—Part 1: Mechanistic model and validation. Geophysics 81 (2), E139–E157.

Peeters, M., Holmes, A., 2014. Review of existing shaly-sand models and introduction of a new method based on dry-clay parameters. Petrophysics 55 (6), 543–553.

Sabouroux, P., Ba, D., 2011. Epsimu, a tool for dielectric properties measurement of porous media: application in wet granular materials characterization. Prog. Electromag. Res. 29, 191–207.

Seleznev, N., Boyd, A., Habashy, T., Luthi, S.M., January 1, 2004. Dielectric Mixing Laws for Fully and Partially Saturated Carbonate Rocks. Society of Petrophysicists and Well-Log Analysts.

Seleznev, N.V., Habashy, T.M., Boyd, A.J., Hizem, M., January 1, 2006. Formation Properties Derived from a Multi-Frequency Dielectric Measurement. Society of Petrophysicists and Well-Log Analysts.

Seleznev, N., Hou, C.-Y., Freed, D., Habashy, T., Feng, L., Fellah, K., Nadeev, A., June 1, 2018. Coherent Interpretation of Wideband Electromagnetic Measurements in the Millihertz to Gigahertz Frequency Range. Society of Petrophysicists and Well-Log Analysts.

Sen, P., 1984. Grain shape effects on dielectric and electrical properties of rocks. Geophysics 49, 586–587. https://doi.org/10.1190/1.1441695.

Simandoux, P., 1963. Dielectric measurements in porous media and application to shaly formation. Rev. Inst. Fr. Petrol 18 (Supplementary Issue), 193–215.

Stroud, D., Milton, G.W., De, B.R., 1986. Analytical model for the di- electric response of brine-saturated rocks. Phys. Rev. B 34, 5145–5153.

Tathed, P., Han, Y., Misra, S., 2018. Hydrocarbon saturation in Bakken Petroleum System based on joint inversion of resistivity and dielectric dispersion logs. Fuel 233, 45–55.

Waxman, M.H., Smits, L.J.M., 1968. Electrical conductivities in oil-bearing shaly sands. Soc. Petrol. Eng. J. 8 (2), 107–122.

Section 2

Mechanistic models for the interpretation of multifrequency electromagnetic data

Chapter 3

Multifrequency conductivity and permittivity of porous material containing non-conductive particles possessing surface conductance

Siddharth Misra[1,2], Yuteng Jin[3]

[1]*Harold Vance Department of Petroleum Engineering, College of Engineering, Texas A&M University, TX, United States;* [2]*Department of Geology and Geophysics, Texas A&M University, TX, United States;* [3]*Harold Vance Department of Petroleum Engineering, Texas A&M University, TX, United States*

1. Interfacial polarization phenomenon because of nonconductive particles possessing surface charge

The negatively charged surface of a nonconductive mineral, such as clay and sand, attracts positively charged ions of the surrounding electrolyte that get attached to the surface because of ionic adsorption, protonation/deprotonation of the hydroxyl groups, and dissociation of other potentially active surface groups. These loosely attached positively charged ions on the surface of nonconductive mineral give rise to interfacial polarization (IFP) in the presence of an externally applied electromagnetic (EM) field. Fig. 3.1 shows the surface-conductance-assisted interfacial polarization (SCAIP) phenomenon in a representative volume of a dilute mixture of uniformly distributed nonconductive spherical inclusions in an electrolyte (host medium). IFP of this representative volume is independent of the direction of the externally applied electric field because of the spherical symmetry. Under the presence of an externally applied electrical field, the positively charged counterion layer will move along the direction of the electrical field. A net positive charge and associated charge separation then accumulates at the solid-liquid interface, which leads to interfacial polarization, referred herein as the SCAIP phenomenon.

Multifrequency Electromagnetic Data Interpretation for Subsurface Characterization
https://doi.org/10.1016/B978-0-12-821439-8.00011-2

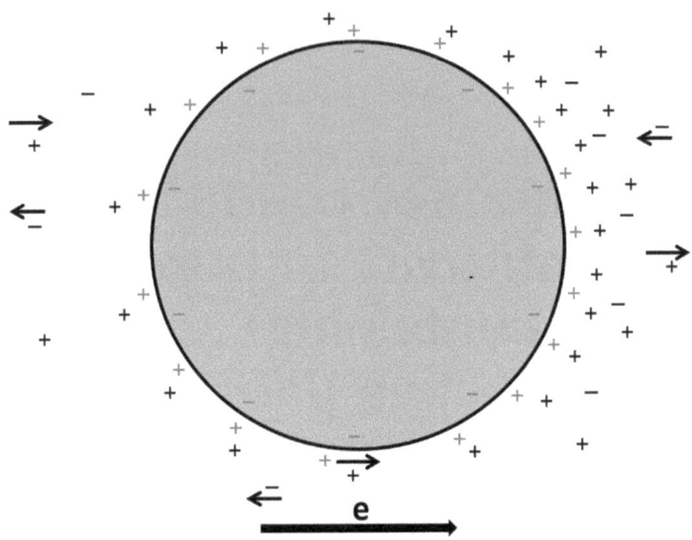

FIGURE 3.1 Cross-section of a nonconductive spherical inclusion possessing surface charge surrounded by an ionic host medium (i.e., electrolyte). The inclusion is negatively charged, surrounded by a positively charged counterion layer, which forms a Gouy-Chapman model. Charge carriers in the ionic host medium are cations, identified by "+" symbol, and anions, identified by "−" symbol. The direction of the externally applied electrical field, **e**, is identified with a bold arrow next to the symbol "**e**". The direction of movement of the charge carriers in the ionic host medium is represented by the arrow next to the symbol of the charge carrier. *Yuteng Jin, Siddharth Misra.*

2. Mechanistic model of interfacial polarization of nonconductive particle possessing surface charge

Different mixing models have been developed to quantify the effects of various IFP phenomena. The model proposed by Schwarz (1962) considers IFP effect around charged nonconductive particles. It assumes a diffusion of counterion layer moving along the surface of the charged particle by calculating the potential outside the counterion layer as a solution of Laplace's equation rather than Poisson's equation. However, this model fails to account for all the bulk diffusion effects. In contrast, Dukhin et al. (1974) concluded that the mechanism behind IFP is the diffusion of ions in the bulk electrolyte around the particle. They were unable to provide analytical expressions for IFP effects in terms of various relaxation parameters because of mathematical complexity caused by nonlinearity of Dukhin et al. (1974) equation. This model, called the standard model in colloidal chemistry, does not consider the existence of a Stern layer with mobile ions. Grosse and Foster (1987) developed an analytical solution of IFP effect by developing a simplified model of charged nonconductive spherical particles in bulk electrolyte. In their model, positive ions from the bulk electrolyte can freely exchange with the positively charged counterion layer while the negative ions are excluded from the counterion layer. This model was generalized in Grosse (1988) by allowing arbitrary charge in nonsymmetrical electrolytes, assuming finite surface conductivity and considering the entire frequency spectrum.

3. Model of interfacial polarization because of nonconductive particle possessing surface charge

3.1 Model description

In this chapter, SCAIP model (Misra et al., 2016; Jin et al., 2019) is used to investigate the IFP phenomena around surface-charge-bearing spherical nonconductive particles. The phenomenological basis of interfacial polarization considered in our work builds on the mechanistic descriptions outlined by Grosse (1988). The negatively charged inclusion, together with its positive counterion layer, essentially behaves as a conductor of positive charge carriers, which allows the positive ions in the host medium to freely exchange with the positive ions in the counterion layer, and as a nonconductor of negative charges. In the presence of an externally applied electric field, this differential treatment of positively and negatively charge carriers leads to charge separation and charge accumulation giving rise to the IFP effects that are dependent on the operation frequency. In the absence of an externally applied electric field, a Gouy-Chapman double layer is assumed around the surface-charge-bearing nonconductive inclusions, where the positive counterion layer is characterized by a finite surface conductivity. We assume the thickness of counterion layer is negligible, which is valid when $a \gg \lambda_D$, where λ_D is the Debye screening length and a is the characteristic length of the inclusion phase.

3.2 Assumptions

To simplify the model, we assume only spherical particles are present in the porous media. The solid matrix comprises spherical nonconductive particles (like sand grain) possessing small surface charge and spherical nonconductive inclusion particle (like clay grain) possessing large surface charge. The pore space is made of spherical oil droplets surrounded by electrolyte. Electrolyte also surrounds the solid particles. All solids and fluids are assumed to have homogeneous, isotropic, and nondispersive electrical properties. Therefore, the frequency dispersion and dielectric enhancement predicted by the SCAIP model solely stems from the SCAIP phenomenon around the negatively charged nonconductive spherical inclusions possessing large surface charge. All the charge carriers are assumed to have unitary charges. Electrolyte is assumed to contain binary, symmetric charge carriers.

3.3 Mechanistic model development

3.3.1 Poisson-Nernst-Planck equation

Both the SCAIP model and PPIP model (perfectly polarized IFP model, to be discussed in the following chapter) are based on the Poisson-Nernst-Planck

(PNP) equations for a dilute solution in a weak electrical field regime. By applying the PNP equations, we analyze the EM response of a representative volume comprising a single, isolated nonconductive inclusion possessing surface charge (as for SCAIP model) or electrically conductive inclusion (as for PPIP model) surrounded by an electrolyte-saturated host medium (Zheng and Wei, 2011). The PNP equation has been used to model electromigration and diffusion of ionic charge carriers in electrolytes (Zheng and Wei, 2011) and that because of holes and electrons in semiconductors (Schmuck and Bazant, 2015). It is based on a mean-field approximation of charge carrier interactions and continuum descriptions of charge concentration and electrostatic potential. PNP equations are used to model the charge dynamics and relaxation in the representative volume containing only two phases: the host medium, denoted by subscript h, and the nonconductive or conductive (to be discussed in the following chapter) particles/inclusions, denoted by subscript i. In our formulation, the host medium can be assumed as a homogeneous mixture of electrolyte and nonconductive matrix or as a pure electrolyte.

At time $t < \mathbf{0}$, it is assumed that there is no external electric field exciting the representative volume and the system is at an equilibrium. At this equilibrium condition, the initial charge carrier densities in both the host and inclusion phases are denoted as $N_{0,j}^{\pm}$, where subscript j takes the form of i for the inclusion phase and h for the host phase. After time $t = 0$, the representative volume experiences a uniform externally applied electric field $E = E_0 e^{\mathrm{i}\omega t}$, where E_0 is the amplitude of the externally applied electric field, i is square root of -1, ω is the angular frequency (rad/s) of the externally applied electric field, and e is Euler's number. Note $\omega = 2\pi f$, where f is frequency (Hz). We assume the negatively charged spherical nonconductive particles and inclusions to be surrounded by a layer of positively charged, conducting counterion layer, which has a surface conductance λ and bears a field-induced surface charge density $\rho e^{\mathrm{i}\omega t}\cos\theta$, where θ is the angle between the normal to the interface and the incident external electric field. Under a weak field approximation, charge carrier densities in host and inclusion phases are perturbed from their equilibrium conditions near the host-inclusion interfaces, resulting in a new linearly approximated charge distribution, expressed as

$$N_j^{\pm}(r,t,\theta) = N_{0,j}^{\pm} + c_j^{\pm}(r)e^{\mathrm{i}\omega t}cos\theta \tag{3.1}$$

such that $\left|c_j^{\pm}\right| \leq N_{0,j}^{\pm}$, $c_j^{\pm}$ is the charge density variation near the host-inclusion interface in medium j because of the externally applied electric field and r is the radial distance along the normal to the interface. Note that in this section, inside the nonconductive inclusion, $c_i^{\pm}(r) = 0$. In addition, there is an absence of charge carriers in the nonconductive inclusion phase, $N_{0,i}^{\pm} = 0$.

Furthermore, the symbol "+" identifies positive-charge carriers such as holes and cations, while the symbol "−" identifies negative-charge carriers such as electrons and anions.

We assume that the characteristic length a of the inclusions phase is far greater than the Debye screening length λ_D. Note that λ_D is a measure of induced charge distribution that forms around an inclusion particle because of surface charges that exist on the inclusion particle in the absence of an externally applied electric field. In other words, λ_D represents a volume outside of which surface charges on an inclusion particle are electrically screened. The characteristic length a is equal to the radius of spherical inclusion. Mathematically, $\lambda_D = \sqrt{\varepsilon_h k_B T / \left(2Z_h^+ Z_h^- q^2 N_{0,h}\right)}$, where ε_h is dielectric permittivity of the host electrolyte, k_B is Boltzmann's constant, T is absolute temperature, $Z_h^\pm$ is charge number of positive and negative charge carriers in the host electrolyte, and q is the elementary charge. The volume fraction of nonconductive particles (for e.g., clays) is assumed to be in the range of 5%−15%. Another simplifying assumption is that all the charge carriers bear unitary charge and that both host and inclusion phases bear binary, symmetric charge carriers. In other words,

$$Z_j^\pm = 1, \mu_h^+ = \mu_h^- = \mu_h, \mu_i^+ = \mu_i^- = \mu_i, N_{0,i}^+ = N_{0,i}^- = N_{0,i}, N_{0,h}^+ = N_{0,h}^- = N_{0,h} \tag{3.2}$$

where $\mu_j^\pm$ is the electrical mobility of positive and negative charge carriers in medium j, and $Z_j^\pm$ is charge number of positive and negative charge carriers in medium j.

The current density of each charge carrier type in the host and inclusion phases is the sum of current density because of drift current and diffusion current. In the absence of generation/recombination reactions, the transport equation representing conservation laws for charge-carrying species can be written as

$$\boldsymbol{j}_j^\pm = \boldsymbol{j}_{j,\text{drift}}^\pm + \boldsymbol{j}_{j,\text{diffusion}}^\pm = qN_j^\pm \mu_j \boldsymbol{e}_j \mp qD_j^\pm \nabla N_j^\pm \tag{3.3}$$

where $\boldsymbol{j}_j^\pm$ is the current density of positive and negative charge carriers, respectively, in medium j, $\boldsymbol{e}_j$ is the net electric field vector in medium j, and $D_j^\pm$ is diffusion coefficient of positive and negative charge carriers, respectively, in medium j. When using the simplifying assumption for electrical mobility of charge carriers, as mentioned in Eq. (3.2), and Einstein's relationship of diffusion coefficient with electrical mobility, namely $D_j = \left(\mu_j \mathrm{k_B} T\right)/q$, we obtain

$$D_h^+ = D_h^- = D_h; D_i^+ = D_i^- = D_i \tag{3.4}$$

By substituting $\boldsymbol{e}_j = -\nabla\phi_j$ into the low-frequency limit of Maxwell's equations (induction neglected) and substituting Eq. (3.4) into Eq. (3.3), we express the charge species conservation condition as

$$\boldsymbol{j}_j^{\pm} = -qN_j^{\pm}\mu_j\nabla\phi_j \mp qD_j\nabla N_j^{\pm} \tag{3.5}$$

where ϕ_j is the electrical potential in medium j. Eq. (3.5) is Nernst–Planck's equation that describes the relationship of the flux of charge-carrying species to its concentration gradient and that to the applied electrical potential gradient in a given medium. Nernst–Planck's equation can alternatively be expressed as

$$\boldsymbol{j}_j^{\pm} = -D_j^{\pm}N_j^{\pm}\nabla\phi_{cj}^{\pm} \tag{3.6}$$

where $\phi_{cj}^{\pm} = \mathrm{k_B}T\ln N_j^{\pm} \pm qZ_j^{\pm}\phi_j$ is the electrochemical potential of charge carriers. The continuity equation for charge carrier density based on mass conservation for each charge carrier type in an incompressible medium without any convective flow can be written as

$$\mp q\frac{\partial N_j^{\pm}}{\partial t} = \nabla\cdot\boldsymbol{j}_j^{\pm} \tag{3.7}$$

By applying Eq. (3.5) to Eq. (3.7), we obtain

$$\frac{\partial N_j^{+}}{\partial t} = \nabla\cdot\left(D_j\nabla N_j^{+} + \mu_j N_j^{+}\nabla\phi_j\right) \tag{3.8}$$

and

$$\frac{\partial N_j^{-}}{\partial t} = \nabla\cdot\left(D_j\nabla N_j^{-} - \mu_j N_j^{-}\nabla\phi_j\right) \tag{3.9}$$

The time derivative of Eq. (3.1) assuming axial symmetry is

$$\frac{\partial N_j^{\pm}}{\partial t} = \mathrm{i}\omega c_j^{\pm} \tag{3.10}$$

where $c_j^{\pm} = c_j^{\pm}(r,t,\theta) = c_j^{\pm}(r)e^{i\omega t}\cos\theta$. We apply Eq. (3.10) to Eqs. (3.8) and (3.9), then we add and subtract Eqs. (3.8) and (3.9) to obtain Eqs. (3.11) and (3.12) expressed as:

$$-\mathrm{i}q\omega d_j = -2qN_{0,j}\mu_j\Delta\phi_j - qD_j\Delta d_j \tag{3.11}$$

and

$$-\mathrm{i}q\omega s_j = -qD_j\Delta s_j \tag{3.12}$$

where $d_j = c_j^{+} - c_j^{-}$ represents net charge density variation, $s_j = c_j^{+} + c_j^{-}$ represents total ion density variation, and Δ (∇^2) is Laplace's operator. Note d_j and s_j are finite everywhere in the representative volume. Inside the nonconductive particles $d_i = s_i = 0$. We obtained Eqs. (3.11) and (3.12) by assuming $d_j\mu_j$ and $s_j\mu_j$ are negligible compared to the other terms as $\left|c_j^{\pm}\right| \leq N_{0,j}^{\pm}$.

Under the influence of an externally applied EM field, the distribution of charge carriers in both media leads to a time-varying electric potential that is expressed as $\phi_j(r,t,\theta) = \phi_j(r)e^{i\omega t}\cos\theta$. Using Gauss's law and Eq. (3.1), we obtain

$$\nabla\cdot\left(\varepsilon_j \boldsymbol{e}_j\right) = P_{f,j} = q\left(N_j^+ - N_j^-\right) = q\left(c_j^+ - c_j^-\right) = qd_j \tag{3.13}$$

where $P_{f,j}$ is the net free charge density in medium j because of charge redistribution in the presence of an externally applied EM field, $\boldsymbol{e}_j$, and $\varepsilon_j = \varepsilon_{r,j}\varepsilon_0$ is the dielectric permittivity of medium j, $\varepsilon_{r,j}$ is the relative permittivity of medium j, and $\varepsilon_0 = 8.854\times10^{-12}\mathrm{F/m}$ is the vacuum permittivity. Eq. (3.13) relates the spatial distribution of electric charge to the time-varying electric field. Assuming both media are linear, isotropic, and homogeneous, and that the electric field can be defined by a scalar electrical potential field, ϕ_j, we obtain

$$\nabla\cdot\left(\varepsilon_j \boldsymbol{e}_j\right) = -\nabla\cdot\left(\varepsilon_j\nabla\phi_j\right) = -\varepsilon_j\Delta\phi_j \tag{3.14}$$

By combining Eqs. (3.13) and (3.14), we obtain an alternate expression of Poisson's equation, expressed as

$$\Delta\phi_j = -\frac{qd_j}{\varepsilon_j} \tag{3.15}$$

Poisson's equation is applied to describe the electric field in terms of the electrical potential, the gradient of which governs electromigration in both media. By substituting Eq. (3.15) into Eq. (3.11), we obtain the PNP equation, given by

$$-iq\omega d_j = 2q^2N_{0,j}\mu_j d_j/\varepsilon_j - qD_j\Delta d_j \tag{3.16}$$

which can be re-written as

$$\Delta d_j = \left(\frac{i\omega}{D_j} + \frac{\sigma_j}{\varepsilon_j D_j}\right)d_j \tag{3.17}$$

where $\sigma_j = 2N_{0,j}\mu_j q$ is the electrical conductivity of medium j. We rewrite Eqs. (3.17) and (3.12) as

$$\Delta d_j = \gamma_j^2 d_j \tag{3.18}$$

where

$$\gamma_j^2 = \left(\frac{i\omega}{D_j} + \frac{\sigma_j}{\varepsilon_j D_j}\right) \tag{3.19}$$

and

$$\Delta s_j = \xi_j^2 s_j \tag{3.20}$$

where

$$\xi_j^2 = \frac{i\omega}{D_j} \tag{3.21}$$

respectively. Eqs. (3.18) and (3.20) are Helmholtz partial differential equation (PDE) which can be solved to obtain distinct analytical expressions of d_j and s_j for the host and inclusion phases, respectively. Eq. (3.18) is inserted into Eq. (3.15) to obtain the following Laplace PDE that can be solved for the electric potential field in the representative volume:

$$\Delta\vartheta_j = 0 \tag{3.22}$$

where

$$\vartheta_j = \phi_j + (qd_j)/\left(\gamma_j^2\varepsilon_j\right) \tag{3.23}$$

3.3.2 Solution of Helmholtz partial differential equations

As mentioned before, inside the nonconductive inclusions, $d_i = s_i = 0$. So, we are solving the Helmholtz PDEs to obtain the distinct analytical expressions of d_h and s_h for the host phase. A sphere of radius equal to a exhibits dipolarizability (dipole moment) in the radial direction. Such an inclusion identifies a grain or vug. To compute the dipolarizability of the representative volume comprising a spherical inclusion in an electrolytic host, Eq. (3.18) can be expressed in spherical coordinates, assuming azimuthal symmetry, axial symmetry, and a separable solution (Young, 2009) for $d_h(r,\theta) = R_h(r)T_h(\theta)$, as

$$\frac{1}{R_h}\frac{\partial}{\partial r}\left(r^2\frac{\partial R_h}{\partial r}\right) - \gamma_h^2 r^2 + \frac{1}{T_h \sin\theta}\frac{\partial}{\partial\theta}\left(\frac{\sin\theta\,\partial T_h}{\partial\theta}\right) = 0 \tag{3.24}$$

and

$$\frac{1}{\sin\theta}\frac{\partial}{\partial\theta}\left(\frac{\sin\theta\,\partial T_h}{\partial\theta}\right) = -n(n+1)T_h \tag{3.25}$$

where n is an integer referring to the order of the standing wave solution. A standing wave solution (Young, 2009) to the above differential equation is

$$T_h = \sum_{n=1}^{\infty}\left[A_{n,h}P_n^0(\cos\theta) + B_{n,h}Q_n^0(\cos\theta)\right] \tag{3.26}$$

where P_n^0 and Q_n^0 are associated Legendre functions of the first and second kind (Weisstein, 2018a), respectively, of n-th order and $A_{n,h}$ and $B_{n,h}$ are unknown complex-valued coefficients of the general solution of the partial differential Eq. (3.25). Substituting Eq. (3.25) in Eq. (3.24), we obtain

$$\frac{\partial}{\partial r}\left(r^2\frac{\partial R_h}{\partial r}\right)-\left[\gamma_h^2 r^2+n(n+1)\right]R_h=0 \tag{3.27}$$

A standing wave solution to the above differential equation is

$$R_h=\sum_{n=1}^{\infty}\left[C_{n,h}i_n(r\gamma_h)+D_{n,h}k_n(r\gamma_h)\right] \tag{3.28}$$

where n is an integer for the standing wave solution (Young, 2009), i_n and k_n are the modified spherical Bessel function of the first and second kind (Weisstein, 2018b), respectively, of n-th order. $C_{n,h}$ and $D_{n,h}$ are unknown complex-valued coefficients of the general solution of the partial differential Eq. (3.27). i_n and k_n can be expressed in terms of modified Bessel function of the first and second kind, respectively, as $i_n(r\gamma_h)=\sqrt{\frac{\pi}{2r\gamma_h}}I_{n+\frac{1}{2}}(r\gamma_h)$ and $k_n(r\gamma_h)=\sqrt{\frac{2}{\pi r\gamma_h}}K_{n+\frac{1}{2}}(r\gamma_h)$, where $I_{n+\frac{1}{2}}$ and $K_{n+\frac{1}{2}}$ are the modified Bessel function of the first and second kind, respectively, of $\left(n+\frac{1}{2}\right)$ -th order. In the analytical expression of our model, the series is reduced to a single term. $n=1$ and $B_{n,h}=0$ by considering the following symmetries of the charge density: (1) axial symmetry, (2) antisymmetry with respect to θ, and (3) uniformity of the external applied field. This simplification is aligned with boundary conditions that cannot be satisfied by other values of n. This reduces Eqs. (3.28) and (3.26) to

$$R_h=C_h i_1(r\gamma_h)+D_h k_1(r\gamma_h) \tag{3.29}$$

and

$$T_h=A_h\cos\theta \tag{3.30}$$

respectively, where C_h, D_h, and A_h and are unknown complex-valued coefficients of the particular solution obtained from Eqs. (3.26) and (3.28). The general representation of $d_h(r,\theta)$ can now be written, by combining Eqs. (3.29) and (3.30), as

$$d_h(r,\theta)=A_h\left[C_h i_1(r\gamma_h)+D_h k_1(r\gamma_h)\right]\cos\theta \tag{3.31}$$

Using the condition that $d_h(r,\theta)$ should be finite at $r \rightarrow \infty$, we obtain a particular solution of d_h for the host phase that can be represented as

$$d_h(r,\theta) = B_{h1}k_1(r\gamma_h)cos\,\theta \tag{3.32a}$$

or

$$d_h(r,\theta) = B_{h1}\left[e^{-r\gamma_h}\left(\frac{1}{r\gamma_h}+\frac{1}{(r\gamma_h)^2}\right)\right]cos\,\theta \tag{3.32b}$$

where B_{h1} is unknown complex-valued coefficient of the particular solution in the host medium obtained from Eq. (3.31). Note when $r \rightarrow \infty$, $d_h(r,\theta) = 0$. Repeat the above procedure, we can obtain a particular solution of s_h for the host phase from Eq. (3.20) that can be represented as

$$s_h(r,\theta) = B_{h2}k_1(r\xi_h)cos\,\theta \tag{3.33a}$$

or

$$s_h(r,\theta) = B_{h2}\left[e^{-r\xi_h}\left(\frac{1}{r\xi_h}+\frac{1}{(r\xi_h)^2}\right)\right]cos\,\theta \tag{3.33b}$$

where B_{h2} is unknown complex-valued coefficient of the particular solution in the host medium.

3.3.3 Solution of Laplace partial differential equation

The Laplacian PDE must be solved to obtain the electric potential field in the representative volume. Assuming azimuthal symmetry and a separable solution for $\vartheta_j(r,\theta,\phi) = R_{\varepsilon j}(r)T_{\varepsilon j}(\theta)$, Eq. (3.22) can be expressed in spherical coordinates as

$$\Delta\vartheta_j = \frac{1}{R_{\varepsilon j}}\frac{\partial}{\partial r}\left(r^2\frac{\partial R_{\varepsilon j}}{\partial r}\right) + \frac{1}{T_{\varepsilon j}\,sin\,\theta}\frac{\partial}{\partial\theta}\left(\frac{sin\,\theta\,\partial T_{\varepsilon j}}{\partial\theta}\right) = 0 \tag{3.34}$$

Assuming axial symmetry, a general solution (Hogg, 2001) to the above PDE can be expressed as

$$\vartheta_j(r,\theta) = \sum_{n=0}^{\infty}\left[A_{n,j}r^n + C_{n,j}r^{-(n+1)}\right]\left[E_{n,j}P_n^0(cos(n\theta)) + F_{n,j}Q_n^0(sin(n\theta))\right] \tag{3.35}$$

where n is an integer and $A_{n,j}$, $C_{n,j}$, $E_{n,j}$, and $F_{n,j}$ are unknown complex-valued coefficients of the general solution of the PDE expressed in Eq. (3.34). For analytical modeling purposes for our model, $n = 1$ and $F_{n,j} = 0$, $A_{0,j} = 0$ and $C_{0,j} = 0$, which ensures remaining terms satisfy the polar angle dependence of the model. Simplified representation of Eq. (3.35) is expressed as

$$\vartheta_j(r,\theta) = \left(A_{1,j}r + C_{1,j}r^{-2}\right)E_{1,j}\,cos\,\theta \tag{3.36a}$$

which can be rewritten using Eq. (3.23) as

$$\phi_j(r,\theta) = \left(A_j r + C_j r^{-2}\right)\cos\theta - \frac{q d_j(r,\theta)}{\gamma_j^2 \varepsilon_j} \quad (3.36b)$$

Using the condition that $d_i = 0$ and ϕ_i should be finite when $r \to 0$, we can obtain $C_i = 0$. So, a standing wave representation of Eq. (3.36b) for the nonconductive inclusion phase is

$$\phi_i(r,\theta) = A_i r \cos\theta \quad (3.37)$$

where A_i is unknown complex-valued coefficient of the particular solution in the nonconductive inclusion phase obtained from Eq. (3.36b). Using the condition when $r \to \infty$, $d_h = 0$, we can obtain $A_h = -E_0$. A standing wave representation of Eq. (3.36b) for the host phase, using Eq. (3.32b), is

$$\phi_h(r,\theta) = \left(-E_0 r + C_h r^{-2}\right)\cos\theta - \frac{q B_{h1}}{\gamma_h^2 \varepsilon_h}\left[e^{-r\gamma_h}\left(\frac{1}{r\gamma_h} + \frac{1}{(r\gamma_h)^2}\right)\right]\cos\theta \quad (3.38)$$

where C_h is unknown complex-valued coefficient of the particular solution in the host obtained from Eq. (3.36b), and E_0 is the amplitude of the externally applied electric field.

3.3.4 Boundary conditions

To obtain an expression for the dipolarizability (dipole moment), following boundary conditions are used (Grosse, 1988):

(a) Continuity of the electric potential at the interface.

$$\phi_i(r=a) = \phi_h(r=a) \quad (3.39a)$$

(b) Discontinuity of the normal component of the displacement current at the interface because of the surface charge distribution on the inclusion phase. This boundary condition is derived from Gauss' Law.

$$\varepsilon_i \left.\frac{\partial \phi_i}{\partial r}\right|_{r=a} - \varepsilon_h \left.\frac{\partial \phi_h}{\partial r}\right|_{r=a} = \rho\cos\theta \quad (3.39b)$$

(c) Continuity of the surface charge density at the host-inclusion interface qualitatively expressed as: Rate of change of surface charge density = normal drift/conduction current at the interface because of potential gradient arising from the external EM field + normal diffusion current because of concentration gradient in the host media at the interface + tangential conduction current because of the potential gradient arising from the surface-charge-bearing inclusion phase. In other words, this boundary condition shows that the time derivative of surface charge density in the counterion layer is equal to the sum of the normal conduction and diffusion current because of potential and concentration difference, separately, from the host medium and the tangential conduction current because of potential from the inclusion phase.

$$-i\omega\rho\cos\theta = -\frac{\sigma_h}{2}\frac{\partial\phi_h}{\partial r}\bigg|_{r=a} - qD_h\frac{\partial c_h^+}{\partial r}\bigg|_{r=a} + \frac{2\lambda}{a}A_i\cos\theta \tag{3.39c}$$

(d) The normal component of the current density of negative ions in the host medium must vanish at the interface because of the assumption that the negative ions are excluded from the counterion layer.

$$j_h^- = -\frac{\sigma_h}{2}\frac{\partial\phi_h}{\partial r}\bigg|_{r=a} + qD_h\frac{\partial c_h^-}{\partial r}\bigg|_{r=a} = 0 \tag{3.39d}$$

(e) Because of the application of the external electric field, we use a simplifying assumption that the relative change of the positive ion density in the counterion layer (which is assumed to be negligibly thin) and that in the host medium must be the equal because the positive ions in the host medium can freely exchange with the ions in the counterion layer.

$$\frac{c_h^+(r=a,\theta)}{N_{0,h}} = \frac{\rho\cos\theta}{\rho_0} \tag{3.39e}$$

where ρ_0 is the initial equilibrium surface change density in the counterion layer, and ρ is the net resultant surface charge density in the counterion layer after the application of electric field.

3.3.5 Solution for the dipolarizability

Using boundary condition Eq. (3.39a), Eqs. (3.37) and (3.38) can be equated on the surface of the sphere of radius equal to a. The resulting equation can be abbreviated as

$$-E_0 a + \frac{C_h}{a^2} - E_h B_{h1} = A_i a \tag{3.40a}$$

where

$$E_h = \frac{q}{\gamma_h^2\varepsilon_h}e^{-a\gamma_h}\left[\frac{1}{a\gamma_h} + \frac{1}{(a\gamma_h)^2}\right] \tag{3.40b}$$

All other parameters/coefficients in the above equation are explained in previous sections. The equation obtained using boundary condition Eq. (3.39b) at the surface of the sphere can be abbreviated as

$$\varepsilon_h\left(E_0+\frac{2C_h}{a^3}-G_hB_{h1}\right)+\varepsilon_iA_i=\rho \tag{3.41a}$$

where

$$G_h=\frac{q}{\gamma_h\varepsilon_h}e^{-a\gamma_h}\left[\frac{1}{a\gamma_h}+\frac{2}{(a\gamma_h)^2}+\frac{2}{(a\gamma_h)^3}\right] \tag{3.41b}$$

Boundary condition Eq. (3.39c) gives us the following abbreviated equation:

$$i\omega\rho=-\frac{\sigma_h}{2}E_0-\frac{\sigma_hC_h}{a^3}+\frac{\sigma_h}{2}G_hB_{h1}-\frac{D_h}{2}\gamma_h^2G_hB_{h1}\varepsilon_h-\frac{D_h}{2}\xi_h^2L_hB_{h2}\varepsilon_h-\frac{2\lambda}{a}A_i \tag{3.42a}$$

where

$$L_h=\frac{q}{\xi_h\varepsilon_h}e^{-a\xi_h}\left[\frac{1}{a\xi_h}+\frac{2}{(a\xi_h)^2}+\frac{2}{(a\xi_h)^3}\right] \tag{3.42b}$$

All other parameters/coefficients in the above two equations are explained in previous sections. Similarly, the equation obtained using boundary condition Eq. (3.39d) can be abbreviated as

$$\frac{\sigma_h}{2}E_0+\frac{\sigma_hC_h}{a^3}-\frac{\sigma_h}{2}G_hB_{h1}=\frac{D_h}{2}\xi_h^2L_hB_{h2}\varepsilon_h-\frac{D_h}{2}\gamma_h^2G_hB_{h1}\varepsilon_h \tag{3.43}$$

For boundary condition Eq. (3.39e), we assume the electrical mobilities in the two regions are the same to rewrite this boundary condition as

$$2qc_h^+(r=a)=\frac{\rho\sigma_h}{\lambda} \tag{3.44}$$

After solving Eqs. (3.40a), (3.41a), (3.42a), (3.43) and (3.44), we obtain the dipolarizability (dipolar field coefficient) of the representative volume comprising a spherical nonconductive inclusion in an electrolytic host as

$$f_{ncond}(\omega)=\frac{C_h}{E_0a^3}=\frac{Q(R+A)-P}{Q(R-2A)+2P} \tag{3.45}$$

where

$$A=\frac{1}{a^2} \tag{3.45a}$$

$$P=\gamma_h^2+\xi_h^2\frac{G}{H}+\frac{2G}{a^2L} \tag{3.45b}$$

$$Q=\frac{1}{iF+1}\left[2-\frac{a^2\xi_h^2}{H}\left(\frac{L}{iF}+E\right)-\frac{2E}{L}\right] \tag{3.45c}$$

$$R=\frac{P}{Q}\left(\frac{iFE+L}{iF+1}\right) \tag{3.45d}$$

$$H=\frac{aL_h}{F_h}, G=\frac{aG_h}{E_h}, L=\frac{2\lambda}{a\sigma_h}, E=\frac{\varepsilon_i}{\varepsilon_h}, F=\frac{\omega\varepsilon_h}{\sigma_h} \tag{3.45e}$$

$$F_h=\frac{q}{\xi_h^2\varepsilon_h}e^{-a\xi_h}\left[\frac{1}{a\xi_h}+\frac{1}{(a\xi_h)^2}\right] \tag{3.45f}$$

3.4 Limitations

A limitation of the PNP equations arises from the omission of the finite volume effect of charge carriers, mutual interactions, and steric effects, effects because of transport of ions in confined channels of the pore system, and correlation effects (Chu and Bazant, 2006). Another limitation arises because the model is developed only for symmetric and binary charge carriers in both the host and inclusion phases. This assumption simplifies the analytical complexity of the PNP formulations. Another drawback of the PNP formulations is that the analysis is performed for materials that contain completely dissociated charge carriers at low-concentration values. Moreover, in this book, unlike Chu and Bazant (2006), we only consider the linear response to weak fields where exact solutions are possible and is closer to the field conditions.

4. Effective medium models of complex conductivity/permittivity of materials exhibiting various polarization phenomena

After obtaining the mathematical expressions of dipolarizability for spherical nonconductive and conductive (to be discussed in the following chapter) inclusion particles surrounded by the electrolyte, we apply the effective-medium theory to determine the effective complex electrical conductivity (σ_{eff}^*) of the mixture (Grosse and Barchini, 1992). For the development of our model, as stated in the previous sections, PNP equations are first used to obtain dipolarizabilities (dipole moment), which are microscopic electrical properties, for the representative volume containing the spherical particle. In the derivations of dipolarizabilities, we neglect multipoles effect because their magnitude decreases with inverse power of distance (Sihvola, 2007). Monopole effects are also neglected since there is zero net charge because of the assumed electroneutrality. The macroscopic electrical properties are then computed using effective-medium formulations based on the theory that a material composed of a mixture of distinct homogeneous media can be seen as a homogeneous material at a sufficiently large observation scale (Giordano, 2003).

Effective medium models are widely used to calculate the frequency-dependent EM properties of materials. Lakhtakia et al. (1998) described the Mossotti–Clausius formalism for estimating the effective permittivity dyadic of a dilute composite containing carbon nanotube inclusions. Lagarkov and Sarychev (1996) developed an effective-medium approximation to calculate an effective dielectric constant of the composites containing elongated conducting inclusions, which has a potential usage in many processes occurring in nature, including dielectric enhancement in grain-saturated porous rocks. The model assumes that each conducting inclussion is embedded in the effective medium. Their model qualitatively agrees with the experimental data conducted by Grimes et al. (2000). Mantese et al. (1996) applied both Maxwell–Garnett and Bruggeman type effective medium theories to the composition and frequency-dependent complex permittivities and permeabilities of ferroelectric/ferromagnetic composites. They find the composites are well described by effective medium theory, while the Bruggemann model is more accurate for higher volume fractions than the Maxwell–Garnett model because the latter assumes that the embedded phase never percolates, which is not true according to microstructural analysis. The authors also mention that both the Maxwell–Garnett and the Bruggeman type effective medium models are limited in their applicability because the microstructure for a composite medium is assumed to be given, while actually the underlying microstructures of real composites are quite complicated and must be described by correlation functions.

On the other hand, for the low-frequency conductivity response of water-saturated packs of glass beads, Leroy et al. (2008) developed a low-frequency polarization model of both the Maxwell–Wagner polarization and the electrochemical polarization of the Stern layer, where the differential effective medium theory at the scale of a representative elementary volume of the porous material is used to account for the Maxwell–Wagner polarization. Their model predictions match well with the experimental data. Pavlin et al. (2002) compared their finite-element method prediction of effective conductivity of cell suspensions with the Maxwell effective medium theory, and they concluded that Maxwell theory is valid also for higher volume fractions.

Also, Stroud (1975) proposed a generalized effective medium approach to the conductivity of an inhomogeneous materials to treat materials consisting of crystallites of arbitrary shape and conductivity tensors of arbitrary symmetry. McLachlan (1989) proposed a generalized effective medium equation of the complex permittivity of a binary mixture by combining the effective medium theory with percolation scaling law. The equation successfully fitted to experimental data on emulsions. Moreover, Sihvola and Kong (1988) generalized the mixing formulas to the arbitrary orientation distribution of the scatterers and anisotropic mixtures as well as to multiphase mixtures where the scatterers can be of different shape and type. They applied their theory to obtain the effective permittivities of show and sea ice.

5. Surface-conductance-assisted interfacial polarization model: mechanistic model of conductivity/permittivity of porous material containing nonconductive particles possessing surface charge

In this book, we obtain the effective electrical properties using a Maxwell–Garnett type effective medium formulation. To meet the requirements of the formulation, the volume fraction of nonconductive and conductive inclusions in the mixture should be less than 20% (Revil et al., 2015). Subsurface water-bearing reservoir rocks have less than 10% volume fraction of conductive mineral inclusions but the volume fraction of nonconductive particles possessing surface charge (e.g., clay and sand) can exceed 20%. Moreover, we invoke the PNP equations in the bulk electrolyte that introduces a decaying length scale, where the Maxwell Garnett formula may become invalid at even lower concentration of inclusion phase (Hou et al., 2018). It is also important to mention this formulation used in the calculation neglects the EM interaction between the inclusions and other components. Because of the assumed dilution of the uniformly distributed inclusion phase, individual elements of the dispersed phase are assumed to be isolated and not in contact with each other. Like other mixing theories, our model includes the assumption that the magnitude of spatial variations of the electric field is smaller than the magnitude of variations in the intrinsic electrical properties and geometrical structures. Moreover, all calculations are performed using a quasi-static assumption that requires the size of heterogeneities to be much smaller than the wavelength of the applied EM field (Cosenza et al., 2009). Also, when dealing with a lossy medium, the skin depth of the EM wave must be considered to avoid strong attenuation of the field amplitudes in the conductive heterogeneities. Most importantly, because of the implementation of the PNP equations, our effective medium formulations unlike other theories (e.g., Giordano, 2003) explicitly accounts for the characteristic lengths of heterogeneities, resulting in a physically consistent way to account for the perturbation because of nonconductive or conductive inclusions.

For a mixture containing only spherical nonconductive particles (such as sand grain, clay grain, and oil droplet), the effective medium formulation of the complex conductivity response can be expressed as

$$\frac{\sigma^*_{\widehat{n},eff} - \sigma^*_h}{\sigma^*_{\widehat{n},eff} + 2\sigma^*_h} = \sum \phi_{ncond} f_{ncond}(\omega) \tag{3.46}$$

where $\sigma^* = \sigma + i\omega\varepsilon$ is a representation of the complex conductivity of a material, $\sigma^*_{\widehat{n},eff}$ is the effective complex conductivity of the geological mixture directed along the $\widehat{n}$ unit vector that can be measured with an externally applied electric field directed along the $\widehat{n}$ unit vector, and σ^*_h is the complex-valued conductivity of the homogenous isotropic host material that surrounds the particles (in our case, the host is electrolyte). ϕ_{ncond} is the volume fraction of the nonconductive phase in the mixture, and $f_{ncond}(\omega)$ is the dipolarizability of nonconductive phase along the direction of $\widehat{n}$ unit vector, along which the externally applied electric field is directed. Section 3 of this chapter describes the derivations and formulation of the dipolarizability of nonconductive spherical particles possessing surface charge. In the above-mentioned formulation, the nonconductive phases include solid particles, such as clay grains, possessing large surface charge as well as the solid particles, such as sand grains, possessing small surface charge and oil droplets without surface charge.

6. Validation of the surface-conductance-assisted interfacial polarization model

In this section, the SCAIP model predictions are compared with the published experimental data. The properties of mixtures used for generating plots are summarized in the corresponding tables, where ϕ is the volume fraction of a specific phase, a is the characteristic length (radius) of spherical particles, D is diffusion coefficient of charge carriers, ε_r is relative permittivity, σ is conductivity, and λ is surface conductance. Property of host medium (electrolyte) and spherical nonconductive inclusion phase is represented with subscripts of h and i, respectively.

Schwan et al. (1962) conducted laboratory investigations of dielectric enhancement and dielectric dispersion of colloidal suspensions of polystyrene spheres of uniform size in an ionic electrolyte. Their experiments considered the frequency range from 10 Hz to several MHz. For modeling purposes, they used a frequency-dependent surface admittance circuit model to explain their laboratory measurements. In that paper, the authors mentioned the need to develop a mechanistic model to predict experimental data. Fig. 3.2 compares the SCAIP model predictions and experimental data from Schwan et al. (1962). SCAIP model predictions are in good agreement with experimental results, for input values that are similar to published ones. Also, we obtain good agreement for the computed $\varepsilon_{r,eff}$ of the suspension with another set of experimental results mentioned in Schwan et al. (1962), as shown in Fig. 3.3. Figs. 3.2 and 3.3 imply that the presence of dispersed polystyrene particles produces drastic dielectric enhancement and dispersion because of IFP phenomena because neither the host nor the inclusion individually possesses dielectric characteristics comparable to that measured in the experiments.

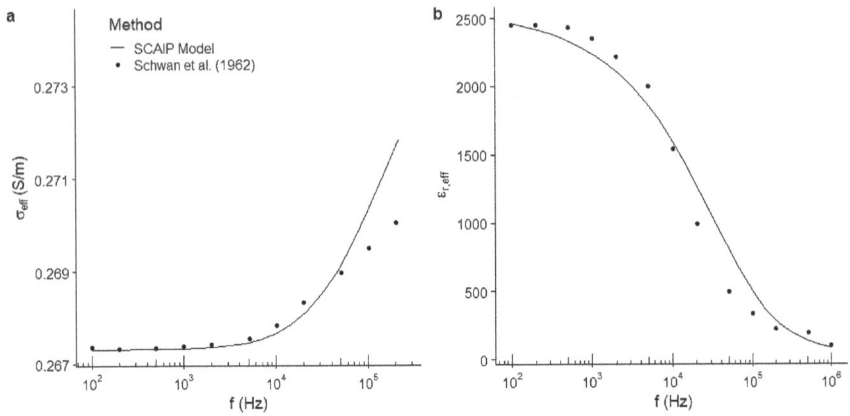

FIGURE 3.2 Comparison of the surface-conductance-assisted interfacial polarization model predictions of (A) σ_{eff} and (B) $\varepsilon_{r,eff}$ against those measured by Schwan et al. (1962). The properties used for generating this plot is shown in Table 3.1. *Yuteng Jin, Siddharth Misra.*

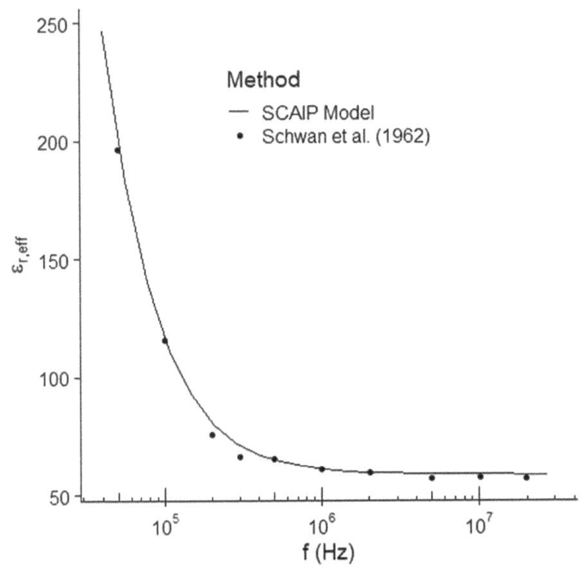

FIGURE 3.3 Comparison of the surface-conductance-assisted interfacial polarization model predictions of $\varepsilon_{r,eff}$ against that measured by Schwan et al. (1962). The properties used for generating this plot are shown in Table 3.2. *Yuteng Jin, Siddharth Misra.*

On the other hand, Fig. 3.2 indicates relatively smaller conductivity dispersion in the order of 1% relative difference between the high- and low-frequency values of conductivity, which is attributed to the absence of high-mobility charge carriers in the inclusion phase.

TABLE 3.1 The known, assumed, and estimated properties used for generating Fig. 3.2.

Property		Known	Assumed	Estimated
Inclusion (polystyrene particles)	ϕ_i (%)	30		
	a_i (μm)	0.094		
	$\varepsilon_{r,i}$		2.5	
	λ_i (S)			6.7×10^{-9}
Host (KCl solution)	D_h (m^2/s)		1.6×10^{-9}	
	$\varepsilon_{r,h}$		78	
	σ_h (S/m)			0.348

TABLE 3.2 The known, assumed, and estimated properties used for generating Fig. 3.3.

Property		Known	Assumed	Estimated
Inclusion (polystyrene particles)	ϕ_i (%)	19.5		
	a_i (μm)	0.2785		
	$\varepsilon_{r,i}$		2.5	
	λ_i (S)			6×10^{-9}
Host (KCl solution)	D_h (m^2/s)			1×10^{-9}
	$\varepsilon_{r,h}$		78	
	σ_h (S/m)	0.125		

TABLE 3.3 The known, assumed, and estimated properties used for generating Fig. 3.4.

Property		Known	Assumed	Estimated
Inclusion (polystyrene particles)	ϕ_i (%)	1		
	a_i (μm)	0.5		
	$\varepsilon_{r,i}$		2	
	λ_i (S)			9×10^{-9}
Host (KCl solution)	D_h (m^2/s)	2×10^{-9}		
	$\varepsilon_{r,h}$		80	
	σ_h (S/m)	0.00554		

TABLE 3.4 The known, assumed, and estimated properties used for generating Fig. 3.5.

Property		Known	Assumed	Estimated
Inclusion (polymer latex balls)	ϕ_i (%)	15.6 or 12.7		
	a_i (nm)	55		
	$\varepsilon_{r,i}$		5	
	λ_i (S)			2.8×10^{-10}
Host (KCl solution)	D_h (m^2/s)			2.3×10^{-10}
	$\varepsilon_{r,h}$		78	
	σ_h (S/m)		0.0147	

TABLE 3.5 Properties used for generating Fig. 3.6, where S1 and S2 represent sand, and C1~C5 represent five different clay types, and the host is an electrolyte of $\varepsilon_{r,h}$ of 70, σ_h of 0.1 S/m, and D_h of 10^{-9} m^2/s.

	ϕ_i (%)	a_i (μm)	$\varepsilon_{r,i}$	λ_i (S)	a_i/λ_i
Sand1 (S1)	70	1000	4	10^{-9}	10^{12}
Sand2 (S2)	80	1000	4	10^{-9}	10^{12}
Clay1 (C1)	10	100	8	10^{-8}	10^{10}
Clay2 (C2)	10	10	8	10^{-8}	10^{9}
Clay3 (C3)	10	100	8	10^{-7}	10^{9}
Clay4 (C4)	10	1	8	10^{-8}	10^{8}
Clay5 (C5)	10	100	8	10^{-6}	10^{8}

Tirado and Grosse (2006) performed broadband dielectric measurements on suspensions of spherical polystyrene particles having high-surface charge distribution in an aqueous solution. IFP phenomenon in this experiment was dominated by surface conductance effects because of high-surface charge of polystyrene particles, similar to the clay grains possessing high-surface charge and high-surface conductivity. Fig. 3.4 shows a good agreement between Tirado and Grosse's (2006) measurements and the SCAIP model predictions.

TABLE 3.6 Properties used for generating Fig. 3.7, where S1 represent sand, and C1 ~ C4 represent four different clay types fully saturated with electrolyte of different conductivities, and of $\varepsilon_{r,h}$ of 70 and D_h of 10^{-9} m^2/s.

	ϕ_i (%)	a_i (μm)	$\varepsilon_{r,i}$	λ_i (S)	σ_h (S/m)
Sand1 (S1)	70	1000	4	10^{-9}	–
Clay1 (C1)	10	10	8	10^{-8}	0.1
Clay2 (C2)	10	10	8	10^{-6}	0.1
Clay3 (C3)	10	10	8	10^{-6}	0.01
Clay4 (C4)	10	10	8	10^{-6}	1

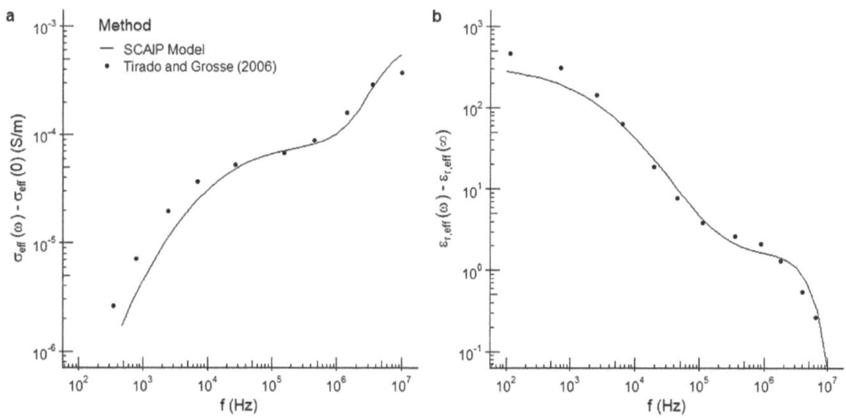

FIGURE 3.4 Comparison of the surface-conductance-assisted interfacial polarization model predictions of (A) change of effective conductivity $(\sigma_{eff}(\omega) - \sigma_{eff}(0))$ and (B) change of effective relative permittivity $(\varepsilon_{r,eff}(\omega) - \varepsilon_{r,eff}(\infty))$ against those measured by Tirado and Grosse (2006). The properties used for generating this plot is shown in Table 3.3. *Yuteng Jin, Siddharth Misra.*

Furthermore, we modeled the experiments performed by Delgado et al. (1998), who worked on identifying laboratory techniques to differentiate surface diffusion mechanism from the volume diffusion mechanism. They carried out dielectric dispersion measurements on suspensions of polymer latex balls in a KCl solution. The SCAIP modeling results matched the experimental results for two different volume fractions of polymer latex balls in the KCl solution, as shown in Fig. 3.5.

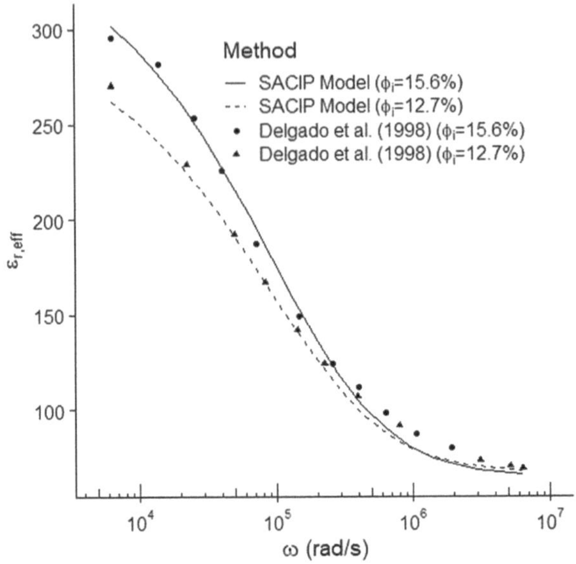

FIGURE 3.5 Comparison of the surface-conductance-assisted interfacial polarization model predictions of $\varepsilon_{r,eff}$ against those measured by Delgado et al. (1998). The properties used for generating this plot are shown in Table 3.4. *Yuteng Jin, Siddharth Misra.*

7. Surface-conductance-assisted interfacial polarization model predictions of multifrequency conductivity and permittivity in 100 Hz to 10 MHz

In this section, the effects of SCAIP phenomenon on low-frequency (around 100 Hz) σ_{eff} predictions and high-frequency (around 10 MHz) $\varepsilon_{r,eff}$ predictions are evaluated. The properties of mixtures used for generating plots are summarized in the corresponding tables, where ϕ_i is the volume fraction of a specific phase, a_i is the characteristic length (radius) of spherical particles, D is diffusion coefficient of charge carriers, ε_r is relative permittivity, σ is conductivity, and λ is surface conductance. Properties of host medium and inclusion phase are represented with a subscript of h and i, respectively.

7.1 The influence of the properties of the nonconductive particles

In Fig. 3.6, curves S1 and S2 act as the references. S1 has porosity of 30%, and S2 has porosity of 20%. Comparing curves S1 against S2, we conclude that the increase in volume fraction of nonconductive particles, like sand grains, results in decrease in porosity leading to decrease of the σ_{eff} and $\varepsilon_{r,eff}$. Comparing curves S1C4 and S1C5 against S1C1, the decrease of the characteristic length or the increase of the surface conductance of clay particles by two orders of magnitude leads to increase of the σ_{eff} and $\varepsilon_{r,eff}$, as shown in Fig. 3.6A and B,

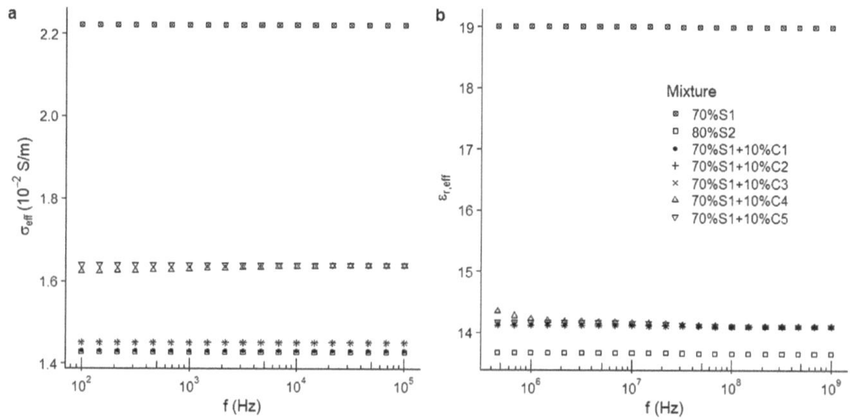

FIGURE 3.6 Comparison of the surface-conductance-assisted interfacial polarization model predictions of (A) σ_{eff} and (B) $\varepsilon_{r,eff}$ of mixtures containing nonconductive spherical sand particles and various types of spherical clay particles (properties listed in Table 3.5) fully saturated with electrolyte of $\varepsilon_{r,h}$ of 70, σ_h of 0.1 S/m, and D_h of 10^{-9} m^2/s. Both clay and sand are nonconductive particles possessing surface charges. Surface conductance of sand particles are negligible compared to that of clay particles. *Yuteng Jin, Siddharth Misra.*

respectively. Moreover, for our model, mixture S1C2 and S1C3 gives the same prediction value for both σ_{eff} and $\varepsilon_{r,eff}$ because the clay particles in both mixtures have the same a_i/λ value and the same relative permittivity. It can also be observed that with the variation in frequency, there are negligible dispersion effects on σ_{eff} and $\varepsilon_{r,eff}$. In conclusion, porosity, surface conductance of clay, and size of clay play an important role in determining σ_{eff} and $\varepsilon_{r,eff}$. Also, σ_{eff} and $\varepsilon_{r,eff}$ of typical sand and clay mixtures are not dependent on the frequency of measurement.

7.2 The influence of the conductivity of pore-filling electrolyte

In Fig. 3.7, curve S1C1 acts as the reference curve. On comparing the curves S1C1 and S1C2 in Fig. 3.7B, it can be concluded that the $\varepsilon_{r,eff}$ exhibits large frequency dependence when the surface conductance of the nonconductive particles are around 10^{-6} S. The σ_{eff} will also increase because of the increase of surface conductance, as shown in Fig. 3.7A. Notably, on comparing the curves S1C2, S1C3, and S1C4, it is evident that an increase in electrolyte conductivity reduces the permittivity dispersion (Fig. 3.7B). At high-electrolyte conductivity, charge transport dominates charge accumulation at the interface, as a result, the dielectric effect at high-electrolyte conductivity is only because of dipolar/orientational polarization of water. Also, as electrolyte conductivity increases, the σ_{eff} will become higher (Fig. 3.7A).

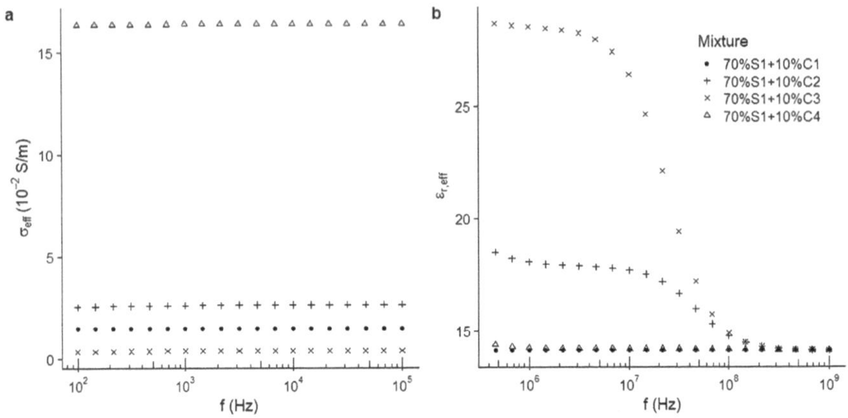

FIGURE 3.7 Comparison of the surface-conductance-assisted interfacial polarization model predictions of (A) σ_{eff} and (B) $\varepsilon_{r,eff}$ of mixtures containing nonconductive spherical sand particles and various types of spherical clay particles fully saturated with electrolyte of different conductivities (properties listed in Table 3.6), and of $\varepsilon_{r,h}$ of 70 and D_h of 10^{-9} m^2/s. Both clay and sand are nonconductive particles possessing surface charges. Surface conductance of sand particles is negligible compared to that of clay particles. *Yuteng Jin, Siddharth Misra.*

7.3 The influence of the volume fractions of oil

In this book, we model oil as nonconductive spherical droplets of specific size (100 μm) uniformly distributed in the mixture surrounded by electrolyte. Oil droplets do not possess any surface charge or surface conductance. An increase in oil saturation results in a decrease in porosity because of the overall increase in the volume fraction of nonconductive spherical particles. Properties of oil droplets are described in Table 3.7. Comparing the curves with reference curve S1 in Fig. 3.8, we conclude that the increase in volume fraction of oil will decrease the σ_{eff} and $\varepsilon_{r,eff}$. Oil is neither conductive nor leads to dielectric effects. Increase in volume fraction of oil leads to decrease in charge transport as well as charge accumulation.

TABLE 3.7 Properties used for generating Fig. 3.8, where S1 represents sand, C1 represents clay, and O1 ~ O3 represent different volume fractions of oil. The host is an electrolyte of $\varepsilon_{r,h}$ of 70, σ_h of 0.1 S/m, and D_h of 10^{-9} m^2/s.

	ϕ_i (%)	a_i (μm)	$\varepsilon_{r,i}$	λ_i (S)
Sand (S1)	70	1000	4	10^{-9}
Clay (C1)	20	10	8	10^{-8}
Oil (O1 ~ 3)	1, 5, 8	100	2	10^{-30}

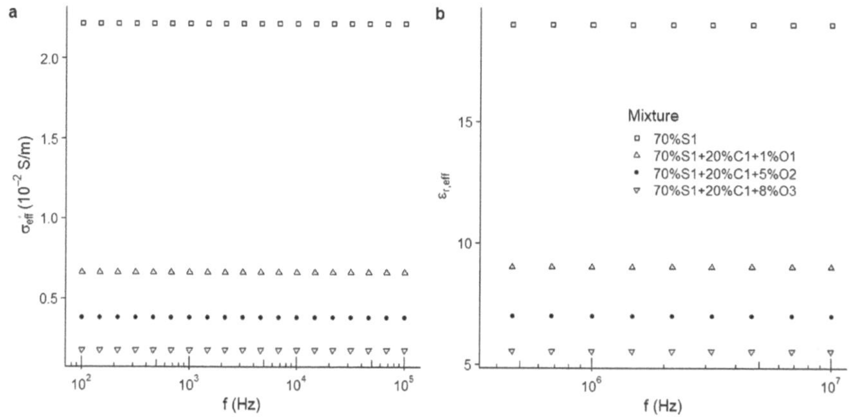

FIGURE 3.8 Comparison of the surface-conductance-assisted interfacial polarization model predictions of (A) σ_{eff} and (B) $\varepsilon_{r,eff}$ of mixtures containing spherical sand and clay particles partially saturated with electrolyte of $\varepsilon_{r,h}$ of 70, σ_h of 0.1 S/m, and D_h of 10^{-9} m^2/s and containing different volume fractions of oil. The properties of sand, clay, and oil are listed in Table 3.7. O1 ~ O3 represent oil volume fractions of 1%, 5%, and 8% (which correspond to oil saturation of 5%, 25%, and 40%, respectively, if porosity is 20%). *Yuteng Jin, Siddharth Misra.*

8. Use of surface-conductance-assisted interfacial polarization model in subsurface/material characterization

The SCAIP model can be used in various fields including colloid and interface science, bioscience, electrochemistry, geophysics, subsurface characterization of water resources systems, geothermal/fossil-fuel energy systems, carbon sequestration, and mining industries to characterize the materials where conductive particles are not present and when the particles that give IFP phenomenon can be represented by surface-charge-bearing nonconductive spherical particles. Such applications include the complex conductivity response of polystyrene balls or polymer latex balls suspend in the electrolyte solution (Schwan et al., 1962; Delgado et al., 1998; Tirado and Grosse, 2006), the oil-water emulsion as mentioned by Hanai et al. (1959) where oil can be treated as nonconductive spherical particles, or even erythrocyte cells uniformly distributed in saline solution (Hanai et al., 1979).

9. Conclusions

We developed a mechanistic model of multifrequency complex conductivity for a homogeneous porous media containing surface-charge-bearing nonconductive particles (e.g., clays and sand) saturated with oil and brine in the frequency range of 100 Hz to 10 MHz. The model accounts for the IFP effects on conductivity and permittivity because of the surface conductance of nonconductive particles. We studied the IFP effects of clays on the effective conductivity (σ_{eff}) in the frequency range of 100 Hz to 100 kHz and on the effective permittivity ($\varepsilon_{r,eff}$) in the frequency range of 0.5 MHz–1 GHz. A decrease in size or an increase in the surface conductance of surface-charge-bearing nonconductive particles, such as clays, uniformly distributed in a porous homogeneous mixture leads to a slight decrease in the σ_{eff} and a drastic increase in the $\varepsilon_{r,eff}$. Notably, clay particles with the same ratio of size to surface conductance and the same relative permittivity (ε_r) have similar effects on the complex conductivity of the mixture. In the frequency range investigated in this chapter, the frequency dispersion of complex conductivity because of the IFP effects of clays are negligible compared to that of conductive particles. Increase in volume fraction of oil leads to decrease in charge transport as well as charge accumulation. At high-electrolyte conductivity, charge transport dominates charge accumulation at the interface; as a result, dielectric permittivity becomes more dispersive at high-electrolyte conductivity. Overall, porosity, surface conductance of clay, and size of clay play an important role in determining the conductivity and permittivity of mixtures containing nonconductive particles possessing surface charges. Interestingly, conductivity and permittivity of typical sand and clay mixtures do not exhibit any dependence on the frequency of measurement in the frequency range of 100 Hz to 10 MHz.

Nomenclature

Acronyms

EM electromagnetic
IFP interfacial polarization
PDE partial differential equations
PNP Poisson-Nernst-Planck
SCAIP surface-conductance-assisted interfacial polarization

Symbols

$\boldsymbol{a}$ characteristic length of inclusion phase (m)
$\boldsymbol{c}$ charge density variation (1/m^3)
$\boldsymbol{d}$ net charge density variation (1/m^3)
$\boldsymbol{D}$ diffusion coefficient of charge carriers (m^2/s)
Δ (∇^2) = Laplace's operator
$\boldsymbol{e}$ Euler's number
$\boldsymbol{e}$ electric field vector
$\boldsymbol{E_0}$ amplitude of the electric field (V)
ε dielectric permittivity (F/m)
ε_0 vacuum permittivity (8.854×10^{-12} F/m)
ε_{eff} effective dielectric permittivity of the mixture (F/m)
ε_r relative permittivity
$\boldsymbol{f}$ frequency (Hz)

$f(\omega)$ dipolarizability (dipolar field coefficient)
i square root of -1
i_n modified spherical Bessel function of the first kind of n th order
I_n modified Bessel function of the first kind of n th order
j current density (A/m^3)
k_B Boltzmann's constant
k_n modified spherical Bessel function of the second kind of n th order
K_n modified Bessel function of the second kind of n th order
λ surface conductance (S)
λ_D Debye screening length (m)
μ electrical mobility [m^2/(V·s)]
n an integer referring to the order of the standing wave solution
N charge carrier density (1/m^3)
ω angular frequency of the electric field (rad/s)
P_f net free charge density (C/m^3)
P_n^0 associated Legendre functions of the first kind of n th order
ϕ electrical potential (V)
ϕ volume fraction (%)
q elementary charge (1.6×10^{-19}°C)
Q_n^0 associated Legendre functions of the second kind of n th order
r radial distance along the normal to the interface (m)
ρ surface charge density (C/m^2)
s total ion density variation (1/m^3)
σ electrical conductivity (S/m)
σ^* complex electrical conductivity (S/m)
σ_{eff} effective electrical conductivity of the mixture (S/m)
σ_{eff}^* effective complex electrical conductivity of the mixture (S/m)
t time (s)
T absolute temperature (K)
θ angle between the normal to the interface and the incident external electric field (°)
Z charge number

Subscripts

0 at time equal to 0 s
eff effective
h host medium
i inclusion phase
j type of medium/phase
n an integer referring to the order
$\hat{n}$ unit vector
ncond nonconductive particles
r relative

Superscripts

\+ positively charged carrier
− negatively charged carrier

References

Chu, K.T., Bazant, M.Z., July 6, 2006. Nonlinear electrochemical relaxation around conductors. Phys. Rev. 74 (1), 011501.

Cosenza, P., Ghorbani, A., Camerlynck, C., Rejiba, F., Guérin, R., Tabbagh, A., October 1, 2009. Effective medium theories for modelling the relationships between electromagnetic properties and hydrological variables in geomaterials: a review. Near Surf. Geophys. 7 (5–6), 563–578.

Delgado, A.V., Arroyo, F.J., González-Caballero, F., Shilov, V.N., Borkovskaya, Y.B., 1998. The effect of the concentration of dispersed particles on the mechanisms of low-frequency dielectric dispersion (LFDD) in colloidal suspensions. Colloids Surf., A Physicochem. Eng. Asp. 140 (1), 139–149.

Dukhin, S.S., Shilov, V.N., Bikerman, J.J., April 1, 1974. Dielectric phenomena and double layer in disperse systems and polyelectrolytes. J. Electrochem. Soc. 121 (4), 154C.

Giordano, S., May 1, 2003. Effective medium theory for dispersions of dielectric ellipsoids. J. Electrost. 58 (1–2), 59–76.

Grimes, C.A., Mungle, C., Kouzoudis, D., Fang, S., Eklund, P.C., March 24, 2000. The 500 MHz to 5.50 GHz complex permittivity spectra of single-wall carbon nanotube-loaded polymer composites. Chem. Phys. Lett. 319 (5–6), 460–464.

Grosse, C., Barchini, R., March 14, 1992. The influence of diffusion on the dielectric properties of suspensions of conductive spherical particles in an electrolyte. J. Phys. Appl. Phys. 25 (3), 508.

Grosse, C., Foster, K.R., May 1987. Permittivity of a suspension of charged spherical particles in electrolyte solution. J. Phys. Chem. 91 (11), 3073–3076.

Grosse, C., June 1988. Permittivity of a suspension of charged spherical particles in electrolyte solution. 2. Influence of the surface conductivity and asymmetry of the electrolyte on the low-and high-frequency relaxations. J. Phys. Chem. 92 (13), 3905–3910.

Hanai, T., Asami, K., Koizumi, N., 1979. Dielectric theory of concentrated suspensions of shell-spheres in particular reference to the analysis of biological cell suspensions. Bull. Inst. Chem. Res., Kyoto Univ. 57, 297–305.

Hanai, T., Koizumi, N., Gotoh, R., November 1, 1959. Dielectric properties of emulsions. Kolloid -Zeitschrift. 167 (1), 41–43.

Hogg, A.J., 2001. Fluid Dynamics Separable Solutions to Laplace's Equation.

Hou, C.Y., Feng, L., Seleznev, N., Freed, D.E., September 1, 2018. Low frequency complex dielectric (conductivity) response of dilute clay suspensions: modeling and experiments. J. Colloid Interface Sci. 525, 62–75.

Jin, Y., Misra, S., Homan, D., Rasmus, J., Revil, A., 2019. Mechanistic model of multi-frequency complex conductivity of porous media containing water-wet nonconductive and conductive particles at various water saturations. Adv. Water Resour. 130, 244–257.

Lagarkov, A.N., Sarychev, A.K., March 1, 1996. Electromagnetic properties of composites containing elongated conducting inclusions. Phys. Rev. B 53 (10), 6318.

Lakhtakia, A., Slepyan, G.Y., Maksimenko, S.A., Gusakov, A.V., Yevtushenko, O.M., January 1, 1998. Effective medium theory of the microwave and the infrared properties of composites with carbon nanotube inclusions. Carbon 36 (12), 1833–1839.

Leroy, P., Revil, A., Kemna, A., Cosenza, P., Ghorbani, A., May 1, 2008. Complex conductivity of water-saturated packs of glass beads. J. Colloid Interface Sci. 321 (1), 103–117.

Mantese, J.V., Micheli, A.L., Dungan, D.F., Geyer, R.G., Baker-Jarvis, J., Grosvenor, J., February 1, 1996. Applicability of effective medium theory to ferroelectric/ferrimagnetic composites with composition and frequency-dependent complex permittivities and permeabilities. J. Appl. Phys. 79 (3), 1655–1660.

McLachlan, D.S., November 1, 1989. The complex permittivity of emulsions: an effective media-percolation equation. Solid State Commun. 72 (8), 831–834.

Misra, S., Torres-Verdín, C., Revil, A., Rasmus, J., Homan, D., March 1, 2016. Interfacial polarization of disseminated conductive minerals in absence of redox-active species—Part 1: mechanistic model and validation. Geophysics 81 (2), E139–E157.

Pavlin, M., Slivnik, T., Miklavcic, D., August 7, 2002. Effective conductivity of cell suspensions. Inst. Electr. Electron. Eng. Trans. Biomed. Eng. 49 (1), 77–80.

Revil, A., Abdel Aal, G.Z., Atekwana, E.A., Mao, D., Florsch, N., August 12, 2015. Induced polarization response of porous media with metallic particles—Part 2: comparison with a broad database of experimental data. Geophysics 80 (5), D539–D552.

Schmuck, M., Bazant, M.Z., June 30, 2015. Homogenization of the Poisson–Nernst–Planck equations for ion transport in charged porous media. SIAM J. Appl. Math. 75 (3), 1369–1401.

Schwan, H.P., Schwarz, G., Maczuk, J., Pauly, H., 1962. On the low-frequency dielectric dispersion of colloidal particles in electrolyte solution. J. Phys. Chem. 66 (12), 2626–2635.

Schwarz, G., December 1962. A theory of the low-frequency dielectric dispersion of colloidal particles in electrolyte solution1, 2. J. Phys. Chem. 66 (12), 2636–2642.

Sihvola, A.H., Kong, J.A., July 1988. Effective permittivity of dielectric mixtures. IEEE Trans. Geosci. Rem. Sens. 26 (4), 420–429.

Sihvola, A., January 1, 2007. Dielectric polarization and particle shape effects. J. Nanomater. 2007 (1), 5.

Stroud, D., October 15, 1975. Generalized effective-medium approach to the conductivity of an inhomogeneous material. Phys. Rev. B 12 (8), 3368.

Tirado, M., Grosse, C., 2006. Dependence of the broad frequency dielectric spectra of colloidal polystyrene particle suspensions on the difference between the counterion and co-ion diffusion coefficients. J. Colloid Interface Sci. 298 (2), 973–981.

Weisstein, E.W., 2018. Legendre Polynomial. http://mathworld.wolfram.com/LegendrePolynomial.html. (Accessed 11 November 2018).

Weisstein, E.W., 2018. Bessel Function. http://mathworld.wolfram.com/BesselFunction.html. (Accessed 11 November 2018).

Young, P., 2009. Helmholtz's and Laplace's Equations in Spherical Polar Coordinates: Spherical Harmonics and Spherical Bessel Functions. Physics 116C Lecture Notes. University of California, Santa Cruz, CA, p. 116 http://physics. ucsc. edu/~ peter C.

Zheng, Q., Wei, G.W., May 21, 2011. Poisson–Boltzmann–Nernst–Planck model. J. Chem. Phys. 134 (19), 194101.

Chapter 4

Multifrequency conductivity and permittivity of porous material containing conductive particles in redox inactive conditions

Yuteng Jin[1], Siddharth Misra[2,3]

[1]*Harold Vance Department of Petroleum Engineering, Texas A&M University, TX, United States;* [2]*Harold Vance Department of Petroleum Engineering, College of Engineering, Texas A&M University, TX, United States;* [3]*Department of Geology and Geophysics, College of Geoscience, Texas A&M University, TX, United States*

1. Interfacial polarization phenomena around conductive particles in redox inactive conditions

Fig. 4.1 shows the perfectly polarized interfacial polarization (PPIP) phenomenon in a representative volume of a dilute mixture of uniformly distributed electrically conductive spherical inclusions in an electrolyte-saturated porous host medium. Because of the assumption of spherical symmetry, such an interfacial polarization (IFP) is independent of the direction of the externally applied electric field. Under the presence of an externally applied electrical field, the charge carriers in the conductive spherical inclusion will move as shown in Fig. 4.1. The charge carriers then accumulate at the solid-liquid interface, which lead to IFP phenomenon (Jin et al., 2019). In this book, only spherical particles are considered. Misra et al. (2016b) also discussed the IFP phenomenon of long rod-like and thin sheet-like particles. Unlike this chapter, Chapter 3 focusses on the IFP phenomenon due to non-conductive particles carrying surface charges and surface conductance. In this chapter, the focus is on the IFP phenomenon due to conductive particles that do not have any surface conductance or intrinsic surface conductance. Chapter 3 described the SCAIP model, whereas Chapter 4 describes PPIP model.

Multifrequency Electromagnetic Data Interpretation for Subsurface Characterization
https://doi.org/10.1016/B978-0-12-821439-8.00007-0

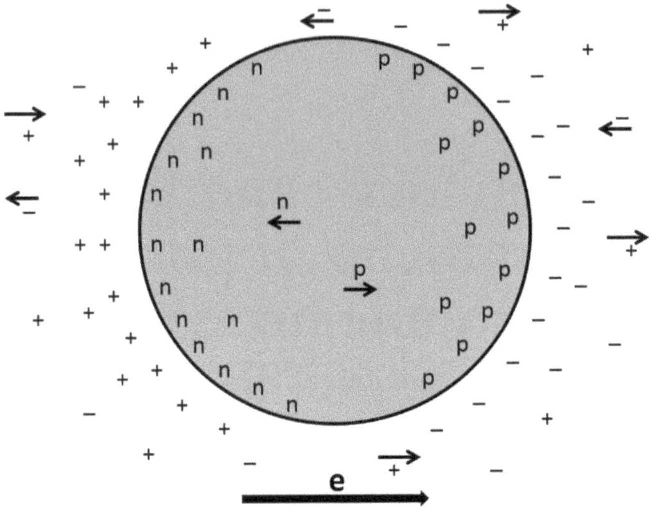

FIGURE 4.1 Cross-section of a perfectly polarized conductive spherical inclusion surrounded by an ionic host medium. Charge carriers in the ionic host medium are cations, identified by "+" symbol, and anions, identified by "-" symbol. Charge carriers in the conductive spherical inclusion are n- and p-charge carriers, identified by symbol "n" and "p," respectively. The direction of the externally applied electrical field, **e**, is identified with a bold arrow next to the symbol "**e**". The direction of movement of the four different types of charge carriers is represented by the arrow next to the symbols of the charge carriers. *Yuteng Jin, Siddharth Misra.*

2. Popular mechanistic models of interfacial polarization due to conductive particles in redox inactive conditions

Garcia et al. (1985) developed a model for conductive spherical particles with insulating shells (for e.g., oxidized surface of pyrite) in a conductive medium, where the diffusive effects play an important role. Grosse and Barchini (1992) improved the previous theory for infinitely conductive spherical particles in bulk electrolyte by considering the ion flow across the interface. Moreover, in comparison to dielectric mixing laws, Tuncer et al. (2001) applied a finite element method on cylinder-like conductive inclusion phase to investigate the dielectric relaxation phenomena. Their result shows the two methods match well at low-inclusion concentrations. However, as the concentration of inclusion increases, mutual interaction of the inclusions becomes significant.

3. Newly proposed mechanistic model of interfacial polarization due to conductive particles in redox inactive conditions

3.1 Model description

In this chapter, PPIP model (Misra et al., 2016a) is presented to investigate the IFP phenomena around conductive particles. The phenomenological basis of IFP considered in our work aligns with the mechanistic descriptions outlined by Revil et al. (2015). Charge carriers in conductive minerals have higher mobility compared to ions in porous geomaterials. In the presence of an externally applied electromagnetic (EM) field, charge carriers in the disseminated electrically conductive inclusions migrate faster and accumulate at the impermeable interfaces. Consequently, electrically conductive inclusions behave as dipoles in the presence of an externally applied electric field. Subsequently, charge carriers in the host medium migrate and accumulate on host-inclusion interfaces under the influence of the externally applied electric field and that of the induced charges in conductive inclusions.

In the absence of an externally applied electric field, a negligible initial surface charge is assumed on electrically conductive inclusions. Thus, there is a negligible double layer around the surface of electrically conductive inclusions, whereby the surface conductance of a conductive inclusion is negligible. Similar assumptions are made in electrochemistry and colloid science with respect to electrochemical relaxation around metallic surfaces (Chu and Bazant, 2006). Also, we assume absence of redox-active species and neglect the influence of pH of pore water (Revil et al., 2015). The host and inclusion phases can be modeled as electrically conductive, insulating, or dielectric material. Also, pore-filling fluid can be modeled as electrically conductive (e.g., brine) or nonconductive material (e.g., oil).

3.2 Assumptions

vTo simplify the model, we assume only spherical particles are present in the porous media. Also, the host, inclusion, and pore-filling fluid are assumed to have homogeneous, isotropic, and nondispersive electrical properties. Therefore, the frequency dispersion and dielectric enhancement predicted by the PPIP model solely stems from the PPIP phenomena around the electrically conductive inclusions. We also assume all the charge carriers bear unitary charge and both host and inclusion phases bear binary, symmetric charge carriers.

3.3 Model development

3.3.1 Poisson–Nernst–Planck equation

Similar to the SCAIP model described in Chapter 3, the PPIP model is also based on the Poisson–Nernst–Planck (PNP) equations for a dilute solution in a weak electrical field regime. The development of the PPIP model is very similar to that of the SCAIP model. For PPIP model development, spontaneous initial accumulation of charges is assumed to be absent on the host-inclusion interfaces. At time $t = 0$, electro-neutrality is assumed throughout the system.

3.3.2 Solution of Helmholtz partial differential equation

Eq. (3.18) as mentioned in Chapter 3 must be solved to obtain an analytical expression for d_j in the host and inclusion phases around the perfectly polarized host-inclusion interface of conductive spherical inclusion. Recall that $d_j = c_j^+ - c_j^-$ represents net charge density variation, where $c_j^{\pm}$ is the charge density variation near the host-inclusion interface in medium j because of the externally applied electric field. A major difference between PPIP model and SCAIP model is due to the existence of non-zero net charge density and that of non-zero total charge-carrier density inside the spherical inclusion in the case of PPIP model. For SCAIP model expression for $d_h(r,\theta)$ for the mixture cotaining conductive sphical inclusion is the same as that in the host for the mixture containing nonconductive spherical inclusion (Eqs. 3.32a or 3.32b). Using the condition that $d_i(r,\theta)$ should be finite at $r \to 0$, we obtain a particular solution for d_i for the mixture containing conductive spherical inclusion that can be represented as

$$d_i(r,\theta) = B_i i_1(r\gamma_i)\cos\theta \tag{4.1a}$$

or (Note to Dr. Misra: i_1() is defined in chapter 3, so not explained here in the main text. But I mentioned it in the nonmenclature section in this chapter 4. This is same for many other parameters like q, r, \theta, …)

$$d_i(r,\theta) = B_i\left[\frac{\cos h(r\gamma_i)}{r\gamma_i} - \frac{\sin h(r\gamma_i)}{(r\gamma_i)^2}\right]\cos\theta \tag{4.1b}$$

where B_i is unknown complex-valued coefficient of the paticular solution in the inclusion phase obtained from Eq. (3.31), substituting the subscript h with i. Note when $r \to 0$, it is assumed that $d_i(r,\theta) = 0$.

3.3.3 Solution of Laplace partial differential equation

Eq. (3.22) as mentioned in Chapter 3 must be solved to obtain the electric potential field in the representative volume. The expression for $\varphi_h(r,\theta)$ for the mixture containing conductive spherical inclusion is the same as that for the mixture containing nonconductive spherical inclusion (Eq. 3.38). Using the condition when $r \to 0$, $d_i = 0$ and φ_i should be finite, we can obtain $C_i = 0$. A standing wave representation of Eq. (3.36b) for the conductive inclusion phase, using Eq. (4.1b), is

$$\varphi_i(r,\theta) = A_i r\cos\theta - \frac{qB_i}{\gamma_i^2\varepsilon_i}\left[\frac{\cosh(r\gamma_i)}{r\gamma_i} - \frac{\sinh(r\gamma_i)}{(r\gamma_i)^2}\right]\cos\theta \tag{4.2}$$

where A_i is unknown complex-valued coefficient of the particular solution in the conductive inclusion phase obtained from Eq. (3.36b).

3.3.4 Boundary conditions

To obtain an expression for the dipolarizability of the conductive particle in an electrolytic host, the following boundary conditions are used (Garcia et al., 1985):

a. Assuming a zero-intrinsic capacitance of the host-inclusion interface, the electric potential must be continuous at the interface.

$$\varphi_i(r=a)=\varphi_h(r=a) \tag{4.3a}$$

b. The normal component of the displacement current must be continuous at the interface. This condition corresponds to the fact that there is no net surface-charge distribution on an electrically conductive inclusion phase.

$$\varepsilon_i \frac{\partial \varphi_i}{\partial r}\bigg|_{r=a} = \varepsilon_h \frac{\partial \varphi_h}{\partial r}\bigg|_{r=a} \tag{4.3b}$$

c. The normal component of the current density must vanish at the interface for both media. This condition expresses the fact that in the absence of transport of charge carriers and exchange of charges along the interface, the diffusive and electromigrative currents in the host must cancel each other at the interface. Our focus is perfectly polarizable or completely blocking interfaces without Faradic processes, wherein fluxes of charge carriers must vanish on both sides of the interface. Note that this boundary condition is used to obtain two equations: one for the outer volume of the sphere in the host medium and the other for the inner volume of the sphere in the inclusion medium.

$$j_j^+ + j_j^- = -2N_{0,j} q \mu_j \frac{\partial \varphi_j}{\partial r}\bigg|_{r=a} - qD_j \frac{\partial d_j}{\partial r}\bigg|_{r=a} = 0 (j=h \text{ or } i) \tag{4.3c}$$

3.3.5 Solution for the dipolarizability

Using boundary condition Eq. (4.3a), Eq. (4.2) and Eq. (3.38) can be equated on the surface of the sphere of radius equal to a. The resulting equation can be abbreviated as

$$-E_0 a + \frac{C_h}{a^2} - E_h B_h = A_i a - F_i B_i \tag{4.4a}$$

where

$$F_i = \frac{q}{\gamma_i^2 \varepsilon_i}\left[\frac{\cosh(a\gamma_i)}{a\gamma_i} - \frac{\sinh(a\gamma_i)}{(a\gamma_i)^2}\right] \tag{4.4b}$$

The equation obtained using boundary condition Eq. (4.3b) at the surface of the sphere can be abbreviated as

$$\varepsilon_h\left(-E_0-\frac{2C_h}{a^3}+G_hB_h\right)=\varepsilon_i(A_i+H_iB_i) \tag{4.5a}$$

where

$$H_i=\frac{q}{\gamma_i\varepsilon_i}\left[\frac{2\cosh(a\gamma_i)}{(a\gamma_i)^2}-\frac{\sinh(a\gamma_i)}{a\gamma_i}-\frac{2\sinh(a\gamma_i)}{(a\gamma_i)^3}\right] \tag{4.5b}$$

Similarly, the equation obtained using boundary condition Eq. (4.3c) at the outer surface of the sphere in the host medium can be abbreviated as

$$C_h=-a^3\left(\frac{E_0}{2}+\frac{\mathrm{i}\omega\varepsilon_hG_hB_h}{2\sigma_h}\right) \tag{4.6}$$

On the other hand, the equation obtained using boundary condition Eq. (4.3c) at the inner surface of the sphere in the inclusion medium can be abbreviated as

$$A_i=\frac{\mathrm{i}\omega\varepsilon_iH_iB_i}{\sigma_i} \tag{4.7}$$

By solving Eqs. (4.4a), (4.5a), (4.6), and (4.7), we obtain the dipolarizability (dipolar field coefficient) of the representative volume comprising a spherical conductive inclusion in an electrolytic host as

$$f_{cond}(\omega)=\frac{C_h}{E_0a^3}=-\frac{1}{2}+\frac{3}{2}\frac{\mathrm{i}\omega}{\left[\frac{2}{a}\frac{\sigma_h}{\varepsilon_h}\frac{E_h}{G_h}-\frac{2}{a}\frac{\sigma_h^*}{\sigma_i^*}\frac{\sigma_i}{\varepsilon_i}\frac{F_i}{H_i}+\mathrm{i}\omega\left(\frac{2\sigma_h^*}{\sigma_i^*}+1\right)\right]} \tag{4.8}$$

where $\sigma_h^*=\sigma_h+\mathrm{i}\omega\varepsilon_h$ is the complex conductivity of the host medium, and $\sigma_i^*=\sigma_i+\mathrm{i}\omega\varepsilon_i$ is the complex conductivity of the inclusion phase. The expressions for E_h, G_h is shown in Chapter 3.

3.4 Limitations

In addition to the limitations associated with the PNP equations as discussed in Chapter 3, PPIP model is reliable for studying the EM response of mixtures containing uniformly distributed conductive particles of characteristic length (a) less than 1 mm, conductivity (σ_i) less than 10^5 S/m, relative permittivity ($\varepsilon_{r,i}$) less than 20, relative magnetic permeability equal to 1, and volume fraction (ϕ_i) less than 20% in the frequency range of 100 Hz to 100 MHz. Beyond these limits, the PPIP model predictions will incur significant discrepancies with measurements because of the skin effect of the inclusion phase. The skin effect is primarily governed by the operating frequency and conductivity of the inclusion phase. PPIP model predictions are physically consistent only when the estimated skin depth is an order of magnitude larger than the characteristic length of the particles, where skin depth is defined by the depth from the surface till which an alternating current flow in the conductive particle. For example, at an operating frequency of 100 kHz, an inclusion phase of conductivity of 10^3 S/m and ε_r of 10 exhibits a skin depth of 50 mm. Therefore, the PPIP model can be used to compute the complex conductivity response of a geological mixture containing such an inclusion phase at 100 kHz only when the characteristic length of the inclusion phase is smaller than 5 mm, which is an order of magnitude smaller than the skin depth of 50 mm.

4. Effective medium theory

Similar to the discussions in Chapter 3, we apply the effective-medium theory to determine the effective complex electrical conductivity (σ_{eff}^*) of the mixture (Grosse and Barchini, 1992) by combining the contributions of dipolarizability for spherical nonconductive and conductive particles surrounded by the electrolyte-saturated host medium. For the development of our model, as stated in the previous sections, PNP equation is first used to obtain dipolarizabilities (dipole moment), which are microscopic electrical properties, for the representative volume containing the spherical inclusion. In the derivations of dipolarizabilities, we neglect multipoles effect because their magnitude decreases with inverse power of distance (Sihvola, 2007). Monopole effects are also neglected because there is zero net charge as a consequence of the assumed electroneutrality. The macroscopic electrical properties are then computed using effective-medium formulations based on the theory that a material composed of a mixture of distinct homogeneous media can be seen as a homogeneous material at a sufficiently large observation scale (Giordano, 2003).

5. Combining the perfectly polarized interfacial polarization (PPIP) model and the surface-conductance-assisted interfacial polarization (SCAIP) model

In this chapter, we obtain the effective electrical properties using a Maxwell–Garnett type effective medium formulation. To meet the requirements of the formulation, the volume fraction of nonconductive and conductive inclusions in the mixture should be less than 20% (Revil et al., 2015). Subsurface water-bearing reservoir rocks have less than 10% volume fraction of conductive mineral inclusions but the volume fraction of nonconductive particles possessing surface charge (e.g., clay and sand) can exceed 20%. Moreover, we invoke the PNP equations in the bulk electrolyte that introduces a decaying length scale, where the Maxwell Garnett formula may become invalid at even lower concentration of inclusion phase (Hou et al., 2018). It is also important to mention this formulation used in the calculation neglects the EM interaction between the inclusions and other components. Because of the assumed dilution of the uniformly distributed inclusion phase, individual elements of the dispersed phase are assumed to be isolated and not in contact with each other. Like other mixing laws, our model includes the assumption that the magnitude of spatial variations of the electric field is smaller than the magnitude of variations in the intrinsic electrical properties and geometrical structures. Moreover, all calculations are performed using a quasi-static assumption that requires the size of heterogeneities to be much smaller than the wavelength of the applied EM field (Cosenza et al., 2009). Also, when dealing with a lossy medium, the skin depth of the EM wave must be considered to avoid strong attenuation of the field amplitudes in the conductive heterogeneities. Most importantly, because of the implementation of the PNP equations, our effective medium formulations unlike other theories (e.g., Giordano, 2003) explicitly accounts for the characteristic lengths of heterogeneities, resulting in a physically consistent way to account for the perturbation because of nonconductive or conductive inclusions.

For a mixture containing only spherical conductive particles, the effective medium formulation of the complex conductivity response can be expressed as

$$\frac{\sigma^*_{\underset{\hat{n},eff}{\mathrm{I}}} - \sigma^*_h}{\sigma^*_{\underset{\hat{n},eff}{\mathrm{I}}} + 2\sigma^*_h} = \sum \phi_{cond} f_{cond}(\omega) \tag{4.9}$$

where $\sigma^* = \sigma + i\omega\varepsilon$ is a representation of the complex conductivity of a material, $\sigma^*_{\underset{\hat{n},eff}{\mathrm{I}}}$ is the effective complex conductivity of the geological mixture directed along the $\hat{n}$ unit vector that can be measured with an externally applied electric field directed along the $\hat{n}$ unit vector, and σ^*_h is the complex-valued conductivity of the homogenous isotropic host material that surrounds the particles. ϕ_{cond} is the volume fraction of the conductive inclusion phase in the mixture. $f_{cond}(\omega)$ is the dipolarizability of conductive inclusion phase along the direction of $\hat{n}$ unit vector, along which the externally applied electric field is directed.

Normally, the PPIP model is combined with the SCAIP model (as described in Chapter 3) for a mixture containing both spherical conductive and nonconductive particles; for example, a pyrite-rich shaly sand containing pyrite grains, clay-sized particles, sand-sized particles, and clay grains. It is desirable to consider the condition where oil is present along with the electrolyte in the pores of the host medium. In this book, we assume that the oil is distributed as spherical droplets in the mixture and can be treated as nonconductive spherical inclusion phase with negligible surface charge. We also assume that the inclusion phases are predominantly water wet. Thus, the effective medium formulation of the complex conductivity response of the PPIP−SCAIP (PS) model can be expressed as

$$\frac{\sigma^*_{\text{I}\atop h,eff} - \sigma^*_h}{\sigma^*_{\text{I}\atop h,eff} + 2\sigma^*_h} = \sum \phi_{cond} f_{cond}(\omega) + \sum \phi_{ncond} f_{ncond}(\omega) + \phi_o f_o(\omega) \quad (4.10)$$

where ϕ_o is the volume fraction of oil in the mixture and $f_o(\omega)$ is the dipolarizability of oil, which takes the same form of $f_{ncond}(\omega)$ in Chapter 3 but with negligible surface charge.

6. Validation of the PPIP−SCAIP (PS) model

In this section, the PPIP−SCAIP (PS) model predictions are compared with the published experimental data. The properties of mixtures used for generating plots are summarized in the corresponding tables, where ϕ is the volume fraction of a specific phase, a is the characteristic length (radius) of spherical particles, D is diffusion coefficient of charge carriers, ε_r is relative permittivity, σ is conductivity, and λ is surface conductance. Property of host medium and inclusion phase are represented with a subscript of h and i, respectively.

Mahan et al. (1986) measured the complex resistivity responses of synthetic samples containing quartz grains of various sizes and conductive sulfide particles of different materials that were fully saturated with electrolytes with or without active ions (e.g., Cu^{2+} ions are active ions with respect to chalcopyrite mineral inclusions). Active ions result in charge transfer across host-inclusion interfaces because of oxidation-reduction reactions. Unlike Mahan et al. (1986) modeling procedure, the PS model takes into account the effects of surface conductance of uniformly distributed 80-μm diameter spherical silica grains that are fully saturated with pore-filling fluid. As shown in Fig. 4.2A and B, the PS model predictions match the Debye-type dispersion response from Mahan et al. (1986) for the complex conductivity response of a mixture of chalcopyrite and silica grains fully saturated with a NaCl solution that contains no active ions.

7. PPIP−SCAIP (PS) model predictions of conductivity and permittivity versus frequency in 100 Hz to 10 MHz

In this section, the effects of PPIP and SCAIP phenomena on low-frequency (100 Hz) σ_{eff} and high-frequency (10 MHz) $\varepsilon_{r,eff}$ prediction are evaluated as a function of the operating frequency of the measurement device or the logging tool. The properties of mixtures used for generating plots are summarized in the corresponding tables, where ϕ_i is the volume fraction of the inclusion

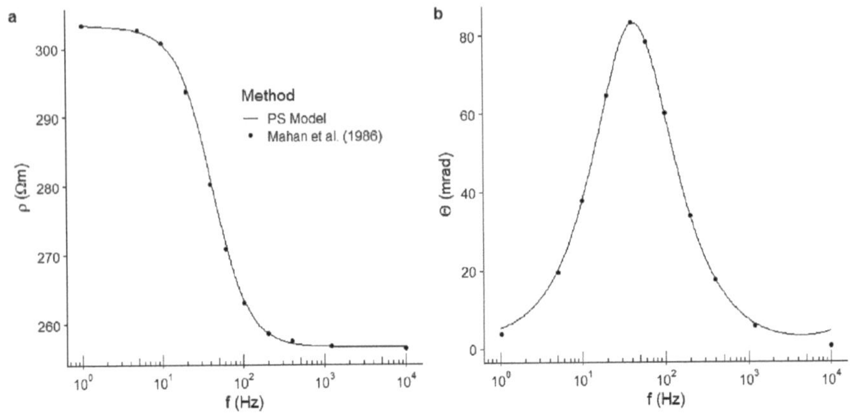

FIGURE 4.2 Comparison of the PS model predictions of (A) resistivity (ρ) and (B) phase angle (Θ) against that measured by Mahan et al. (1986). The properties used for generating this plot is shown in Table 4.1. *Yuteng Jin, Siddharth Misra.*

TABLE 4.1 The known, assumed, and estimated properties used for generating Fig. 4.2.

Property		Known	Assumed	Estimated
Inclusion (pyrite)	ϕ_i (%)			2.23
	a_i (μm)		138	
	$\varepsilon_{r,i}$		10	
	D_i (m^2/s)		1×10^{-9}	
	σ_i (S/m)		1000	
Inclusion (sand)	ϕ_i (%)			65.7
	a_i (μm)		40	
	$\varepsilon_{r,i}$		2	
	λ_i (S)			2.58×10^{-8}
Host (NaCl solution)	D_h (m^2/s)			2.56×10^{-11}
	$\varepsilon_{r,h}$		80	
	σ_h (S/m)	0.00943		

phase, a_i is the characteristic length (radius) of spherical inclusion, D is diffusion coefficient of charge carriers, ε_r is relative permittivity, σ is conductivity, and λ is surface conductance. Properties of host medium and inclusion phase are represented with a subscript of h and i, respectively.

7.1 The influence of the properties of the conductive particles

The mixtures studied in Fig. 4.3 contain 70% volume fraction of sand and 10% volume fraction of uniformly distributed pyrite. In comparison with the reference curve S1, we can conclude that the presence of conductive particles increases the $\varepsilon_{r,eff}$ and decreases σ_{eff} with the lowering of frequency. There is an increase in σ_{eff} at high frequency. The results show that a higher relative permittivity or conductivity of conductive inclusions, like mixture S1P2 or S1P4, increases σ_{eff} and decreases $\varepsilon_{r,eff}$ of mixture, while a mixture with higher diffusion coefficient like S1P3 decreases σ_{eff} and increases $\varepsilon_{r,eff}$. Moreover, for our model, mixtures S1P1 and S1P5 give the same prediction for both σ_{eff} and $\varepsilon_{r,eff}$ because both mixtures have the same σ/D value and the same ε_r. Mixture S1P2 and S1P4 also give the same prediction for both σ_{eff} and $\varepsilon_{r,eff}$ because both mixtures have the same $\sigma\varepsilon_r$ and D. It's also worth to notice that the variation in σ_{eff} prediction only occurs for frequency between 1 and 100 kHz, whereas the variation in $\varepsilon_{r,eff}$ prediction only occurs for frequency lower than about 5 MHz.

7.2 The influence of the properties of the mixture of conductive and nonconductive particles

The PS model predictions of multifrequency complex conductivity of mixtures containing both nonconductive and conductive inclusions/particles

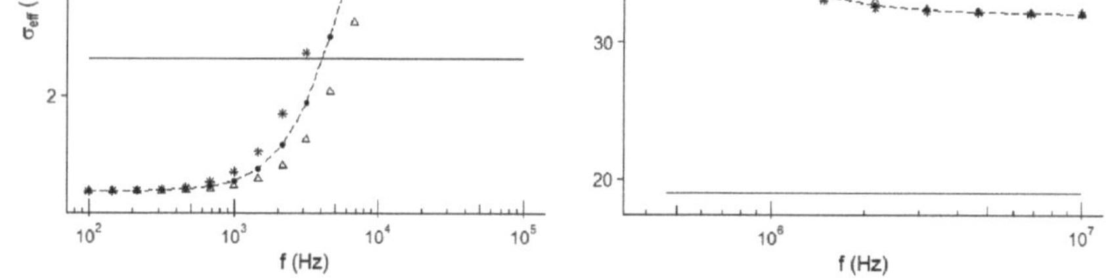

FIGURE 4.3 Comparison of the PS model predictions of (A) σ_{eff} and (B) $\varepsilon_{r,eff}$ of mixtures containing surface-charge-bearing nonconductive spherical sand particles and various types of conductive spherical pyrite particles (properties listed in Table 4.2) fully saturated with electrolyte of $\varepsilon_{r,h}$ of 70, σ_h of 0.1 S/m and D_h of 10^{-9} m^2/s. Characteristic lengths of pyrite and sand particles are assumed to be 200 and 1000 μm, respectively. The surface conductance λ of the sand is 10^{-9} S. *Yuteng Jin, Siddharth Misra.*

TABLE 4.2 Properties used for generating Fig. 4.3, where S1 represents sand, P1 ~ P5 represent five different pyrite types, and the host is an electrolyte of $\varepsilon_{r,h}$ of 70, σ_h of 0.1 S/m and D_h of 10^{-9} m^2/s.

	ϕ_i (%)	D_i (m^2/s)	$\varepsilon_{r,i}$	σ_i (S/m)	σ_i/D_i	$\sigma_i\varepsilon_{r,i}$
Sand (S1)	70	–	4	–	–	–
Pyrite1 (P1)	10	5×10^{-5}	12	500	10^7	6×10^3
Pyrite2 (P2)	10	5×10^{-5}	24	500	10^7	1.2×10^4
Pyrite3 (P3)	10	10^{-4}	12	500	5×10^6	6×10^3
Pyrite4 (P4)	10	5×10^{-5}	12	1000	2×10^7	1.2×10^4
Pyrite5 (P5)	10	10^{-4}	12	1000	10^7	1.2×10^4

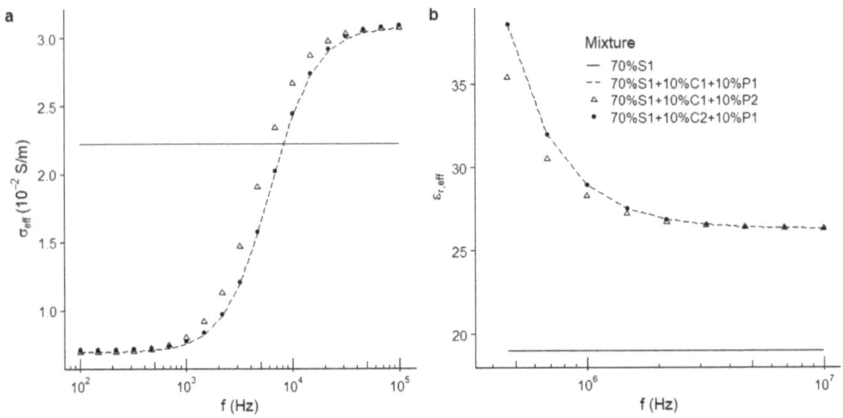

FIGURE 4.4 Comparison of the PS model predictions of (A) σ_{eff} and (B) $\varepsilon_{r,eff}$ of mixtures containing nonconductive spherical sand particles, various types of nonconductive, surface-charge-bearing clay particles, and various types of conductive spherical pyrite particles (properties listed in Table 4.3) fully saturated with electrolyte of $\varepsilon_{r,h}$ of 70, σ_h of 0.1 S/m, and D_h of 10^{-9} m^2/s. P1 is less conductive than P2. C1 has less surface conductance than C2. *Yuteng Jin, Siddharth Misra.*

are evaluated in this section. The resulting plots are shown as Fig. 4.4. Curve S1 acts as reference curve representing a clean mixture of sand and electrolyte without any clay and pyrite particles. The presence of uniformly distributed surface-charge-bearing nonconductive and conductive inclusion particles will decrease the σ_{eff} to a value lower than that of clean formation at low frequency (Fig. 4.4A). This is because at low frequency, the charge carriers quickly reach the equilibrium distribution around the interfaces under the influence of a time-varying electric field, so that the polarized conductive and nonconductive particles act as insulators because of the interface that does not allow charge

TABLE 4.3 Properties used for generating Fig. 4.4, where S1 represents sand, C1 and C2 represent two different clay types, and P1 and P2 represent two different pyrite types. The host is an electrolyte of $\varepsilon_{r,h}$ of 70, σ_h of 0.1 S/m, and D_h of 10^{-9} m^2/s.

	ϕ_i (%)	a_i (μm)	D_i (m^2/s)	$\varepsilon_{r,i}$	σ_i (S/m)	λ_i (S)
Sand (S1)	70	1000	–	4	–	10^{-9}
Clay1 (C1)	10	100	–	8	–	10^{-8}
Clay2 (C2)	10	100	–	8	–	10^{-7}
Pyrite1 (P1)	10	200	5×10^{-5}	12	500	–
Pyrite2 (P2)	10	200	5×10^{-5}	12	1000	–

migration between the particle and host electrolyte. Hence, the reduction in the net charge transport. For frequencies over 1 kHz, the σ_{eff} increases as frequency increases, which will become higher than that of the clean mixture if the frequency is high enough. σ_{eff} saturates for frequencies over 100 kHz. The presence of both nonconductive and conductive inclusion phases will increase the $\varepsilon_{r,eff}$ (Fig. 4.4B) because the PPIP and SCAIP phenomena leads to larger charge accumulation around the interfaces, and hence higher net EM energy storage. The $\varepsilon_{r,eff}$ decreases as frequency increases and become stable for frequencies over around 5 MHz. When both conductive and nonconductive spherical inclusions exist in the mixture, the interfacial polarization effects of nonconductive inclusions are negligible compared to that of conductive inclusions, which is illustrated by the overlap of S1C1P1 and S1C2P1. In other words, the physical properties of conductive inclusions dominate the predictions.

7.3 The influence of the characteristic lengths of conductive and nonconductive particles

In this section, the PS model predictions of complex conductivity for mixtures containing both nonconductive and conductive inclusions/particles of various sizes are evaluated. The resulting plots are shown as Fig. 4.5. Curve S1 is the reference curve. The characteristic length of surface-charge-bearing nonconductive particle seems have little effect on model predictions when both conductive and nonconductive particles are present in the mixture (as observed from the overlap between S1C1P1 and S1C2P1). For conductive inclusions, a

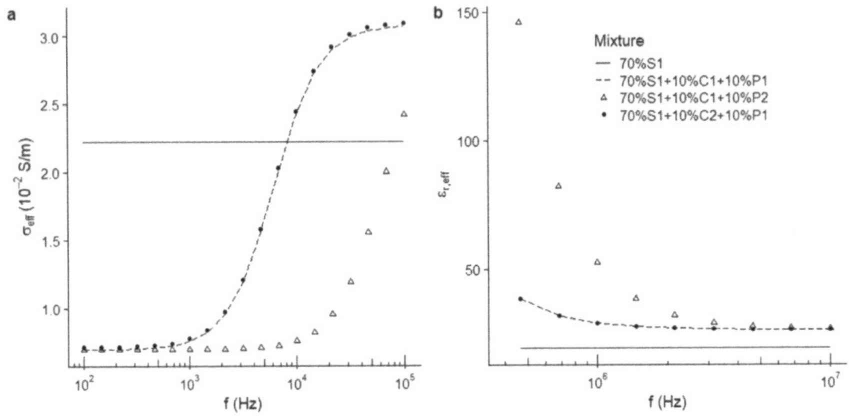

FIGURE 4.5 Comparison of the PS model predictions of (A) σ_{eff} and (B) $\varepsilon_{r,eff}$ of mixtures containing spherical sand, clay, and pyrite particles of various sizes (properties listed in Table 4.4) and fully saturated with electrolyte of $\varepsilon_{r,h}$ of 70, σ_h of 0.1 S/m and D_h of 10^{-9} m^2/s. C1 is larger than C2. P1 is larger than P2. *Yuteng Jin, Siddharth Misra.*

TABLE 4.4 Properties used for generating Fig. 4.5, where S1 represents sand, C1 and C2 represent two different clay types, and P1 and P2 represent two different pyrite types and the host is an electrolyte of $\varepsilon_{r,h}$ of 70, σ_h of 0.1 S/m, and D_h of 10^{-9} m^2/s.

	ϕ_i (%)	a_i (μm)	D_i (m^2/s)	$\varepsilon_{r,i}$	σ_i (S/m)	λ_i (S)
Sand (S1)	70	1000	–	4	–	10^{-9}
Clay1 (C1)	10	100	–	8	–	10^{-8}
Clay2 (C2)	10	10	–	8	–	10^{-8}
Pyrite1 (P1)	10	200	5×10^{-5}	12	500	–
Pyrite2 (P2)	10	20	5×10^{-5}	12	500	–

smaller characteristic length will shift the frequency dispersion of σ_{eff} and $\varepsilon_{r,eff}$ response to higher frequencies (i.e., shift the curves toward right in both the plots). Therefore, σ_{eff} reduces and $\varepsilon_{r,eff}$ increases for mixtures containing conductive particles of smaller characteristic lengths.

7.4 The influence of the volume fractions of conductive and nonconductive particles

The effects of volume fraction of nonconductive and conductive particles are evaluated in this section. Fig. 4.6 shows the result, where Curve S1 is the

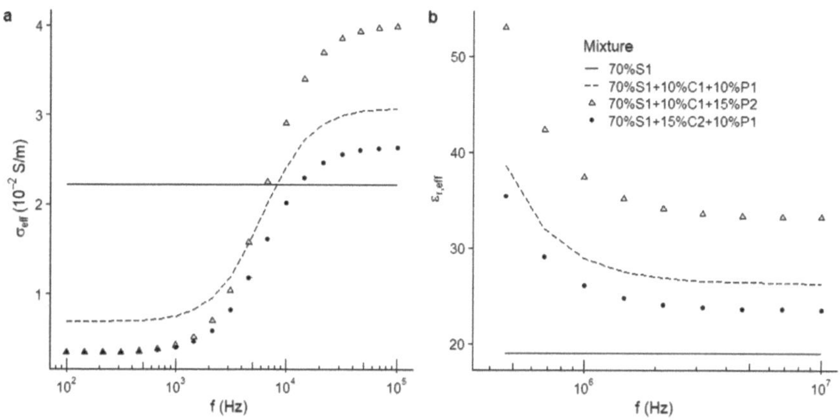

FIGURE 4.6 Comparison of the PS model predictions of (A) σ_{eff} and (B) $\varepsilon_{r,eff}$ of mixtures containing spherical sand, clay, and pyrite particles at various volume fractions of the inclusion phase (Table 4.5) and fully saturated with electrolyte of $\varepsilon_{r,h}$ of 70, σ_h of 0.1 S/m, and D_h of 10^{-9} m^2/s. C2 has higher concentration of clays as compared to C1. P2 has higher concentration of pyrites as compared to P1. *Yuteng Jin, Siddharth Misra.*

reference curve. As the volume fraction of particles increases, the porosity decreases. At low frequency, both the conductive and surface-charge bearing nonconductive particles act as insulators because of charge accumulation at the interface. Consequently, the conductivity of mixture decreases with increase in volume fraction of the particles. This is evident in the S1C2P1 and S1C1P2 curves exhibiting the same values at low frequency. Comparing the curves S1C1P1 and S1C2P1 in Fig. 4.6A, an increase in volume fraction of surface-charge-bearing nonconductive particles shift the curve downward, as demonstrated in Section 7.1 in Chapter 3. On the other hand, comparing the curves S1C1P1 and S1C1P2, an increase in the volume fraction of conductive particles causes a steep rise (a high rate of increase) in the frequency dispersion of σ_{eff}. For frequencies lower than 1 kHz, σ_{eff} values are constant. Comparison of S1C1P1 and S1C1P2 curves in Fig. 4.6B indicates an increase in volume fraction of conductive particles shifts the curve to higher values because of the effect of IFP, whereas comparison of S1C1P1 and S1C2P1 curves indicates an increase in the volume fraction of nonconductive particles shifts the frequency dispersion curve of $\varepsilon_{r,eff}$ to lower values as demonstrated in Section 7.1 in Chapter 3.

7.5 The influence of the conductivity of pore-filling electrolyte

The effects of conductivity of electrolyte are evaluated in this section. As for σ_{eff} prediction, we can clearly observe that the increase of electrolyte conductivity shifts the dispersion to higher frequencies (right shift).

TABLE 4.5 Properties used for generating Fig. 4.6, where S1 represents sand, C1 and C2 represent clay with different volume fractions, and P1 and P2 represent pyrite with different volume fractions, and the host is an electrolyte of $\varepsilon_{r,h}$ of 70, σ_h of 0.1 S/m, and D_h of 10^{-9} m^2/s.

	ϕ_i (%)	a_i (μm)	D_i (m^2/s)	$\varepsilon_{r,i}$	σ_i (S/m)	λ_i (S)
Sand (S1)	70	1000	–	4	–	10^{-9}
Clay1 (C1)	10	100	–	8	–	10^{-8}
Clay2 (C2)	15	100	–	8	–	10^{-8}
Pyrite1 (P1)	10	200	5×10^{-5}	12	500	–
Pyrite2 (P2)	15	200	5×10^{-5}	12	500	–

The low-frequency dispersion can hardly be observed for electrolyte with conductivity value higher than 1 S/m (Fig. 4.7C). When the electrolyte conductivity is increased by one order of magnitude, the corresponding σ_{eff} prediction also increases by one order of magnitude, similar to Archie's law. As for $\varepsilon_{r,eff}$ prediction, it can also be concluded that $\varepsilon_{r,eff}$ is positively related to the conductivity of electrolyte because a greater charge carrier concentration in electrolyte leads to greater charge accumulation around the interface. Moreover, the conductive particles dominate the prediction of $\varepsilon_{r,eff}$ for frequencies below 5 MHz, especially with the conductivity of the electrolyte is high.

8. Use of PPIP–SCAIP model in subsurface/material characterization

The PS model can be used in various fields, including colloid and interface science, bioscience, electrochemistry, geophysics, subsurface characterization of water resources systems, geothermal/fossil-fuel energy systems, carbon sequestration and mining industries to characterize the materials containing conductive particles in redox inactive conditions. Typical applications include the complex conductivity response of chalcopyrite/pyrite grains uniformly distributed in quartz grain/porous sand matrix and fully saturated with electrolyte without active ions (Mahan et al., 1986; Abdel Aal et al., 2014).

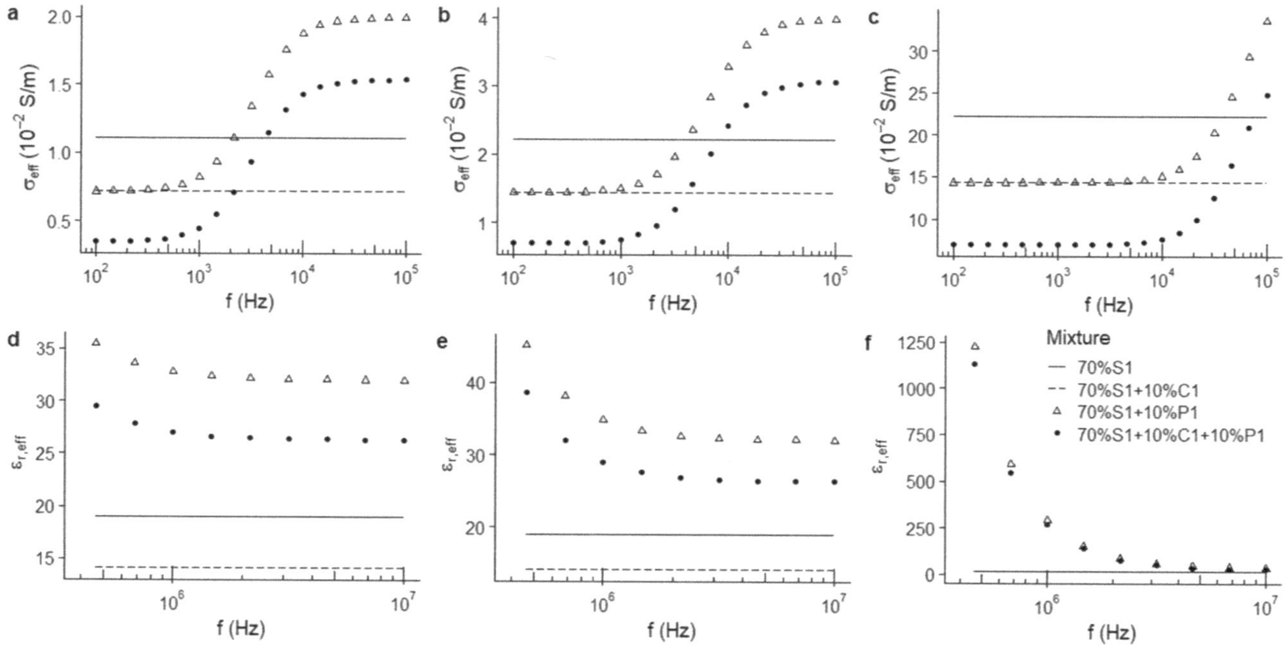

FIGURE 4.7 Comparison of the PS model predictions of (A) σ_{eff} and (B) $\varepsilon_{r,eff}$ of mixtures containing spherical sand, clay, and pyrite particles (properties listed in Table 4.6) and fully saturated with electrolyte of various conductivity σ_h, $\varepsilon_{r,h}$ of 70, and D_h of 10^{-9} m^2/s (A), (B), (C) represent σ_{eff} prediction, and (D), (E), (F) represent $\varepsilon_{r,eff}$ prediction for mixtures fully saturated with 0.05, 0.1, and 1 S/m electrolyte, respectively. *Yuteng Jin, Siddharth Misra.*

TABLE 4.6 Properties used for generating Fig. 4.7, where S1 represents sand, C1 represents clay, and P1 represents pyrite, and the host is an electrolyte of $\varepsilon_{r,h}$ of 70 and D_h of 10^{-9} m^2/s.

	ϕ_i (%)	a_i (μm)	D_i (m^2/s)	$\varepsilon_{r,i}$	σ_i (S/m)	λ_i (S)
Sand (S1)	70	1000	–	4	–	10^{-9}
Clay (C1)	10	100	–	8	–	10^{-8}
Pyrite (P1)	10	200	5×10^{-5}	12	500	–

9. Conclusions

The chapter presents a mechanistic model of multifrequency conductivity and permittivity of a homogeneous fluid-filled of porous media containing surface-charge-bearing nonconductive particles (e.g., clays and sand) and conductive mineral particles (e.g., pyrite). The porous media is saturated with hydrocarbon and water. PPIP and SCAIP models are combined into PS model, which accounts for the IFP effects due to the surface conductance of clays and that due to the conductive particles on the conductivity and permittivity of porous materials.

The IFP effects of clays and conductive minerals on the effective conductivity are studied in the frequency range of 100 Hz to 100 kHz and that on the effective permittivity are studied in the frequency range of 0.5 MHz to 1 GHz. A decrease in size or an increase in the surface conductance of surface-charge-bearing nonconductive particles (such as clays) uniformly distributed in a porous homogeneous mixture leads to a slight decrease in the effective conductivity and an increase in the effective permittivity. The effects of clay particles with the same ratio of size to surface conductance and the same relative permittivity on the complex conductivity of the mixture are similar. In the frequency windows mentioned above, the frequency dispersion of complex conductivity because of the IFP effects for clays are negligible compared to conductive particles.

The presence of conductive particles increases the effective permittivity and decreases the effective conductivity at lower frequencies. A higher relative permittivity or conductivity of conductive particles increases effective conductivity and decreases effective permittivity of mixture. An opposite trend is observed with an increase in the diffusion coefficient of conductive particles. Notably, the effects of conductive particles with the same ratio of particle conductivity to diffusion coefficient and the same relative permittivity or those with the same value of conductivity times permittivity and diffusion coefficient on the complex conductivity of the mixture are similar. For conductive inclusions, a smaller size shifts the frequency dispersion of effective conductivity and effective permittivity response to higher frequencies. Therefore, effective conductivity reduces and effective permittivity increases for mixtures containing conductive particles of smaller characteristic lengths. The low-frequency dispersion can hardly be observed for electrolyte with conductivity value higher than 1 S/m. Effective permittivity is positively related to the conductivity of electrolyte because a greater charge carrier concentration in electrolyte leads to greater charge accumulation around the interface. Moreover, the conductive particles dominate the prediction of effective permittivity for frequencies below 5 MHz, especially when the conductivity of the electrolyte is high. Overall, for the conductive particles studied in this paper, the frequency dispersion in effective conductivity occurs for frequency between 1 and 100 kHz, whereas the dispersion in effective permittivity occurs for frequency lower than about 5 MHz.

Nomenclature

Acronyms

EM electromagnetic
IFP interfacial polarization
PNP Poisson–Nernst–Planck
PPIP perfectly polarized interfacial polarization
PS PPIP–SCAIP
SCAIP surface-conductance-assisted interfacial polarization

Symbols

$\boldsymbol{a}$ characteristic length of inclusion phase (m)
$\boldsymbol{c}$ charge density variation ($1/m^3$)
$\boldsymbol{d}$ net charge density variation ($1/m^3$)
$\boldsymbol{D}$ diffusion coefficient of charge carriers (m^2/s)
$\boldsymbol{e}$ electric field vector
ε dielectric permittivity (F/m)
ε_{eff} effective dielectric permittivity of the mixture (F/m)
ε_r relative permittivity
$\boldsymbol{f}$ frequency (Hz)
$\boldsymbol{f}(\omega)$ dipolarizability (dipolar field coefficient)
$\boldsymbol{j}$ current density (A/m^3)
$\mathbf{i}$ square root of -1
$\boldsymbol{i_n}$ modified spherical Bessel function of the first kind of nth order
λ surface conductance (S)
μ electrical mobility [$m^2/(V\cdot s)$]
$\boldsymbol{N}$ charge carrier density ($1/m^3$)
ω angular frequency of the electric field (rad/s)
φ electrical potential (V)
ϕ volume fraction (%)
$\boldsymbol{q}$ elementary charge (1.6×10^{-19} C)

r radial distance along the normal to the interface (m)
ρ resistivity (Ω·m)
σ electrical conductivity (S/m)
σ^* complex electrical conductivity (S/m)
σ_{eff} effective electrical conductivity of the mixture (S/m)
σ^*_{eff} effective complex electrical conductivity of the mixture (S/m)
θ angle between the normal to the interface and the incident external electric field (°)
Θ phase angle (mrad)

Subscripts

0 at time equal to 0 s
cond conductive particles
eff effective
h host medium
i inclusion phase
j type of medium/phase
$\hat{n}$ unit vector
ncond nonconductive particles
o oil
r relative

Superscripts

+ positively charged carrier
- negatively charged carrier

References

Abdel Aal, G.Z., Atekwana, E.A., Revil, A., September, 2014. Geophysical signatures of disseminated iron minerals: a proxy for understanding subsurface biophysicochemical processes. J. Geophys. Res.: Biogeosci. 119 (9), 1831–1849.

Chu, K.T., Bazant, M.Z., July 6, 2006. Nonlinear electrochemical relaxation around conductors. Phys. Rev. 74 (1), 011501.

Cosenza, P., Ghorbani, A., Camerlynck, C., Rejiba, F., Guérin, R., Tabbagh, A., October 1, 2009. Effective medium theories for modelling the relationships between electromagnetic properties and hydrological variables in geomaterials: a review. Near Surf. Geophys. 7 (5–6), 563–578.

Garcia, A., Barchini, R., Grosse, C., September 14, 1985. The influence of diffusion on the permittivity of a suspension of spherical particles with insulating shells in an electrolyte. J. Phys. Appl. Phys. 18 (9), 1891.

Giordano, S., May 1, 2003. Effective medium theory for dispersions of dielectric ellipsoids. J. Electrost. 58 (1–2), 59–76.

Grosse, C., Barchini, R., March 14, 1992. The influence of diffusion on the dielectric properties of suspensions of conductive spherical particles in an electrolyte. J. Phys. Appl. Phys. 25 (3), 508.

Hou, C.Y., Feng, L., Seleznev, N., Freed, D.E., September 1, 2018. Low frequency complex dielectric (conductivity) response of dilute clay suspensions: modeling and experiments. J. Colloid Interface Sci. 525, 62–75.

Jin, Y., Misra, S., Homan, D., Rasmus, J., Revil, A., August 1, 2019. Mechanistic model of multi-frequency complex conductivity of porous media containing water-wet nonconductive and conductive particles at various water saturations. Adv. Water Resour. 130, 244–257.

Mahan, M.K., Redman, J.D., Strangway, D.W., August 1, 1986. Complex resistivity of synthetic sulphide bearing rocks. Geophys. Prospect. 34 (5), 743–768.

Misra, S., Torres-Verdín, C., Revil, A., Rasmus, J., Homan, D., March 1, 2016. Interfacial polarization of disseminated conductive minerals in absence of redox-active species—Part 1: mechanistic model and validation. Geophysics 81 (2), E139–E157.

Misra, S., Torres-Verdín, C., Revil, A., Rasmus, J., Homan, D., March 1, 2016. Interfacial polarization of disseminated conductive minerals in absence of redox-active species—Part 2: effective electrical conductivity and dielectric permittivityInterfacial polarization due to inclusions. Geophysics 81 (2), E159–E176.

Revil, A., Abdel Aal, G.Z., Atekwana, E.A., Mao, D., Florsch, N., August 12, 2015. Induced polarization response of porous media with metallic particles—part 2: comparison with a broad database of experimental data. Geophysics 80 (5), D539–D552.

Sihvola, A., January 1, 2007. Dielectric polarization and particle shape effects. J. Nanomater. 2007 (1), 5.

Tuncer, E., Gubański, S.M., Nettelblad, B., June 15, 2001. Dielectric relaxation in dielectric mixtures: application of the finite element method and its comparison with dielectric mixture formulas. J. Appl. Phys. 89 (12), 8092–8100.

Chapter 5

Effects of wettability of conductive and nonconductive particles on the multifrequency electromagnetic response of porous material

Siddharth Misra[1,2], **Yuteng Jin**[3]
[1]*Harold Vance Department of Petroleum Engineering, College of Engineering, Texas A&M University, TX, United States;* [2]*Department of Geology and Geophysics, Texas A&M University, TX, United States;* [3]*Harold Vance Department of Petroleum Engineering, Texas A&M University, TX, United States*

1. Wettability of conductive and nonconductive particles and its influence on electromagnetic properties of fluid-filled porous materials

Conductive minerals such as graphite, pyrite, and rutile are common in the subsurface, especially in hydrocarbon reservoir. Nonconductive minerals such as clay minerals can exhibit surface conductance. Consequently, the bulk conductivity and surface conductanceof minerals influence the electromagnetic (EM) field propagating through the geological materials. Kerogen, commonly present in mudrocks, on excessive maturation generates graphite (Ujiié, 1978). In the deep subsurface, high temperature and pressure mature the kerogen through diagenesis, catagenesis, and metagenesis processes. Graphitization of kerogen occurs when the maturation goes beyond the metagenesis stage (Spötl et al., 1998). During the graphitization, the structure of kerogen undergoes progressive rearrangements to form ordered carbon structure and increased aromatization. Consequently, the molecules in kerogen neatly stack as sheets. The final stable structure of such molecules under high pressure and temperature constitutes graphite (Tissot and Welte, 2013). Thus, graphite can be found in hydrocarbon-bearing rocks, sedimentary rocks, and shales (Winchell, 1911; Clark, 1921; Bustin et al., 1995). Other conductive minerals, such as pyrite and

Multifrequency Electromagnetic Data Interpretation for Subsurface Characterization
https://doi.org/10.1016/B978-0-12-821439-8.00008-2

rutile, are generated together with kerogen/hydrocarbon production and are common in shales and mudrocks (Laughrey et al., 2011; Bernard et al., 2013).

Wettability is defined as the tendency of one fluid to spread on, or adhere to, a solid surface in the presence of other immiscible fluids (Crain, 2002). An understanding of wettability of conductive/nonconductive minerals is required for solving various geophysical, fluid distribution, fluid flow, and material-characterization problems. Chakarov et al. (1995) mentioned that water can adsorb on the graphite surface by forming hydrogen-bonded aggregates. As discussed by Kozbial et al. (2014), highly ordered pyrolytic graphite is intrinsically water wet, which was overserved when measuring the water contact angle within 10 seconds after graphite exfoliation. The authors observed the wettability of graphite gradually change to oil wet (or intermediate wet) after exposure to the ambient air. Kocabag et al. (1990) conducted experiments to investigate wettability of pyrite. They concluded that pyrite is oleophilic when its surface is free of oxidation, but it is hydrophilic with respect to a gas bubble. The oleophilicity will increase because of the formation of sulfur by oxidation at acidic $pH < 4$, while the oleophilicity will decrease with oxidation at neutral and alkaline solutions because of the formation of $Fe(OH)_2$ and $Fe(OH)_3$ on the surface.

Wettability of conductive/nonconductive particles governs the preferential spreading of fluids on the surface of the particles that influences the interfacial polarization (IFP) phenomena and charge transport/accumulation around the particles. Consequently, wettability of conductive/nonconductive particles influences the EM properties of fluid-filled porous materials especially when the fluids have distinct conductivity, permittivity, and wettability (Misra and Han, 2016; Han et al., 2017). Wettability of minerals influences the EM characterization of geomaterials and geological formations. Misra et al. (2021) and Misra et al. (2019) developed methodologies to estimate the wettability of a reservoir and access its organic maturity by processing the multifrequency electromagnetic measurements. To investigate the effect of wettability on dielectric properties, Garrouch and Sharma (1994) conducted a series of experiments on brine-saturated Berea sandstone and Ottawa sand-bentonite packs. They concluded that dielectric constant is not significantly affected by wettability for fully-brine-saturated rocks in the absence of any conductive inclusions. Furthermore, Capaccioli et al. (2000) experimentally investigated the wettability effect on electrical responses of partially saturated porous material. Their findings show that the change from strongly water wet to strongly oil wet will affect connectivity and shape of water phase, which led to a smaller charge transport contribution at low frequencies and smaller dielectric strength as well as shorter characteristic times of Maxwell–Wagner–Sillars relaxation at higher frequency. Bona et al. (2002) performed experiments to study the influence of wettability on the electrical properties of porous material. They concluded that at low frequencies, charge transport is the dominant mechanism, and the governing factor is the connectivity of the water phase, whereas at high frequencies, the shape factor of the water phase is the controlling parameter. Bona et al. (1998), (1999) observed that strongly water-wet and strongly oil-wet glass filter and Berea sandstone samples show markedly different electrical responses. Moss et al. (2002) suggested that complex resistivity measurements have the potential as a noninvasive technique to evaluate reservoir wettability. Al-Ofi et al. (2018) observed linear correlations between wettability and dielectric dispersion parameters. In addition, Nguyen et al. (1999) observed that the dielectric permittivity of oil-wet sand is smaller than that of the water-wet sand at low-water saturation, which might be caused by the presence of an insulating oil film along the surface of the sand grains. On the contrary, the dielectric permittivity of oil-wet sand becomes much larger than that of the water-wet sand at higher water saturation. Revil et al. (2011) found that with the increase of oil saturation of the sand partially saturated with wetting oil, both phase magnitude and resistivity will decrease.

Most of the EM log interpretation methodology assume fully water-wet grains (Bücker et al., 2018; Han and Misra, 2018; Tathed et al., 2018; Zhdanov et al., 2018). This assumption is not valid for porous geological materials filled with two or more fluids having distinct conductivity, permittivity, and wettability. There are limited in-depth investigation of the effects of wettability of grains such as clay, sand, and pyrite on the conductivity and permittivity of the mixture at various operating frequency ranging from 1 Hz to 1 GHz (Jin et al., 2020a,b). Wettability influences the IFP phenomena that affects the frequency-dependent electrical properties (e.g., conductivity and permittivity); as a result, wettability can be estimated by analyzing the EM measurements. In this chapter, we developed a new mechanistic model to quantify the effects of wettability of conductive particles (e.g. pyrite) and nonconductive particles (e.g. clay) on the multifrequency complex conductivity/permittivity of fluid-filled porous geological materials as a function of the fluid saturation and the operating frequency of the externally applied EM field. The model allows the wettability to range from strongly wet to intermediate wet with respect to the two pore-filling fluids. The mechanistic model described in this chapter is a novel model for the EM-based characterization of geological formations, geomaterials, and porous materials with a special emphasis on hydrogeological applications and upstream oil and gas exploration. For purposes of simplifying the reading experience, the two pore-filling fluids having distinct conductivity, permittivity, and wettability are assumed to be oil and water.

2. Wettability model for spherical conductive and nonconductive particles in two immiscible fluids

For developing the mechanistic model of the effects of wettability of conductive/nonconductive particles on the multifrequency conductivity/permittivity of fluid-filled porous materials, we first need to develop a mechanistic model of wetting angle of a conductive/nonconductive spherical particle in a mixture of two immiscible fluids. This requires us to solve the Young–Laplace equation for the spherical particle in the mixture of two fluids (e.g., oil and water) having a certain proportion of the two fluids, referred as the fluid saturations.

In this chapter, wettability refers to the differential spreading of the water phase and oil phase on the surface of a spherical particle inside a mixture of the two immiscible fluids. Young–Laplace equation is used to model this differential spreading. Young–Laplace equation is a nonlinear partial differential equation that describes the capillary pressure difference across the interface between two static fluids around a solid surface because of surface tension and interfacial tension. Young–Laplace equation relates the pressure difference to the shape of fluid interface around the solid interface to determine the shape of the oil-water interface (meniscus) at equilibrium by applying appropriate boundary conditions. In doing so, we obtain the expression of wetting angle of the spherical particle as a function of contact angle of the particle and properties of the fluids (e.g., saturation and density). In this chapter, contact angle is measured in the water phase where the oil–water interface meets the particle surface. Contact angle is measured in the water phase as the angle between the tangent to the surface and tangent to the interface. Contact angle depends on the wettability. A water-wet particle exhibits low values of contact angle indicating preferential spreading of the water phase on the particle, whereas an oil-wet particle exhibits large values of contact angle indicating preferential spreading of the oil phase on the particle.

Mechanistic modeling of the wetting angle enables the calculation of particle surface areas covered by the two fluids. Following that, Poisson–Nernst–Planck equation is invoked to quantify dipolarizability of the two differentially wetted surfaces (as discussed in Chapters 3 and 4). This improves the PPIP, SCAIP, and the PS model proposed by Misra et al. (2016a) and Jin et al. (2019) by introducing the expression of surface areas covered by the two immiscible fluids.

2.1 Model description

Fig. 5.1 shows the representative volume assumed for developing the mechanistic model of wetting angle of spherical particle in a mixture of two immiscible fluids (Jin et al. 2020b). The two fluids are chosen to be oil and water to simplify the reading experience. Oil and water have distinct conductivity, permittivity, and wettability. The wetting phase will try to surround the solid particle to satisfy the contact angle with respect to the non-wetting phase. The movement of the oil-water interface around the particle (based on the wettability of the particle) generates a wetting angle, which represents the degree of exposure and preference of the particle to the water phase. The IFP phenomena because of such solid particles are entirely governed by the extent to which the solid particle is surrounded by water versus oil, which is governed by the oil saturation and wettability of the solid particle. In Fig. 5.1, the wetting angle ϕ represents the degree of exposure of the particle to the water phase, whereas the contact angle θ is a fixed property of the particle for a

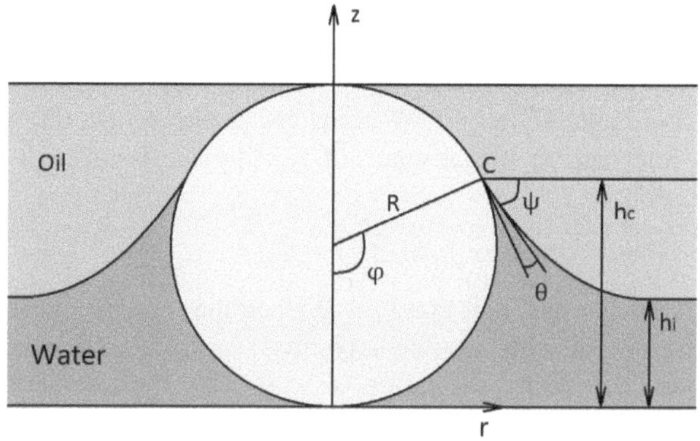

FIGURE 5.1 Cross-section of the representative volume assumed for developing the mechanistic model of wetting angle of a conductive/nonconductive spherical particle in a mixture of two fluids. The representative volume contains a spherical particle suspended in an oil-water medium, where the point C denotes the point where the oil-water interface contacts the particle surface; θ is the contact angle of spherical particle; ϕ is the wetting angle; ψ is the angle between oil-water interface and the horizon (r-axis) at point C; R is the radius of spherical particle; h_i is the uniform height of oil-water interface in the absence of wetting of the particle (far-field height); h_c is the height where the oil-water interface contacts the particle surface, such that $h_c = R(1 - \cos\phi)$; and $h(r)$ is the height of oil-water interface at any distance r away from the vertical axis z. *Yuteng Jin, Siddharth Misra.*

given wetting/non-wetting phase. For example, the contact angle of n-Heptane on polytetrafluoroethylene in the presence of air is fixed to be 21 degrees (Fox and Zisman, 1950). It should be noted that θ is hard to measure, and the particles like graphite and pyrite are not spherical in shape. Limitations and assumptions of this study are listed in a later section of the chapter.

2.2 Assumptions

At the level of representative volume, we assume the following:

- The high-density conductive fluid layer (e.g., water) lies below the low-density nonconductive fluid layer (e.g., oil). The particle can get preferentially wetted by either the condutive or the non-conductive fluid, which are immisible fluids.
- The two fluid layers have a common smoothly varying interface that is affected because of the presence of the spherical particle.
- The spherical particle suspends at the interface and one of the two immiscible fluids preferentially wet the spherical particle.
- The location of the wetting/non-wetting phase interface far away from the spherical particle is proportional to the fluid saturation; this assumes that particles are far away from each other and have no influence on each other, i.e. the fluid layers and the wetting/non-wetting phase interface are spread over horizontal distances that are orders of magnitude larger than the particle size.

2.3 Model development

2.3.1 Far-field height of the interface

We assume the far-field height of oil-water interface or the wetting/non-wetting fluid interface in the absence of wetting, h_i, has linear relationship with oil saturation:

$$h_i = 2R(1 - S_o) \tag{5.1}$$

where S_o is oil saturation, such that $h_i = 0$ when the representative volume has 100% saturation of oil (the low-density fluid) and $h_i = 2R$ when the representative volume has 100% saturation of water (the high density fluid). This assumption is valid when the layers of water and oil are assumed to be spread over horizontal distances that are orders of magnitude larger than the radius of the sphere. S_o can be related to the volume fraction of oil in the material, ϕ_o, using the following equation:

$$S_o = \frac{\phi_o}{\phi} \tag{5.2}$$

where ϕ denotes porosity of the porous material. The two-phase fluid fills the entire porosity.

2.3.2 Young–Laplace equation for computing the shape of the interface around a spherical particle due to preferential-wettting conditions

When two liquids contact a solid particle, the wetting phase will spread on the surface of particle, which changes the shape of the interface of the two liquids. The shape of the fluid-fluid interface (meniscus) at equilibrium is described by Young–Laplace equation (Cavallaro, 2012):

$$\Delta p = 2H\gamma \tag{5.3}$$

where Δp is Laplace pressure defined as the pressure difference between the inside and the outside of the curved surface that forms the boundary between the oil and water phases. H is mean curvature of the meniscus surface, and γ is interfacial tension between oil and water. Laplace pressure can be expressed as

$$\Delta p = (\rho_w - \rho_o)g[h(r) - h_i] \tag{5.4}$$

where ρ_w and ρ_o are densities of water and oil, respectively, and g denotes gravitational acceleration. The mean curvature can be expressed as a surface divergence (Kralchevsky et al., 1994):

$$2H = \nabla \cdot \left[\frac{\nabla h}{\sqrt{1 + (\nabla h)^2}} \right] \tag{5.5}$$

Bond number in this chapter is defined by $B_o = \frac{(\rho_w - \rho_o)gR^2}{\gamma}$, which is the ratio between gravity force and surface tension force. For large Bond number, the size of spherical particle is much larger compared to the capillary length (Champmartin et al., 2016). Assume typical values: $\rho_o = 800\text{kg}/\text{m}^3$, $\rho_w = 1000\text{kg}/\text{m}^3$, $g = 9.8\text{N}/\text{kg}$, $R = 200\mu\text{m}$, and $\gamma = 0.05\text{N}/\text{m}$, we can calculate $B_o = 0.0016 \ll 1$. Such a small Bond number represents negligible gravity force, and the mean curvature remains constant everywhere on the interface, which renders a small slope assumption. Under this condition, we have $(\nabla h)^2 \ll 1$; consequently, the mean curvature simplifies to (Cavallaro, 2012):

$$2H = \nabla \cdot \nabla h = \Delta h \tag{5.6}$$

Since the meniscus surface is axisymmetric, the mean curvature can be further simplified under cylindrical coordinates as

$$2H = \frac{1}{r}\frac{\partial}{\partial r}\left(r\frac{\partial h}{\partial r}\right) = h'' + \frac{h'}{r} \tag{5.7}$$

where h' and h'' represents $\frac{dh}{dr}$ and $\frac{d^2h}{dr^2}$, respectively. Substituting Eqs. (5.4) and (5.7) into Eq. (5.3), the Young–Laplace equation becomes:

$$\frac{(\rho_w - \rho_o)g[h(r) - h_i]}{\gamma} = h'' + \frac{h'}{r} \tag{5.8}$$

Following dimensionless variables need to be defined to normalize Eq. (5.8):

$$\widehat{r} = \frac{r}{L_c}, G(\widehat{r}) = \widehat{h}(r) - \widehat{h}_i, \ \widehat{h} = \frac{h}{L_c} \tag{5.9}$$

where L_c is capillary length, defined by

$$L_c = \sqrt{\frac{\gamma}{(\rho_w - \rho_o)g}} \tag{5.10}$$

Substituting Eqs. (5.9) and (5.10) into Eq. (5.8), we obtain a modified Bessel differential equation (Fiegel et al., 2005):

$$G'' + \frac{G'}{\widehat{r}} - G = 0 \tag{5.11}$$

where G' and G'' represent $\frac{dG}{d\widehat{r}}$ and $\frac{d^2G}{d\widehat{r}^2}$, respectively. The solution of the differential Eq. (5.11) consists of modified Bessel functions of the first and second kind of order 0, which is shown below:

$$G = c_1 I_0(\widehat{r}) + c_2 K_0(\widehat{r}) \tag{5.12}$$

2.3.3 Boundary conditions

Two boundary conditions are required to solve Eq. (5.12):

- The height of oil-water interface at infinite distance from the vertical axis z along the direction of horizontal axis r, $h(r)|_{r\to\infty}$, is equal to h_i, which gives

$$\lim_{\widehat{r}\to\infty} G=0 \tag{5.13a}$$

- The height of oil-water interface at distance $r=R\sin\phi$ is h_c, which gives

$$G(\widehat{r}=\sqrt{B_o}\sin\phi)=\widehat{h}_c-\widehat{h}_i \tag{5.13b}$$

2.3.4 Determination of wetting angle

Based on boundary condition expressed as Eq. (5.13a), $c_1=0$. Based on boundary condition expressed as Eq. (5.13b), $c_2=\frac{\widehat{h}_c-\widehat{h}_i}{K_0(\sqrt{B_o}\sin\phi)}$, where K_0 is modified Bessel function of the second kind of order 0. Substitute c_1, c_2, and Eq. (5.9) into Eq. (5.12) to obtain the expression for the shape of the oil-water interface around the spherical particle at low Bond number:

$$\widehat{h}=\widehat{h}_i+\frac{\widehat{h}_c-\widehat{h}_i}{K_0(\sqrt{B_o}\sin\phi)}K_0(\widehat{r}) \tag{5.14}$$

At point C, we have

$$\tan\psi=-\frac{dh}{dr}\bigg|_{r=R\sin\phi}=-\frac{d\widehat{h}}{d\widehat{r}}\bigg|_{\widehat{r}=\sqrt{B_o}\sin\phi}=\frac{\widehat{h}_c-\widehat{h}_i}{K_0(\sqrt{B_o}\sin\phi)}K_1(\sqrt{B_o}\sin\phi) \tag{5.15}$$

where K_1 is modified Bessel function of the second kind of order 1. The angle ψ can thereby calculated as

$$\psi=\tan^{-1}(\tan\psi)=\tan^{-1}\left[\frac{\widehat{h}_c-\widehat{h}_i}{K_0(\sqrt{B_o}\sin\phi)}K_1(\sqrt{B_o}\sin\phi)\right] \tag{5.16}$$

The three angles, θ, ϕ, and ψ, can be related using equation

$$180=\theta+\phi+\psi \tag{5.17}$$

Substituting Eq. (5.16) into Eq. (5.17), we have

$$\phi = 180 - \theta - \tan^{-1}\left[\frac{\widehat{h}_c - \widehat{h}_i}{K_0(\sqrt{B_o}\sin\phi)} K_1(\sqrt{B_o}\sin\phi)\right] \tag{5.18a}$$

It is hard to obtain an analytical solution for wetting angle ϕ from Eq. (5.18a) because it is a transcendental equation. Eq. (5.18a) can be solved using numerical root-finding methods to solve function $f(\phi)$ obtained by modifying Eq. (5.18a).

$$f(\phi) = 180 - \phi - \theta - \tan^{-1}\left[\frac{\widehat{h}_c - \widehat{h}_i}{K_0(\sqrt{B_o}\sin\phi)} K_1(\sqrt{B_o}\sin\phi)\right] \tag{5.18b}$$

Newton–Raphson method is used to find ϕ that leads to $f(\phi) = 0$. To ensure that the equation has a unique solution, we examine the monotonicity of the function by plotting $f(\phi)$, as shown in Fig. 5.2. The plot shows that $f(\phi)$ is a monotonic function in the range of $0° < \phi < 180°$.

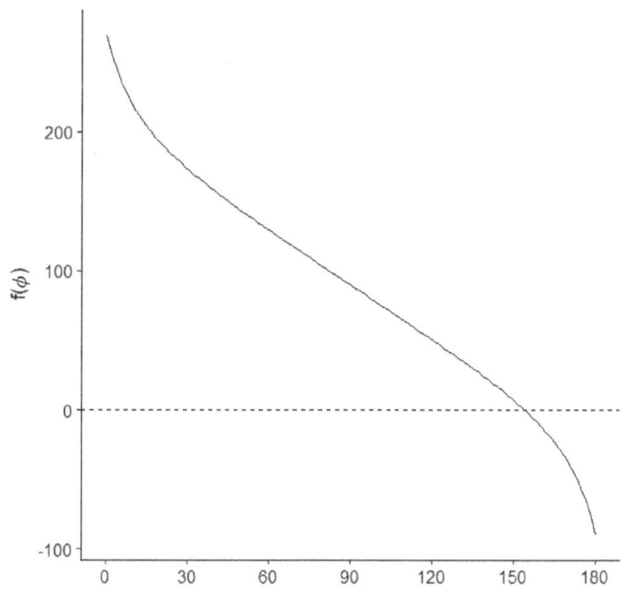

FIGURE 5.2 The plot of $f(\phi)$ in the range of $0° < \phi < 180°$. This plot is generated assuming following parameters: $\rho_o = 800\text{kg/m}^3$, $\rho_w = 1000\text{kg/m}^3$, $g = 9.8\text{N/kg}$, $R = 200\mu\text{m}$, $\gamma = 0.05\text{N/m}$, $\theta = 0°$, and $S_o = 0.5$ defining the representative volume shown in Fig. 5.1. *Yuteng Jin, Siddharth Misra.*

2.3.5 Fractions of surface area of particle covered by water and oil

After calculating the wetting angle, the fraction of surface area of the particle covered by water (p_w) and that by oil (p_o) can be calculated using equation for the curved surface area of a spherical cap based on the wetting angle:

$$p_w = \frac{A_w}{A_s} = \frac{2\pi R^2(1 - \cos\phi)}{4\pi R^2} = \frac{1 - \cos\phi}{2} \tag{5.19}$$

and

$$p_o = \frac{A_o}{A_s} = \frac{2\pi R^2(1 + \cos\phi)}{4\pi R^2} = \frac{1 + \cos\phi}{2} \tag{5.20}$$

respectively, where A_s is the surface area of the spherical particle, A_w and A_o is the surface area of the spherical particle (i.e., curved surface area of a spherical cap) covered by water and oil, respectively.

2.4 Limitations

Wetting angle estimation requires an assumption that the oil-water interface is smoothly varying with small rate of change in the height of interface $h(r)$ as a function of radial distance r. This requires that the horizontal distance r over which the interface deviates from the equilibrium height h_i is smaller than the capillary length L_c. If $r > L_c$, gravitational effects become significant and cause the interface height to decay exponentially (Cavallaro, 2012).

3. Improved PS model: mechanistic model of conductivity/permittivity of porous material containing conductive and nonconductive particles of any wettability immersed in two immiscible fluids

As discussed in the previous two chapters, in the presence of externally applied EM field, conductive particles (Chapter 4) and surface-charge-bearing nonconductive particles (Chapter 3) surrounded by pore-filling fluid give rise to interfacial polarization (IFP) phenomena that strongly influence the charge transport and charge accumulation. Under an external EM field generated by EM measurement tool/device, counterion cloud and diffused-charge distribution develops around the fluid-solid interfaces giving rise to the IFP effects. When the equilibrium state is perturbed because of the variation of the EM field, there is a delay in the counterion cloud and diffused-charge distribution reaching the new equilibrium state leading to frequency dispersion of effective electrical conductivity σ_{eff} and effective relative dielectric permittivity $\varepsilon_{r,eff}$ of the fluid-filled porous material containing conductive particles (e.g., pyrite particles) and surface-charge-bearing nonconductive particles (e.g., sand and clay particles).

The IFP phenomenon around strongly water-wet conductive particle is described by perfectly polarized interfacial polarization (PPIP) model as discussed in Chapter 4 and the IFP phenomenon around strongly water-wet surface-charge-bearing nonconductive particles is described by surface-conductance-assisted interfacial polarization (SCAIP) model as discussed in Chapter 3. In this chapter, we formally integrate PPIP model and SCAIP model into the PS model and improve the previously discussed PS model (designed for only strongly water-wet particles) by accounting for the effects of wettability of conductive/nonconductive particles on the IFP phenomena. Few predictions of PS model and comparison between the PPIP and SCAIP model predictions are presented in Chapter 4. First, a new mechanistic model is invoked to quantify the wetting angle because of the differential/preferential spread of wetting and nonwetting fluid around the spherical particles of a specific wettability, which can range from strongly oil-wet to strongly water-wet. After estimating the wetting angle, the dipolarizability of the particle for any wettability condition can be calculated by adopting a simple mixing law. Eqs. (5.19) and (5.20) express the fraction of surface area of the particle covered by water (p_w) and that covered by oil (p_o) as a function of oil saturation and contact angle.

In this chapter, we assume a fluid-filled porous material comprising spherical intermediately wet graphite (a conductive particle), spherical water-wet sand (a non-conductive particle with very low surface conductance), spherical intermediately wet clay (a non-conductive particle with substantial surface conductance), and spherical oil droplet (a non-conductive particle with no surface conductance) at any water saturation. According to the constituents of the mixture, the improved PS model expresses the effective complex conductivity of the porous material directed along the $\hat{n}$ unit vector as

$$\begin{aligned}\frac{\sigma^*_{\hat{n},eff} - \sigma^*_h}{\sigma^*_{\hat{n},eff} + 2\sigma^*_h} &= \phi_g p_{g,w} f_{g,w}(\omega) + \phi_g p_{g,o} f_{g,o}(\omega) + \phi_s f_{s,w}(\omega) \\ &\quad + \phi_c p_{c,w} f_{c,w}(\omega) + \phi_c p_{c,o} f_{c,o}(\omega) + \phi_o f_{o,w}(\omega)\end{aligned} \tag{5.21}$$

where $\sigma^*_{eff} = \sigma_{eff} + i\omega\varepsilon_{eff}$ is effective complex conductivity, σ_{eff} is effective conductivity, ε_{eff} is effective permittivity, $\omega = 2\pi f$ is the angular frequency, σ^*_h is the complex conductivity of host electrolyte/brine, ϕ_i is the volume fraction of inclusion i in the porous material, and $f_{i,h}(\omega)$ is the dipolarizability of inclusion i when surrounded by host phase h. The subscript i takes the form g, s, c, and o for graphite, sand, clay, and oil, respectively. The subscript h takes the form w and o for water and oil, respectively. For example, $f_{g,w}$ is dipolarizability of water-wet graphite and $f_{c,o}$ is dipolarizability of oil-wet clay. When the inclusion is strongly water-wet conductive particle, the dipolarizability is calculated using Eq. (4.8). When the inclusion is strongly water-wet nonconductive particle, the dipolarizability is calculated using Eq. (3.45).

Notably, oil droplets are considered as nonconductive particles without any surface charge or surface conductance, which makes modeling any fluid saturation in the porous material possible. Oil droplet is fully surrounded by water. $p_{i,h}$ is the fraction of surface area of the inclusion i covered by host phase h. The use of fractional surface area of the particle covered by water (p_w) and that covered by oil (p_o) in Eq. (5.21) ensures that the IFP effect of the spherical particle at any wettability can be accurately quantified for various saturations and operating frequencies.

4. Improved PS model predictions of conductivity and permittivity versus frequency

The effect of wettability of conductive/nonconductive particle and oil saturation on effective conductivity (σ_{eff}) and effective relative permittivity ($\varepsilon_{r,eff}$) predictions of the mixture is evaluated in this section. The properties of host mediums and inclusions used for generating the plots are summarized in the corresponding tables, where γ is interfacial tension between oil and water, ϕ_j is the volume fraction of a specific phase, a_j is the characteristic length (radius) of spherical particles, D_j is diffusion coefficient of charge carriers, $\varepsilon_{r,j}$ is relative permittivity, σ_j is conductivity, λ_j is surface conductance, θ_j is contact angle, and ρ_j is density. The subscript j takes the form of i for the inclusion phase (i.e., conductive/nonconductive particles) and h for the host phase (i.e., brine/electrolyte or oil). In this chapter, spherical oil blobs/droplets are considered as the inclusion phase when implementing the effective medium model of complex conductivity of the fluid-filled porous material. However, when estimating the wetting angle because of the preferential spreading of oil versus water on the conductive/nonconductive particle, oil phase is considered as the host layer around the conductive/nonconductive particle.

4.1 The porous material containing intermediately wet graphite

In the following four cases, we assume a porous material containing spherical intermediately wet graphite, spherical water-wet sand, spherical water-wet clay, and spherical oil droplets at any water saturation.

4.1.1 The influence of the contact angle of conductive particle

In this section, we investigate the influence of the contact angle of spherical conductive (graphite) particle on the improved PS model predictions of σ_{eff} and $\varepsilon_{r,eff}$ at the frequency range of 100 Hz to 10 MHz. In this chapter, contact angle is the angle measured in the water phase where the oil–water interface meets the particle surface. Contact angle of 0 degrees represents strongly water-wet particle and contact angle of 180 degrees represents strongly oil-wet particle. A water-wet particle exhibits smaller contact angle whereas the oil-wet particle exhibits higher contact angle. When the conductive particle (graphite) is oil wet, the net IFP of the mixture will reduce because of the reduction in the surface area of the conductive particle covered less by the conductive electrolyte. Fig. 5.3 indicates that the effect of wettability of conductive particles on conductivity and permittivity is negligible for frequencies greater than 0.1 and 1 MHz, respectively. Such a frequency threshold is a function of size of the conductive particles, host conductivity, and many other factors. As the frequency of the EM field increases, the occurrence of IFP phenomena, which require large scale reorientation of charge carriers and charge separations, become more difficult; consequently, at high frequency, only small-scale polarization phenomena (such as orientation/dipolar polarization) exist because such polarization phenomena can rapidly respond to the fast variations in EM field at high frequencies. In other words, at high frequencies, large-scale IFP phenomena fade away and only small-scale orientation/dipolar polarization phenomena exist. Wettability alteration being an interfacial process primarily influences the large-scale IFP phenomena as compared to the small-scale polarization phenomena. Large-scale IFP phenomena are dominant at lower frequencies. Wettability (contact angle) of conductive particles primarily governs the large-scale IFP at the interface of conductive particle and the surrounding host fluid; consequently, at high frequencies beyond the above-mentioned frequency thresholds, both conductivity and permittivity become insensitive to the variations in wettability.

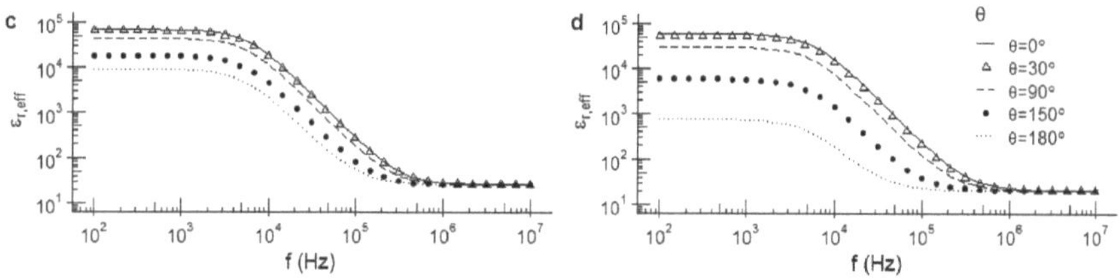

FIGURE 5.3 Comparison of the improved PS model predictions of (A and B) σ_{eff} and (C and D) $\varepsilon_{r,eff}$ of porous materials containing conductive graphite particle having various contact angles (0–180 degrees), sand grains and clay grains partially saturated with electrolyte and (A and C) 1% and (B and D) 9% volume fractions of oil, respectively, which correspond to oil saturations of 10% and 90%, respectively, given 10% porosity assumed for this case. The properties of host media are listed in Table 5.1. The properties of graphite, sand, clay, and oil as inclusions are listed in Table 5.2. *Yuteng Jin, Siddharth Misra.*

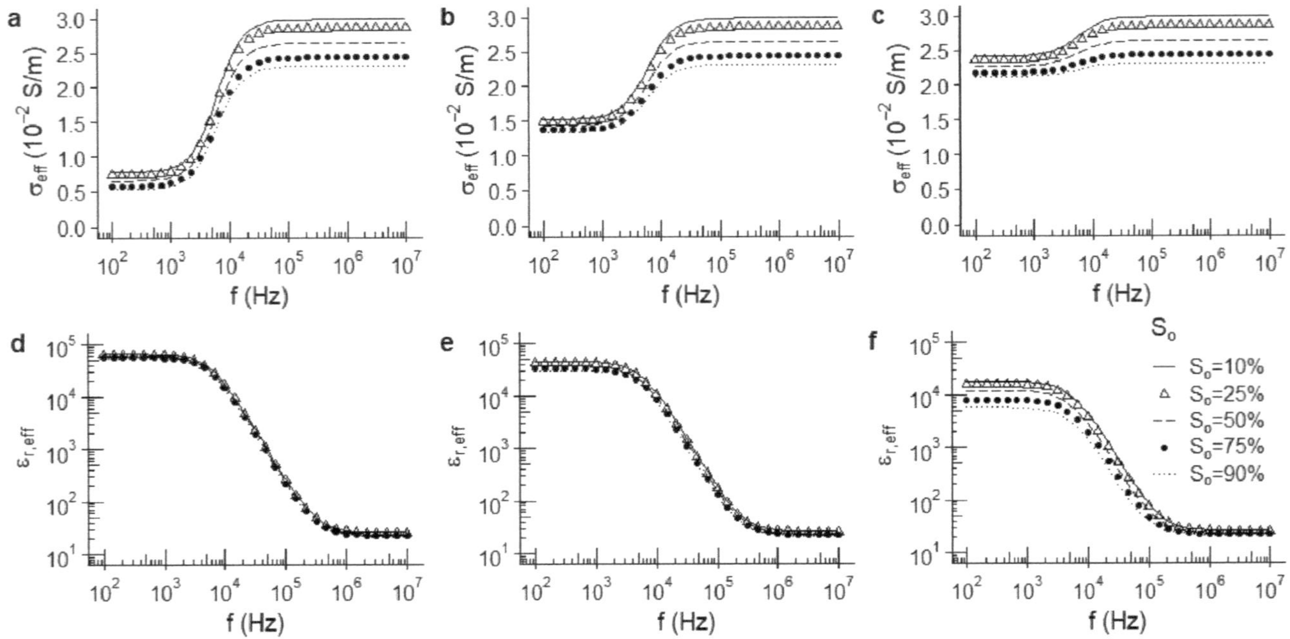

FIGURE 5.4 Comparison of the improved PS model predictions of (A, B and C) σ_{eff} and (D, E and F) $\varepsilon_{r,eff}$ of fluid-filled porous materials containing conductive graphite particles having contact angles of (A and D) 30 degrees, (B and E) 90 degrees or (C and F) 150 degrees, sand grains and clay grains partially saturated with electrolyte/brine and oil at various volume fractions. The properties of host media are listed in Table 5.1. The properties of graphite, sand, clay, and oil as inclusion are listed in Table 5.3. Oil volume fractions of 1%, 2.5%, 5%, 7.5%, and 9% corresponds to oil saturations of 10%, 25%, 50%, 75%, and 90%, respectively, given 10% porosity assumed for this case. *Yuteng Jin, Siddharth Misra.*

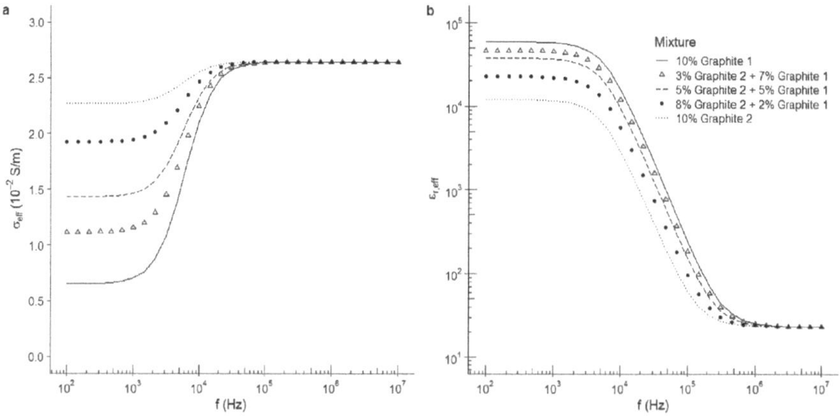

FIGURE 5.5 Comparison of the improved PS model predictions of (A) σ_{eff} and (B) $\varepsilon_{r,eff}$ of fluid-filled porous materials containing various volume fractions of water-wet graphite with contact angle of 30 degrees (Graphite 1), various volume fractions of oil-wet graphite with contact angle of 150 degrees (Graphite 2), sand grains and clay grains partially saturated with electrolyte and oil at 5% volume fraction of oil (which correspond to oil saturation of 50% given 10% porosity assumed for this case). The total volume fraction of graphite in each case is kept constant at 10%. The properties of host media are listed in Table 5.1. The properties of graphite, sand, clay, and oil as the inclusion phases are listed in Table 5.4. *Yuteng Jin, Siddharth Misra.*

For the lower frequencies, Fig. 5.3 indicates that as the conductive particles (graphite) become more oil wet, the frequency dependence of conductivity becomes negligible and that of permittivity decreases. This decrease in frequency dependence of conductivity and permittivity is because of reduction in IFP around the conductive particles when the conductive particles are preferentially wetted (covered) by oil instead of water. Notably, with increase in oil saturation, the sensitivity of permittivity dispersion to wettability increases (Fig. 5.3C and D); in other words, wettability of conductive particles has more influence on the permittivity at higher oil saturations. Wettability effects of conductive particles cannot be neglected when dealing with conductivity and permittivity measurements at low frequencies, especially for higher oil saturations. At higher oil saturations, permittivity dispersion measurement can serve as a better indicator of wettability as compared to conductivity dispersion measurement.

At low frequency, the charge carriers reach the equilibrium distribution around the particle interface, so that the polarized particles act as insulators, which lead to lower σ_{eff} and higher $\varepsilon_{r,eff}$ (Misra et al., 2016b; Jin et al., 2019). Counterintuitively, low-frequency conductivity decreases with increase in water wetness of conductive particles, indicating that at higher water-wet state of conductive particles, more charge carriers are involved in the charge polarization and accumulation as compared to charge transport. This is further supported by the drastic increase in the low-frequency permittivity when the

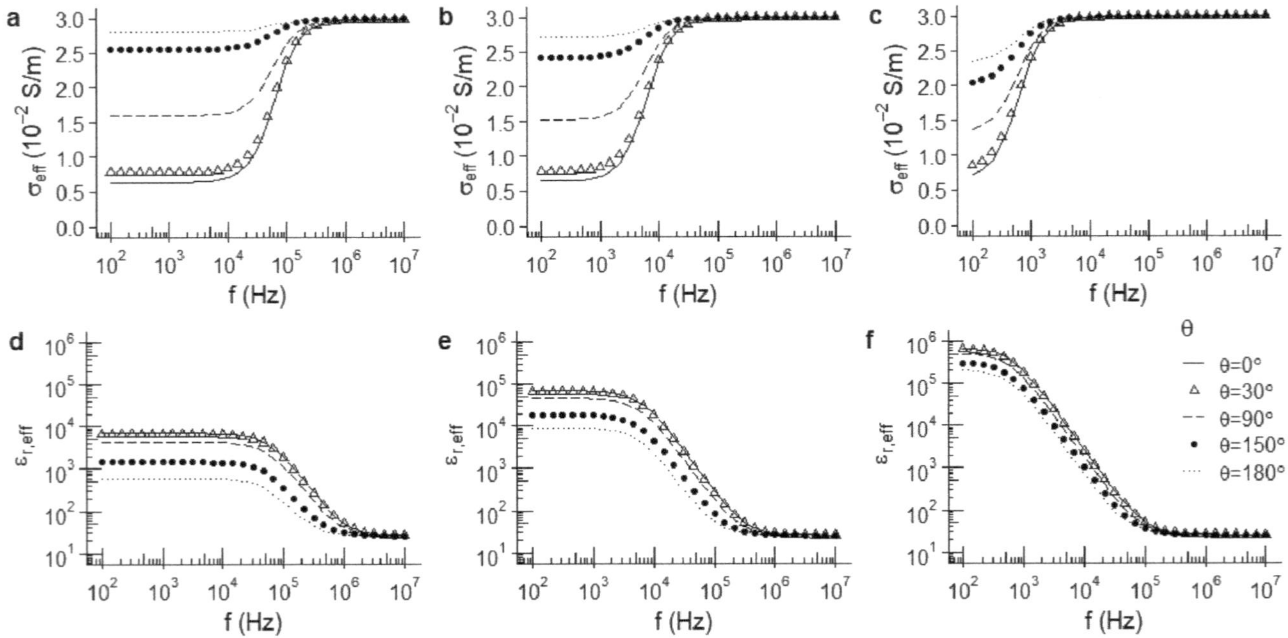

FIGURE 5.6 Comparison of the improved PS model predictions of (A, B and C) σ_{eff} and (D, E and F) $\varepsilon_{r,eff}$ of porous materials containing graphite (conductive) particle of radius of (A and D) 20 μm, (B and E) 200 μm and (C and F) 2000 μm having various contact angles (0–180 degrees), sand grains, and clay grains partially saturated with electrolyte and 1% volume fractions of oil, which corresponds to 10% oil saturation, given 10% porosity assumed for this case. The properties of host are listed in Table 5.1. The properties of graphite, sand, clay, and oil as inclusion phases are listed in Table 5.5. *Yuteng Jin, Siddharth Misra.*

conductive particles are strongly water wet as compared to oil wet scenarios. This is also a consequence of the assumption that the conductive and nonconductive particles do not touch/interact with each other. The result shows the ratio of high-frequency σ_{eff} at 90% to that at 10% oil saturation is around 1.29, while the ratio of high-frequency $\varepsilon_{r,eff}$ at 10% to that at 90% oil saturation is around 1.23. Moreover, for strongly water-wet conductive particle, the ratio of low-frequency σ_{eff} at 10% and 90% oil saturation is around 2.22. For strongly oil-wet conductive particle, the ratio of low-frequency $\varepsilon_{r,eff}$ at 10% and 90% oil saturation is around 11.96.

4.1.2 The influence of the wetting fluid saturation

In this section, we investigate the influence of the oil saturation of the fluid-filled porous material on the improved PS model predictions of σ_{eff} and $\varepsilon_{r,eff}$ at the frequency range of 100 Hz to 10 MHz. The frequency-dependent dispersions in conductivity and permittivity decrease with the increase in oil saturation because of the reduction in IFP phenomena arising from the reduction in charge carriers in the porous material. Low-frequency conductivity of the mixture is more sensitive to contact angle (wettability) of the conductive particle as compared to the oil saturation, whereas the high-frequency conductivity is more sensitive to oil saturation as compared to the contact angle of the conductive particle. Compared to low-frequency permittivity, high-frequency permittivity is much less sensitive to oil saturation. Compared to high-frequency conductivity, low-frequency conductivity is much less sensitive to oil saturation. Also, by comparing the rate of change

TABLE 5.1 Properties of host used for generating Figs. 5.3–5.6.

	D_h (m^2/s)	$\varepsilon_{r,h}$	σ_h (S/m)	ρ_h (kg/m^3)	γ (N/m)
Water	10^{-9}	70	0.1	1000	0.05
Oil	10^{-15}	2	10^{-30}	800	

TABLE 5.2 Properties of inclusions used for generating Fig. 5.3.

	ϕ_i (%)	a_i (μm)	D_i (m^2/s)	$\varepsilon_{r,i}$	σ_i (S/m)	λ_i (S)	θ_i (°)
Graphite	10	200	5×10^{-5}	12	500	–	0–180
Sand	70	1000	–	4	–	10^{-9}	–
Clay	10	100	–	8	–	10^{-8}	–
Oil	1 or 9	100	–	2	–	10^{-30}	–

TABLE 5.3 Properties of inclusions used for generating Fig. 5.4.

	ϕ_i (%)	a_i (μm)	D_i (m^2/s)	$\varepsilon_{r,i}$	σ_i (S/m)	λ_i (S)	θ_i (°)
Graphite	10	200	5×10^{-5}	12	500	–	30, 90 or 150
Sand	70	1000	–	4	–	10^{-9}	–
Clay	10	100	–	8	–	10^{-8}	–
Oil	1–9	100	–	2	–	10^{-30}	–

among curves in Fig. 5.3A and C and Fig. 5.4A and D, we can conclude that in our model, the effect of oil saturation is less than the effect of contact angle (i.e., the contact angle plays a primary effect and oil saturation plays a secondary effect).

4.1.3 The influence of the volume fraction of water-wet and oil-wet conductive particles

In this section, we investigate the influence of the volume fractions of water-wet graphite particles (referred as Graphite 1) and oil-wet graphite particles (referred as Graphite 2) on improved PS model predictions of σ_{eff} and $\varepsilon_{r,eff}$ at the frequency range of 100 Hz to 10 MHz. In this case, the total volume fraction of graphite particles (Graphite 1 and Graphite 2) in the fluid-filled porous material is fixed at 10% while varying the proportion of Graphite 1 and Graphite 2. Fig. 5.5 shows the frequency dispersions of conductivity and permittivity and their sensitivities to the change in wettability of the graphite particles. As the volume fraction of water-wet graphite decreases with the simultaneous increase in volume fraction of oil-wet graphite, the frequency dispersion reduces because more graphite surface is covered by oil resulting in lower IFP phenomena. High-frequency conductivity and permittivity are not sensitive to the wettability of the graphite particles.

TABLE 5.4 Properties of inclusions used for generating Fig. 5.5.

	ϕ_i (%)	a_i (μm)	D_i (m^2/s)	$\varepsilon_{r,i}$	σ_i (S/m)	λ_i (S)	θ_i (°)
Graphite 1	0–10	200	5×10^{-5}	12	500	–	30
Graphite 2	10–0	200	5×10^{-5}	12	500	–	150
Sand	70	1000	–	4	–	10^{-9}	–
Clay	10	100	–	8	–	10^{-8}	–
Oil	5	100	–	2	–	10^{-30}	–

4.1.4 The influence of the radius of conductive particles

In this section, we investigate the influence of the radius of spherical conductive (graphite) particle on the improved PS model predictions of σ_{eff} and $\varepsilon_{r,eff}$ at the frequency range of 100 Hz to 10 MHz. Compare Fig. 5.6D, E and F, we can see that the difference between low- and high-frequency permittivity increases as grain size increases. As the grain size increases, the low-frequency permittivity increases because more charge carriers are stored because of IFP. The high-frequency permittivity remains the same because IFP effect is not obvious. In the meantime, the curves move toward left as grain size increases. Compare Fig. 5.6A, B and C, we can see that as grain size increases, the curves move toward left as well. At low frequency, the conductivity of the mixture containing water-wet graphite remains the same at first, then increases. The conductive grain size has no effect on high-frequency conductivity.

4.2 The porous material containing intermediately wet clay

In the following two cases, we assume a porous material containing spherical water-wet sand, spherical intermediately wet clay and spherical oil droplets at any water saturation. These cases do not contain conductive particles.

4.2.1 The influence of the contact angle of nonconductive particle

In this section, we investigate the influence of the contact angle of spherical nonconductive (clay) particle on the improved PS model predictions of σ_{eff} and $\varepsilon_{r,eff}$ at the frequency range of 100 Hz to 1 GHz. Fig. 5.7 shows the sensitivity of the improved PS model responses of σ_{eff} and $\varepsilon_{r,eff}$ of fluid-filled porous material containing spherical oil droplets, spherical water-wet sand and spherical intermediately wet clay grains to the contact angles of clay.

TABLE 5.5 Properties of inclusions used for generating Fig. 5.6.

	ϕ_i (%)	a_i (μm)	D_i (m^2/s)	$\varepsilon_{r,i}$	σ_i (S/m)	λ_i (S)	θ_i (°)
Graphite	10	20, 200 or 2000	5×10^{-5}	12	500	–	0–180
Sand	70	1000	–	4	–	10^{-9}	–
Clay	10	100	–	8	–	10^{-8}	–
Oil	1	100	–	2	–	10^{-30}	–

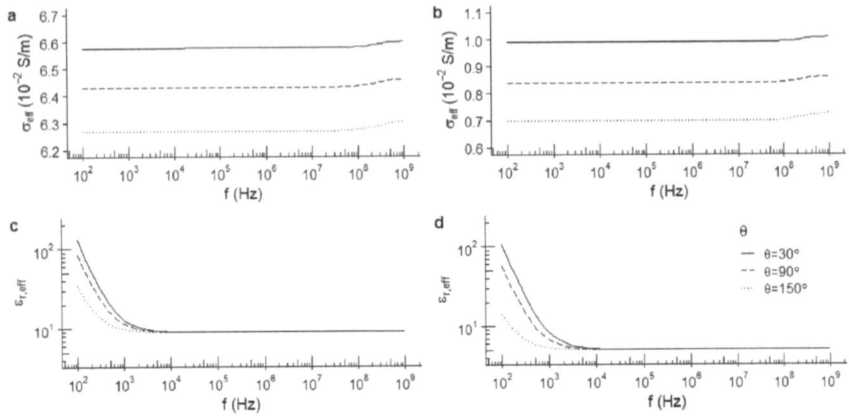

FIGURE 5.7 Comparison of the improved PS model responses of (A and B) σ_{eff} and (C and D) $\varepsilon_{r,eff}$ of fluid-filled porous materials containing spherical oil droplets, spherical water-wet sand and spherical intermediately wet clay grains having various contact angles (30–150 degrees) at (A and C) 10% or (B and D) 90% oil saturation. The properties of host media are listed in Table 5.6. The properties of sand, clay, and oil as inclusion phases are listed in Table 5.7. *Yuteng Jin, Siddharth Misra.*

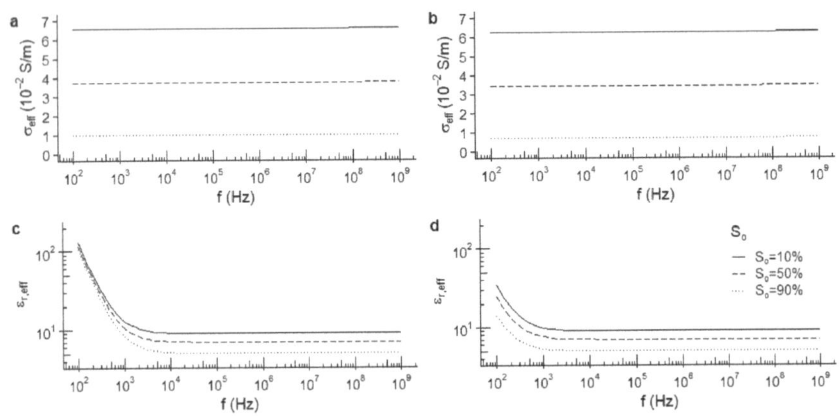

FIGURE 5.8 Comparison of the improved PS model responses of (A and B) σ_{eff} and (C and D) $\varepsilon_{r,eff}$ of fluid-filled porous materials containing spherical oil droplets, spherical water-wet sand and spherical intermediately wet clay with contact angle of (A and C) 30 degrees or (B and D) 150 degrees at various oil saturations (10%–90%). The properties of host media are listed in Table 5.6. The properties of sand, clay, and oil as inclusion phases are listed in Table 5.8. *Yuteng Jin, Siddharth Misra.*

The contact angle is measured in the water phase where the oil–water interface meets the surface of clay. A large contact angle denotes an oil-wet clay whereas a small contact angle denotes a water-wet clay. As shown in Fig. 5.7, no frequency dispersion of electrical conductivity occurs until around 10 MHz, while permittivity shows frequency dispersion before 10 kHz. The frequency dispersion of conductivity is negligible. As contact angle increases (clay becomes more oil wet) the frequency dispersion of permittivity reduces. This is because the surface of clay is covered more by oil, which has much less charge carriers than water and the IFP in the fluid phase is impeded. As contact angle decreases, the conductivity increases because the surface conductance of clay assists charge transport. At low frequency, the charge carriers quickly reach the equilibrium distribution, so that the particles behave like insulators, which lead to lower σ_{eff} and higher $\varepsilon_{r,eff}$. At higher water saturation, there is a greater effect of change in wettability on the conductivity. For low-contact angle, wettability does not much affect the permittivity response even when saturation varies from 10% to 90%. Wettability has negligible effect on high-frequency permittivity, whereas conductivity is affected by wettability in the entire frequency range. Interestingly, frequency dispersion of conductivity is visible for frequencies greater than 100 MHz and that of permittivity is visible for frequencies lower than 10 kHz.

4.2.2 The influence of the wetting fluid saturation

In this section, we investigate the influence of the oil saturation of the fluid-filled porous material on the improved PS model predictions of σ_{eff} and $\varepsilon_{r,eff}$ at the frequency range of 100 Hz to 1 GHz. Fig. 5.8 shows the sensitivity of the improved PS model responses of σ_{eff} and $\varepsilon_{r,eff}$ of fluid-filled porous material containing spherical oil droplets, spherical water-wet sand and spherical intermediately wet clay grains to the oil saturations. As shown in Fig. 5.8, frequency dispersion of electrical conductivity is negligible at the

TABLE 5.6 Properties of host used for generating Figs. 5.7 and 5.8.

	D_h (m²/s)	$\varepsilon_{r,h}$	σ_h (S/m)	ρ_h (kg/m³)	γ (N/m)
Water	10^{-9}	70	1	1000	0.05
Oil	10^{-15}	2	10^{-30}	800	

TABLE 5.7 Properties of inclusions used for generating Fig. 5.7.

	ϕ_i (%)	a_i (μm)	$\varepsilon_{r,i}$	λ_i (S)	θ_i (°)
Sand	70	1000	4	10^{-9}	—
Clay	20	100	8	10^{-6} or 10^{-30}	30–150
Oil	1 or 9	100	2	10^{-30}	—

TABLE 5.8 Properties of inclusions used for generating Fig. 5.8.

	ϕ_i (%)	a_i (μm)	$\varepsilon_{r,i}$	λ_i (S)	θ_i (°)
Sand	70	1000	4	10^{-9}	–
Clay	20	100	8	10^{-6} or 10^{-30}	30 or 150
Oil	1–9	100	2	10^{-30}	–

frequency range of 100 Hz to 1 GHz, while permittivity shows frequency dispersion at frequency range of 100 Hz to 10 kHz. As oil saturation increases, both σ_{eff} and $\varepsilon_{r,eff}$ reduce, which has the similar trend as predicted by the well-known Archie's law and Complex Refractive Index model. Permittivity has negligible sensitivity to saturation at low frequencies for preferentially water-wet clays. Increase in preferential oil-wetness of clays leads to higher sensitivity of permittivity. Saturation change has greater effect on conductivity as compared to wettability change.

5. Conclusions

A novel mechanistic model was developed to quantify the effects of wettability (in terms of contact angle) of conductive and nonconductive particles on the multifrequency complex conductivity of fluid-filled porous materials. For a fluid-filled porous material, conductivity increases and permittivity decreases with the increase in frequency of measurement. At high frequencies, large-scale interfacial polarization (IFP) phenomena fade away and only small-scale orientation/dipolar polarization phenomena exist. Wettability alteration being an interfacial process primarily influences the large-scale IFP phenomena as compared to the small-scale polarization phenomena. Wettability (contact angle) of conductive particles primarily governs the large-scale IFP at the interface of conductive particle and the surrounding host fluid; consequently, at high frequencies beyond certain frequency thresholds, both conductivity and permittivity are insensitive to the variations in wettability. Low-frequency conductivity of the mixture is more sensitive to contact angle (wettability) of the conductive particle as compared to the oil saturation, whereas the high-frequency conductivity is more sensitive to oil saturation as compared to the contact angle of the conductive particle. Compared to low-frequency permittivity, high-frequency permittivity is much less sensitive to oil saturation.

The wettability effects of conductive particles cannot be neglected when dealing with conductivity and permittivity measurements at low frequencies, especially at high oil saturations. Wettability effects can be neglected when dealing with conductivity and permittivity measurements at frequencies greater than 0.1 and 1 MHz, respectively. Frequency dispersions of conductivity and permittivity decrease with increase in oil wetness of the conductive particle. Counterintuitively, low-frequency conductivity decreases with increase in water wetness of conductive particles, indicating that at higher water-wet state of conductive particles, more charge carriers are involved in the charge polarization and accumulation as compared to charge transport. Contrary to high-frequency conductivity, low-frequency conductivity is more sensitive to contact angle of the conductive particle as compared to the oil saturation. The effect of oil saturation is less than the effect of contact angle on the frequency dependence of conductivity and permittivity (i.e., the contact angle has a primary effect and oil saturation has a secondary effect).

On the other hand, regarding to the wettability effects of surface-charge-bearing nonconductive particles, such as clays, the newly developed model shows that the frequency dispersions of multifrequency conductivity of porous material are generally negligible for low frequencies. Interestingly, frequency dispersion of conductivity is visible for frequencies greater than 100 MHz and that of permittivity is visible for frequencies lower than 10 kHz. Both the multifrequency conductivity and permittivity decreases with the decrease in wetness of the clay with respect to the conductive fluid and with the decrease in saturation of the conductive fluid. Unlike conductive particles, the frequency-dependent complex conductivity of porous material containing surface-charge-bearing nonconductive particles is more sensitive to change in fluid saturation as compared to change in the wetness (i.e., contact angle). Saturation change has greater effect on conductivity as compared to wettability change. At higher water saturation, there is a greater effect of change in wettability on the conductivity. For low-contact angle, wettability does not much affect the permittivity response even when saturation varies from 10% to 90%. Wettability has negligible effect on high-frequency permittivity, whereas conductivity is affected by wettability in the entire frequency range.

Nomenclature

EM electromagnetic
IFP interfacial polarization
PPIP model perfectly polarized interfacial polarization model
PS model PPIP−SCAIP model
SCAIP model surface-conductance-assisted interfacial polarization model
$\boldsymbol{a}$ characteristic length of inclusion phase (m)
$\boldsymbol{A_o}$ surface area of particle covered by oil (m^2)
$\boldsymbol{A_s}$ surface area of particle (m^2)
$\boldsymbol{A_w}$ surface area of particle covered by water (m^2)
$\boldsymbol{B_o}$ Bond number
$\boldsymbol{C}$ the point where fluid interface contact the particle
$\boldsymbol{D_j}$ diffusion coefficient of charge carriers of medium j (m^2/s)
$\varepsilon_{r,eff}$ effective relative dielectric permittivity of the mixture (F/m)

$\varepsilon_{r,j}$ relative permittivity of medium j
f frequency (Hz)
$f_j(\omega)$ dipolarizability (dipolar field coefficient) of medium j
$f(\phi)$ a function of wetting angle ϕ
g gravitational acceleration (N/kg)
G dimensionless form of $h - h_i$
γ interfacial tension between oil and water (N/m)
$h(r)$ height of oil-water interface at any distance r away from the vertical axis z (m)
$\widehat{h}$ dimensionless form of h
h_c height where the oil-water interface contacts the particle surface (m)
h_i height of oil-water interface in the absence of wetting (far-field height) (m)
H mean curvature of the meniscus surface (m^{-1})
I_0 modified Bessel function of the first kind of order 0
K_0 modified Bessel function of the second kind of order 0
K_1 modified Bessel function of the second kind of order one
L_c capillary length (m)
λ surface conductance (S)
ω angular frequency of the electric field (rad/s)
Δp Laplace pressure (Pa)
p_o proportion of particle surface that covered by oil (%)
p_w proportion of particle surface that covered by water (%)
ϕ wetting angle (°)
ϕ porosity of the porous media (%)
ϕ_j volume fraction of medium j in the mixture (%)
ϕ_o volume fraction of oil in the mixture (%)
ψ angle between oil-water interface and the horizon (r-axis) at contact point (°)
r distance from vertical axis z (m)
$\widehat{r}$ dimensionless form of r
R radius of particle (m)
ρ_o density of oil (kg/m^3)
ρ_w density of water (kg/m^3)
S_o oil saturation (%)
σ_{eff} effective electrical conductivity of the mixture (S/m)
$\sigma^*_{\widehat{n},eff}$ effective complex electrical conductivity of the mixture along the $\widehat{n}$ unit vector (S/m)
σ_j electrical conductivity of medium j (S/m)
σ^*_j complex electrical conductivity of medium j (S/m)
θ contact angle (°)

References

Al-Ofi, S.M., Dyshlyuk, E., Sauerer, B., Valori, A., Ali, F., Abdallah, W., August 16, 2018. Correlating dielectric dispersion data and wettability Index of a carbonate rock. In: SPE Kingdom of Saudi Arabia Annual Technical Symposium and Exhibition. Society of Petroleum Engineers.

Bernard, S., Wirth, R., Schreiber, A., Bowen, L., Aplin, A.C., Mathia, E.J., Schulz, H.M., Horsfield, B., Aplin, A., Mathia, E., 2013. FIB-SEM and TEM investigations of an organic-rich shale maturation series from the lower Toarcian Posidonia Shale, Germany: Nanoscale pore system and fluid-rock interactions. Electron microscopy of shale hydrocarbon reservoirs. AAPG Memoir 102, 53–66.

Bona, N., Ortenzi, A., Capaccioli, S., April 1, 2002. Advances in understanding the relationship between rock wettability and high-frequency dielectric response. J. Petrol. Sci. Eng. 33 (1–3), 87–99.

Bona, N., Rossi, E., Capaccioli, S., 1999. Electrical measurements in the 100Hz-10GHz for efficient rock wettability evaluation. Int. Symp. Oilfield Chem. 50720–50721 (Society of Petroleum Engineers).

Bona, N., Rossi, E., Venturini, C., Capaccioli, S., Lucchesi, M., Rolla, P.A., November 1, 1998. Characterization of rock wettability though dielectric measurements. Rev. Inst. Fr. Petrol 53 (6), 771–783.

Bücker, M., Orozco, A.F., Kemna, A., 2018. Electrochemical polarization around metallic particles—Part 1: the role of diffuse-layer and volume-diffusion relaxation. Geophysics 83 (4), E203–17.

Bustin, R.M., Ross, J.V., Rouzaud, J.N., October 1, 1995. Mechanisms of graphite formation from kerogen: experimental evidence. Int. J. Coal Geol. 28 (1), 1–36.

Capaccioli, S., Lucchesi, M., Casalini, R., Rolla, P.A., Bona, N., May 7, 2000. Influence of the wettability on the electrical response of microporous systems. J. Phys. Appl. Phys. 33 (9), 1036.

Cavallaro Jr., M., 2012. Assembly of Rod-like Particles at Fluid Interfaces. The Johns Hopkins University.

Chakarov, D.V., Oesterlund, L., Kasemo, B., April 1995. Water adsorption and coadsorption with potassium on graphite (0001). Langmuir 11 (4), 1201–1214.

Champmartin, S., Ambari, A., Le Pommelec, J.Y., May 12, 2016. New procedure to measure simultaneously the surface tension and contact angle. Rev. Sci. Instrum. 87 (5), 055105.

Clark, T.H., April 1, 1921. The origin of graphite. Econ. Geol. 16 (3), 167–183.

Crain, E.R., 2002. Crain's Petrophysical Handbook. Spectrum 2000 Mindware Limited.

Fiegel, J., Jin, F., Hanes, J., Stebe, K., November 15, 2005. Wetting of a particle in a thin film. J. Colloid Interface Sci. 291 (2), 507–514.

Fox, H.W., Zisman, W.A., December 1, 1950. The spreading of liquids on low energy surfaces. I. polytetrafluoroethylene. J. Colloid Sci. 5 (6), 514–531.

Garrouch, A.A., Sharma, M.M., June 1994. The influence of clay content, salinity, stress, and wettability on the dielectric properties of brine-saturated rocks: 10 Hz to 10 MHz. Geophysics 59 (6), 909–917.

Han, Y., Misra, S., April 6, 2018. Joint petrophysical inversion of multifrequency conductivity and permittivity logs derived from subsurface galvanic, induction, propagation, and dielectric dispersion measurements. Geophysics 83 (3), D97–D112.

Han, Y., Misra, S., Simpson, G., June 17, 2017. Dielectric Dispersion Log Interpretation in Bakken Petroleum System. InSPWLA 58th Annual Logging Symposium. Society of Petrophysicists and Well-Log Analysts.

Jin, Y., Misra, S., Homan, D., Rasmus, J., May 15, 2020b. Effects of wettability of conductive particles on the multifrequency conductivity and permittivity of fluid-filled porous material. Fuel 268, 117411.

Jin, Y., Misra, S., Homan, D., Rasmus, J., Revil, A., August 1, 2019. Mechanistic model of multi-frequency complex conductivity of porous media containing water-wet nonconductive and conductive particles at various water saturations. Adv. Water Resour. 130, 244–257.

Jin, Y., Misra, S., Homan, D., 2020. Estimation of In-Situ Wettability Using Multi-Frequency Electromagnetic Measurements. In: *Unconventional Resources Technology Conference, 20–22 July 2020*. Unconventional Resources Technology Conference (URTeC), pp. 4181–4199.

Kocabag, D., Shergold, H.L., Kelsall, G.H., October 1, 1990. Natural oleophilicity/hydrophobicity of sulphide minerals, II. Pyrite. Int. J. Miner. Process. 29 (3–4), 211–219.

Kozbial, A., Li, Z., Sun, J., Gong, X., Zhou, F., Wang, Y., Xu, H., Liu, H., Li, L., August 1, 2014. Understanding the intrinsic water wettability of graphite. Carbon 74, 218–225.

Kralchevsky, P.A., Eriksson, J.C., Ljunggren, S., April 15, 1994. Theory of curved interfaces and membranes: mechanical and thermodynamical approaches. Adv. Colloid Interface Sci. 48, 19–59.

Laughrey, C.D., Ruble, T.E., Lemmens, H., Kostelnik, J., Butcher, A.R., Walker, G., Knowles, W., 2011. AV black shale diagenesis—insights from integrated high-definition analyses of post-mature Marcellus formation rocks, Northeastern Pennsylvania. In: Annual Convention and Exhibition. AAPG.

Misra, S., Han, Y., August 3, 2016. Petrophysical interpretation of multi-frequency electromagnetic measurements in clay-and conductive-mineral-rich mudrocks. In: Unconventional Resources Technology Conference, San Antonio, Texas, 1-3 August 2016, pp. 543–556 (Society of Exploration Geophysicists, American Association of Petroleum Geologists, Society of Petroleum Engineers).

Misra, S., Homan, D., Jin, Y., Rasmus, J., 2021. Systems and Methods for Determining Properties of Porous, Fluid-Filled Geological Formations Based on Multi-Frequency Measurements. U.S. Patent Application 16/940,492, filed February 4, 2021.

Misra, S., Rasmus, J., Homan, D., Torres-Verdin, C., 2019. Subsurface Estimation of Level of Organic Maturity. U.S. Patent 10,386,529, issued August 20, 2019.

Misra, S., Torres-Verdín, C., Revil, A., Rasmus, J., Homan, D., March 1, 2016. Interfacial polarization of disseminated conductive minerals in absence of redox-active species—Part 1: mechanistic model and validationInterfacial polarization of conductive inclusions. Geophysics 81 (2), E139–E157.

Misra, S., Torres-Verdín, C., Revil, A., Rasmus, J., Homan, D., March 1, 2016. Interfacial polarization of disseminated conductive minerals in absence of redox-active species—Part 2: effective electrical conductivity and dielectric permittivityInterfacial polarization due to inclusions. Geophysics 81 (2), E159–E176.

Moss, A.K., Jing, X.D., Archer, J.S., April 1, 2002. Wettability of reservoir rock and fluid systems from complex resistivity measurements. J. Petrol. Sci. Eng. 33 (1–3), 75–85.

Nguyen, B.L., Bruining, J., Slob, E.C., January 1, 1999. Effects of wettability on dielectric properties of porous media. In: SPE Annual Technical Conference and Exhibition. Society of Petroleum Engineers.

Revil, A., Schmutz, M., Batzle, M.L., 2011. Influence of oil wettability upon spectral induced polarization of oil-bearing sands. Geophysics 76 (5), A31–A36.

Spötl, C., Houseknecht, D.W., Jaques, R.C., June 1, 1998. Kerogen maturation and incipient graphitization of hydrocarbon source rocks in the Arkoma Basin, Oklahoma and Arkansas: a combined petrographic and Raman spectrometric study. Org. Geochem. 28 (9–10), 535–542.

Tathed, P., Han, Y., Misra, S., 2018. Hydrocarbon saturation in upper Wolfcamp shale formation. Fuel 219, 375–388.

Tissot, B.P., Welte, D.H., November 11, 2013. Petroleum Formation and Occurrence. Springer Science & Business Media.

Ujiié, Y., March 1978. Kerogen maturation and petroleum genesis. Nature 272 (5652), 438.

Winchell, A.N., April 1, 1911. A theory for the origin of graphite as exemplified in the graphite deposit near Dillon, Montana. Econ. Geol. 6 (3), 218–230.

Zhdanov, M.S., Burtman, V., Endo, M., Lin, W., 2018. Complex resistivity of mineral rocks in the context of the generalised effective-medium theory of the induced polarisation effect. Geophys. Prospect. 66 (4), 798–817.

Section 3

Inversion-based interpretation of multifrequency electromagnetic data

Chapter 6

Unified deterministic inversion of multifrequency electromagnetic measurements using relaxation models

Yifu Han[1], Siddharth Misra[2,3]
[1]*Schlumberger Technology Corporation, Beijing, China;* [2]*Harold Vance Department of Petroleum Engineering, College of Engineering, Texas A&M University, TX, United States;* [3]*Department of Geology and Geophysics, Texas A&M University, TX, United States*

1. Introduction

Electromagnetic (EM) properties, such as electrical conductivity, dielectric permittivity, or magnetic permeability, exhibit frequency dependence, i.e., dispersion (Marshall and Madden, 1959; Wong, 1979). Frequency dependence of EM property is because of various polarization phenomena, such as membrane polarization (Marshall and Madden, 1959), concentration polarization (Šumi, 1965), induced polarization (Seigel, 1959), Maxwell–Wagner polarization (Schurr, 1964), surface conductance assisted interfacial polarization (Jin et al., 2019), bulk conductivity assisted interfacial polarization (Jin et al., 2020), and dipolar/orientation polarization (Fuller and Ward, 1970). Most polarization phenomena develop in the presence of external EM field (Misra et al., 2016). There are several experimental (Misra et al., 2015) and field-scale studies (Misra and Han, 2016; Han and Misra, 2017, 2018a) on the frequency dispersions of electrical conductivity and dielectric permittivity of geological formations/materials. Recently, there has been a push toward understanding the polarization phenomena in the presence of conductive minerals and their effects on EM properties (Misra et al., 2016; Revil et al., 2015, 2017; Tathed et al., 2018a).

Multifrequency measurement of a dispersive EM property is commonly analyzed for purposes of material characterization. Such an analysis requires inversion/interpretation of the multifrequency measurement based on a relaxation model, such as Cole–Cole model or Pelton's model. A relaxation model quantifies the multifrequency electrical conductivity, dielectric permittivity, or magnetic permeability in terms of various empirical parameters. Such models

Multifrequency Electromagnetic Data Interpretation for Subsurface Characterization
https://doi.org/10.1016/B978-0-12-821439-8.00003-3

describe a measurable spectral property of a perturbed physical system returning to equilibrium using descriptive empirical parameters specific to the system and the property being measured. For example, electrical conductivity and dielectric permittivity depend on water saturation, connate water salinity, porosity, tortuosity, wettability, clay content and distribution, conductive mineral content and distribution, grain sizes, grain texture, and pore size distribution along with various polarization phenomena. However, a relaxation model represents the frequency dependence of an EM property using a relatively simple mathematical formulation dependent on three to six model parameters, which combine/integrate the physical factors governing the frequency dependence.

EM-based material characterization involves estimation of the empirical/physical parameters of a relaxation model that are unique to the material with respect to the measurement and the measured physical process. Relaxation models have been used for characterization of soil (Lesmes and Friedman, 2005), mineral (Wong, 1979), hydrocarbon reservoirs (Revil et al., 2013), groundwater (Binley et al., 2005), subsurface contaminant, and geological formations. The dispersive EM measurement can be processed to estimate hydraulic conductivity (Binley et al., 2005), pore or grain size distribution (Revil and Florsch, 2010), and mineral content (Misra and Han, 2016; Han and Misra, 2018a; Gurin et al., 2015). Kemna et al. (2012) summarized the near surface applications of spectral induced polarization (SIP) method that correlates the mechanism of the induced polarization with geomaterial properties. Liu et al. (2016) evaluated SIP signatures of more than 50 ore samples using seven phenomenological models. To process a dispersive EM measurement, a deterministic or stochastic inversion algorithm can be applied to iteratively minimize the difference between the measurements and the numerical predictions of a relaxation model for a set of relaxation model parameters, which in turn is determined based on the mathematical formulation of the inversion scheme.

This chapter presents a unified inversion scheme that can be coupled to any type of relaxation models to independently process the multifrequency measurement of any EM property for purposes of improved EM-based material characterization. The proposed inversion scheme is applied to multifrequency complex permittivity, complex conductivity, complex impedance, and complex resistivity measurements. The proposed deterministic-inversion method can estimate up to seven relaxation-model parameters while exhibiting convergence for random initializations of the relaxation-model parameters within 3-orders of magnitude variations. The proposed inversion method implements a bounded Levenberg–Marquardt algorithm with tunable damping parameter and its iterative adjustment factor. Interestingly, these tunable inversion-specific parameters are fixed in all the cases presented in this chapter, irrespective of the type of EM property and the type of relaxation model. Notably, jump-out step and jump-back-in step are implemented as automated methods in the inversion scheme to prevent the inversion from getting trapped around local minima and to honor physical bounds of model parameters. The proposed inversion scheme can be easily used to process various types of EM measurements without major modifications to the inversion scheme.

2. Relaxation models

Cole–Cole model is the most frequently used relaxation model, which was formulated to quantify the multifrequency complex permittivity response of polar liquids and dielectrics. Cole–Cole model (Cole and Cole, 1941) is expressed as

$$\varepsilon^* = \varepsilon_{r\infty} + \frac{\varepsilon_{r0} - \varepsilon_{r\infty}}{1 + (i\omega\tau)^{1-\alpha}}, \tag{6.1}$$

where ε_{r0} and $\varepsilon_{r\infty}$ are the low-frequency and high-frequency permittivity values, respectively, τ is the central relaxation time, $\omega = 2\pi f$ is the angular frequency of EM measurement, f is the frequency of EM measurement, and exponent α controls the width of the loss peak, such that $0 < \alpha \leq 1$. For porous material characterization, τ is typically between 10^{-3} s and 10^3 s (Kemna, 2005). Another widely used relaxation model is the Cole-Davidson model that was proposed for certain fluids exhibiting asymmetric dispersion spectra. The Cole-Davidson model (Davidson and Cole, 1951) is written as

$$\varepsilon^* = \varepsilon_{r\infty} + \frac{\varepsilon_{r0} - \varepsilon_{r\infty}}{(1 + i\omega\tau)^{\beta}}, \tag{6.2}$$

where β is referred as shape parameter describing an asymmetric distribution of the loss peak about the relaxation time, such that $0 < \beta \leq 1$ (Mashimo et al., 1991). The Cole–Cole model and Cole-Davidson model reduce to Debye model for $\alpha = 0$ and $\beta = 1$, respectively. Both the models, expressed in Eqs. (6.1) and (6.2), constitute special case of the Havriliak–Negami model that accounts for the broadness and asymmetry of dielectric dispersion curves. The Havriliak–Negami model (Havriliak and Negami, 1966) can be expressed as

$$\varepsilon^* = \varepsilon_{r\infty} + \frac{\varepsilon_{r0} - \varepsilon_{r\infty}}{\left[1 + (i\omega\tau)^{1-\alpha}\right]^{\beta}}, \tag{6.3}$$

where α and β are shape characteristics of the dielectric dispersion curve, such that $0 < \alpha,\ \beta \leq 1$. Dias (2000) analyzed several models, such as Cole–Cole model, used to describe induced polarization effects and grouped them together under a common circuit-analog representation. Eq. (6.3) can be modified and applied as

$$\varepsilon^* = \varepsilon_{r\infty} + \frac{\Delta\varepsilon_r}{[1 + (i\omega\tau)^{\beta}]^{\alpha}}, \tag{6.4}$$

where $\Delta\varepsilon_r = \varepsilon_{r0} - \varepsilon_{r\infty}$ is the relaxation increment, and the positions of α and β are switched with respect to Eq. (6.3). The relaxation time of the mixture in the measured frequency range can be in the order of picoseconds. To avoid singular values of the Jacobian matrix during inversion iterations, we can invert the logarithmic value of the relaxation time; consequently, for purposes of inversion, Eq. (6.4) can be modified to

$$\varepsilon^* = \varepsilon_{r\infty} + \frac{\Delta\varepsilon_r}{\left[1 + \left(i\omega 10^{\log\,\tau}\right)^{\beta}\right]^{\alpha}}. \tag{6.5}$$

Pelton et al. (1978) modified the Cole–Cole model to obtain an impedance model to analyze complex impedance measurements over 10^{-2} Hz–10^{5} Hz made on 26 mineral deposits. The modified model can be expressed as

$$Z^* = R_0\left[1 - m\left(1 - \frac{1}{1 + (i\omega\tau)^c}\right)\right], \tag{6.6}$$

where Z^* is the measured complex impedance of the mineral samples, R_0 is the DC (low frequency) resistance, m is the chargeability, and $c = 1 - \alpha$ is the Cole–Cole model exponent. The value of the Cole–Cole model exponent c is typically in the range 0.1–0.6 for mineral-bearing rocks (Pelton et al., 1978). The model can be formulated in terms of resistivity as

$$\rho^* = \rho_0\left[1 - m\left(1 - \frac{1}{1 + (i\omega\tau)^c}\right)\right], \tag{6.7}$$

where ρ^* is the measured complex resistivity and ρ_0 is the low-frequency resistivity. The low-frequency resistivity values of sedimentary rocks are typically between 10 and 10,000 Ωm with most values less than 1000 Ωm (Loke, 2001). The model also can be formulated in terms of complex conductivity, as developed by Tarasov and Titov (2013) as

$$\sigma^* = \sigma_0\left[1 + m\left(\frac{(i\omega\tau)^c}{1 + (i\omega\tau)^c(1 - m)}\right)\right], \tag{6.8}$$

where σ^* is the measured complex conductivity and σ_0 is the low-frequency conductivity. The Cole–Cole model resistivity and conductivity formulations are not exactly equivalent. Therefore, it is to be noted, one should not use the Cole–Cole model conductivity formulation on resistivity data that was transformed to conductivity or vice-versa (Tarasov and Titov, 2013; Macnae, 2015). The Cole–Cole model resistivity formulation is characterized by a single peak in the complex resistivity phase angle spectrum and the location of peak along the frequency axis is related directly to the relaxation time constant τ. A phase angle spectrum with a single peak is expected for induced polarization response of rocks with a unimodal distribution of length scales (pore-size distribution or unimodal grain) (Chen et al., 2008) or a single type of polarization phenomenon (Han et al., 2017). However, phase angle spectra with more than one phase peak, such as bimodal distributions (Leroy et al., 2008), can be observed for more complex distributions of length scales, wherein different peaks reflect relaxation processes at different scales (Chen et al., 2008). A multiple Cole–Cole model for complex resistivity is expressed as

$$\rho^* = \rho_0 \left[1 - \sum_{l=1}^{L} m_l \left(1 - \frac{1}{1 + (i\omega\tau_l)^{c_l}} \right) \right], \tag{6.9}$$

where L is the number of Cole–Cole models that are required to compositely match with complex resistivity measurements. The typical value for L is between 1 and 3, depending on the number of relaxation scales in the polarization phenomena. The symbols m_l, τ_l, and c_l represent chargeability, relaxation time, and Cole–Cole exponent value for the l-th dispersion term in the multiple Cole–Cole model, respectively. Dual Cole–Cole model ($L = 2$) is often encountered in practice either to describe an induced polarization response with two relaxation domains or to describe a single relaxation induced polarization response contaminated by inductive and/or capacitive coupling associated with the measurement layout (Pelton et al., 1978; Kemna et al., 1999).

3. Previous studies on the deterministic-inversion-based interpretation of multifrequency electromagnetic measurements

For purposes of EM-based characterization of subsurface-earth materials, multifrequency EM measurement can be inverted using a specific relaxation model to estimate the relaxation-model parameters by finding the best match between the relaxation-model predictions and the measurements (Revil et al., 2013). In doing so, the estimated parameters of the relaxation model serve as characteristics/identifiers of the material under investigation (Revil, 2012). Deterministic inversion of multifrequency EM measurements using a relaxation model commonly implements Gauss-Newton algorithm (GNA) or Levenberg–Marquardt algorithm (LMA) (Kemna et al., 1999). Roy (1999) proposed a modified LMA-based inversion of moderately noisy induced-polarization data for purposes of imaging. The adaptive damping parameter in the algorithm depends on the noise. The proposed method exhibits better efficiency and robustness compared to the traditional LMA and simulated-annealing based inversion (discussed in detail in Chapter 10). Kemna (2005) proposed an inversion of multifrequency complex resistivity based on Tikhonov approach, which involves a univariate line search to find an optimum value of the regularization parameter. Cole–Cole model is fitted to the multifrequency complex resistivity at each depth in the mineral-bearing formation by means of a nonlinear least-squares inversion procedure, using simple Marquardt-regularization.

Kemna et al. (1999) developed an inversion methodology coupled with a finite-element EM forward model and another inversion methodology coupled with Cole–Cole relaxation model to jointly invert the resistivity magnitude and resistivity phase for purposes of delineating subsurface hydrocarbon contamination at a jet fuel depot and for imaging sulfide deposit in a metamorphic formation. Slater et al. (2006) conducted induced-polarization measurements on kaolinite-, iron-, and magnetite-sand mixtures for various weight fractions of the disseminated mineral constituents. The SIP measurements on the magnetite and iron mixtures were fitted using the Cole–Cole model. The global polarization magnitude obtained from the Cole–Cole modeling of the induced polarization measurements exhibited a single, near-linear dependence on the ratio of surface area to pore volume. This relationship between SIP measurements and the ratio of surface area to pore volume was first identified by Börner and Schön (1991).

Revil et al. (2013) developed a polarization model to predict the complex conductivity of clayey materials as function of the porosity, cation exchange capacity, and salinity of the pore water for binary electrolytes. Using the forward model, multifrequency complex conductivity measurements were inverted to estimate the parameters of Cole–Cole model. They observed relationships between the relaxation time and pore throat size, and between Cole–Cole exponent c and pore-size-distribution. Zisser et al. (2010) examined the relationship between hydraulic permeability and low-frequency (10 mHz–100 Hz) electrical properties of low-permeability sandstones from a tight gas reservoir. First, they computed relaxation-time distribution from frequency-domain induced polarization data using a modification of Debye decomposition procedure. The measured spectral response was inverted into relaxation-time distributions assuming 1000 logarithmically spaced relaxation times. They observed a strong power-law correlation between median relaxation time distribution and rock permeability. Such a Debye decomposition describes the SIP signature using linear superposition of large number of elementary Debye polarization terms. In doing so, SIP signatures are described in terms of relaxation-time distributions. Debye decomposition can be applied to model any number of polarization peaks and requires regularization to achieve a stable solution (Weigand and Kemna, 2016). Nordsiek and Weller (2008) also developed a new fitting approach based on Debye decomposition to fit the SIP signatures.

Xiang et al. (2001) inverted the complex impedance using Cole–Cole model. The proposed inversion algorithm consists of two parts: (1) multifold least-squares estimation that uses an assumed parameter for the frequency dependence of the Cole–Cole model, c, to estimate other Cole–Cole model parameters, namely the DC resistivity, $\rho_0(c)$, polarizability or chargeability, $m(c)$, and relaxation time, $\tau(c)$ from the SIP data, and (2) search the optimal value of c using the golden section minimization algorithm. Xiang et al. (2001) reported a reduced dependence of the optimal solution on the initial guesses. Cao et al. (2005) inverted complex resistivity data using an improved genetic algorithm to overcome the instability and nonconvergence issues associated with the estimation of Cole–Cole model parameters based on linear inversion theory.

Several published methods for deterministic inversion of multifrequency EM measurements were found to be highly sensitive to the initial guesses of the unknown relaxation-model parameters (Chen et al., 2008), such that unique estimates of the relaxation-model parameters are generated for initial guesses lying in close vicinity to the optimal solution (Roy, 1999). Notably, we found several publications on the deterministic inversion of multifrequency EM measurements did not report the model sensitivity to initial guesses of the unknown relaxation-model parameters. Moreover, for several published methods, when there is no prior information about the range of values for the unknown model parameters, it is harder to obtain reliable estimations (Kozhevnikov and Antonov, 2012), unless narrow ranges are assumed for the unknown parameters. Several published inversion methods for multifrequency EM measurements that implement regularization parameters or adaptive parameters in the mathematical formulation of the inversion scheme do not mention the chosen values of such parameters, for instance, the initial values of damping parameter λ and rate of change factor ν used in the LMA-based inversion (Lazović et al., 2014). LMA-based inversion scheme is not a global search algorithm; it is inevitably sensitive to initial guesses (Han and Misra, 2018b; Tathed et al., 2018b).

To address the above-mentioned shortcomings in published methods, we develop a unified inversion scheme that can be coupled to any type of relaxation model to process laboratory multifrequency measurement of any EM property for purposes of EM-based material characterization. We implement a bounded Levenberg–Marquardt nonlinear inversion algorithm based on tunable damping parameter and its iterative adjustment factor, which are fixed in all the cases presented in this chapter irrespective of the type of measured property and the type of relaxation model. The inversion method can invoke jump-back-in and jump-out sequences to facilitate better exploration of the model-parameter space. These features ensure that the inversion scheme is universal and converges for initial guesses with uncertainty within 3-orders of magnitude. In doing so, the inversion scheme is insensitive to user inputs, biases, and errors.

Estimation of the parameters of relaxation model as an inverse problem is well studied in technical literature. Most of the publications in this area focus on applications of different inversion methods on one particular form of dispersive EM property and relaxation model. There are also some open-source inversion codes (e.g., Weigand and Kemna, 2016; Cockett et al., 2015; Bérubé et al., 2017) for laboratory or field-induced polarization measurements. In this chapter, we propose a unified inversion scheme that works well for various dispersive EM properties and relaxation models. Furthermore, we demonstrate that fixed initial values of damping parameter and its iterative adjustment factor significantly improve the speed, convergence, and consistency of parameter estimation. The proposed model is successfully applied to six different parameter-estimation cases ranging from four to seven unknown parameters. We want to point out that, in the original papers or researches, the data/signatures corresponding to these cases were mostly fitted using manual methods or inversion methods with less emphasis on the reliability and robustness of the inversion scheme or the accuracy of the inversion-derived parameters. Our focus is to test the accuracy of the inversion-derived estimates while achieving low-computational cost and avoiding local minima. Inversion algorithms tend to get trapped in local minima; consequently, the inversion-derived parameters can have large inaccuracy and uncertainty. To that end, we perform a methodical deterministic inversion that is automated to a large extent making the proposed inversion code free from user bias, inputs, and errors. Our goal is to automate the inversion scheme to get out of local minima and to effectively honor the upper and lower bounds of each inversion-derived parameters. In Section 5 of this chapter, we present an extensive study on the robustness of the proposed inversion algorithm when applied to synthetic data containing various amount of Gaussian noise. There are few tunable parameters in the proposed inversion scheme such as damping parameter and its iterative adjustment factor, the initial values, and ranges of unknown parameters. Although the ideal choice of damping parameter and its iterative adjustment factor are specific to the parameter-estimation problem and available data, these parameters are fixed in all the cases presented in this chapter.

4. Proposed unified deterministic inversion scheme

4.1 Nonlinear inversion as an error minimization problem

The nonlinear inverse problem discussed in this chapter involves the estimation of relaxation-model parameters that generate least misfit between the measured and modeled multifrequency EM measurements. The measurements used in this chapter are the real part and imaginary part (or magnitude and phase) of complex permittivity, complex conductivity, complex impedance, or complex resistivity of geological samples/formations measured at multiple frequencies. A bounded Levenberg Marquardt algorithm is implemented to find the solution of this nonlinear inverse problem by simultaneously matching the real part and imaginary part (or magnitude and phase) of the measured data with the relaxation model predictions for all the measurement frequencies. Fig. 6.1 illustrates the flowchart of the inversion scheme.

In the proposed inversion scheme, a relaxation model $R(\mathbf{m}, \omega)$ is first chosen to model the frequency-dispersive complex permittivity, complex conductivity, complex impedance, or complex resistivity at a specific angular frequency ω. The relaxation model is described using a specific l-sized relaxation-model parameter vector $\mathbf{m}$. An n-sized modeled data vector

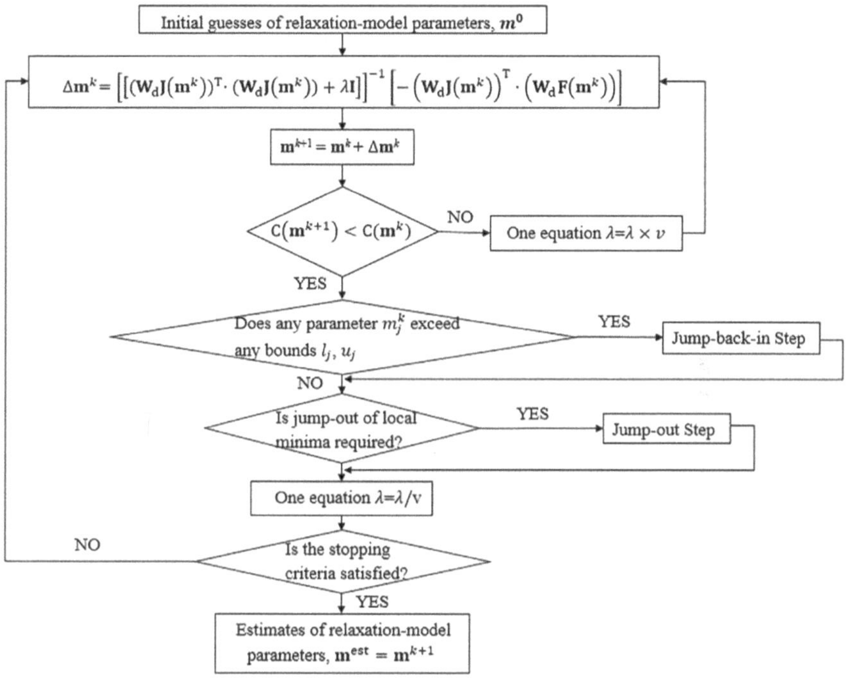

FIGURE 6.1 Flowchart of the newly proposed unified inversion scheme. *Adapted from Han Y, Misra S. A unified inversion scheme to process multifrequency measurements of various dispersive electromagnetic properties. J. Appl. Geophys. April, 2018;151:23–39.*

$\mathbf{D}^{\mathrm{mod}}(\mathbf{m})$ is built as a collection of relaxation model predictions for the n frequencies of measurements and a specific relaxation-model parameter vector $\mathbf{m}$, such that $\mathbf{D}^{\mathrm{mod}}(\mathbf{m}) = [R(\mathbf{m}, \omega_1); R(\mathbf{m}, \omega_2); \ldots\ldots; R(\mathbf{m}, \omega_n)]$. Following that, the measured data $\mathbf{D}^{\mathrm{meas}}$ is matched with modeled data vector $\mathbf{D}^{\mathrm{mod}}(\mathbf{m})$ to find the vector $\mathbf{m}$ comprising the estimates of l relaxation-model parameters that quantitatively characterize the material under investigation. To that end, the inversion algorithm iteratively computes the l-sized estimated model-parameter vector $\mathbf{m}^{\mathrm{est}}$ comprising the estimates of l relaxation-model parameters, such that the n-sized modeled data vector $\mathbf{D}^{\mathrm{mod}}(\mathbf{m}^{\mathrm{est}})$ best matches the measured data vector $\mathbf{D}^{\mathrm{meas}}$. The inversion algorithm iteratively adjusts the vector of relaxation-model parameters $\mathbf{m}$ to obtain the inverted relaxation-model parameter vector $\mathbf{m}^{\mathrm{est}}$ that minimizes the error between the measured data vector $\mathbf{D}^{\mathrm{meas}}$ and the modeled data vector $\mathbf{D}^{\mathrm{mod}}(\mathbf{m}^{\mathrm{est}})$. This requires minimizing quadratic cost function $C(\mathbf{m})$ at each iteration defined as

$$C(\mathbf{m}) = \| \operatorname{real}\left[\mathbf{D}^{\text{mod}}(\mathbf{m}^{\text{est}}) - \mathbf{D}^{\text{meas}}\right] \|_2^2 + w \cdot \| \operatorname{imag}\left[\mathbf{D}^{\text{mod}}(\mathbf{m}^{\text{est}}) - \mathbf{D}^{\text{meas}}\right] \|_2^2, \tag{6.10}$$

where $\| \|_2^2$ is the square of the L_2 norm (or Euclidean length) and w is the n-sized weight factor which is expressed as

$$w = 10^a, \tag{6.11}$$

where a is a positive integer computed as the ratio of magnitude of the real part to that of the imaginary part of the measured EM property. The weight factor w in Eq. (6.10) is used to normalize the errors in real part and imaginary part to ensure that the inversion scheme honors both the errors without getting biased by either conductivity or permittivity. Our extensive sensitivity study indicates that the weight factor should be applied when the real part and imaginary part are more than two orders of magnitude different. In this unified inversion scheme, the n-sized vector $\mathbf{D}$ is replaced with vectors $\boldsymbol{\varepsilon}$, $\boldsymbol{\sigma}$, $\mathbf{z}$, or $\boldsymbol{\rho}$ when the measurement is complex permittivity, complex conductivity, complex impedance, or complex resistivity, respectively. For purposes of minimization, we define an n-sized cost function vector $\mathbf{F}(\mathbf{m})$. Individual components of the vector $\mathbf{F}(\mathbf{m})$ are referred as $f_i(\mathbf{m}^k)$, where superscript k is used as the iteration count for the inversion scheme, $\mathbf{m}^k$ is the l-sized relaxation-model parameter vector computed at the k-th iteration, and subscript i denotes one of the n measurement frequencies. $\mathbf{F}(\mathbf{m}^k)$ is then expressed as

$$\mathbf{F}(\mathbf{m}^k) = \begin{bmatrix} f_1(\mathbf{m}^k) \\ f_2(\mathbf{m}^k) \\ \cdot \\ \cdot \\ f_n(\mathbf{m}^k) \end{bmatrix}, \tag{6.12}$$

where $f_i(\mathbf{m}^k)$ is expressed in the form of the cost function formulation as

$$f_i(\mathbf{m}^k) = \left[\operatorname{real}\left(\mathbf{D}^{\text{mod}}(\mathbf{m}^k) - \mathbf{D}^{\text{meas}}\right)\right]_i^2 + \text{w} \cdot \left[\operatorname{imag}\left(\mathbf{D}^{\text{mod}}(\mathbf{m}^k) - \mathbf{D}^{\text{meas}}\right)\right]_i^2, \tag{6.13}$$

In this scheme, we perform real-valued inversion of complex-valued data (Seleznev et al., 2017) to reduce the computational cost of array and matrix manipulation. This scheme can be improved by using complex-valued inversion or using real-valued inversion with stacked real and imaginary part (Han et al., 2019; Weigand and Kemna, 2016), which is implemented in cases in following chapters. Based on the Levenberg–Marquardt nonlinear inversion method (Aster et al., 2005), the inversion scheme can be expressed as

$$\left[\left(\mathbf{W}_{\mathrm{d}}\mathbf{J}\left(\mathbf{m}^k\right)\right)^{\mathrm{T}}\cdot\left(\mathbf{W}_{\mathrm{d}}\mathbf{J}\left(\mathbf{m}^k\right)\right)+\lambda\mathbf{I}\right]\cdot\Delta\mathbf{m}^k=-\left(\mathbf{W}_{\mathrm{d}}\mathbf{J}\left(\mathbf{m}^k\right)\right)^{\mathrm{T}}\cdot\left(\mathbf{W}_{\mathrm{d}}\mathbf{F}\left(\mathbf{m}^k\right)\right), \tag{6.14}$$

where $\mathbf{J}(\mathbf{m}^k)$ is the $n\times l$ Jacobian matrix of $\mathbf{F}(\mathbf{m}^k)$, I is $l\times l$ identity matrix, superscript T is matrix transpose operator, $\Delta\mathbf{m}^k$ is the l-sized correction vector that determines the direction toward the convergence of relaxation-model parameter estimation. $\mathbf{W}_{\mathrm{d}}$ is the $n\times n$ data weight matrix and is the inverse of the data covariance matrix (Torres-Verdin et al., 2000) describing the estimated variance for each particular measurement and the estimated correlation between measurements or the estimated uncertainties in the data because of noise contamination. If data error estimates of the measurement are known, then the weight matrix should be applied. When the measured data varies by many orders of magnitude along the frequency and data normalization is not performed, weight matrix should also be applied. λ is the damping parameter, which is adaptively adjusted by the iterative adjustment factor v during the course of the iteration to ensure convergence. Based on the damping parameter, the LMA interpolates between Gauss-Newton and gradient descent methods. Using $\Delta\mathbf{m}^k$ computed in Eq. (6.14), we can calculate the successive estimates for the relaxation-model parameters as

$$\mathbf{m}^{k+1}=\mathbf{m}^k+\Delta\mathbf{m}^k, \tag{6.15}$$

such that all elements of the l-sized vector $\mathbf{m}^{k+1}$ lie within the predefined lower and upper limits for each of the relaxation-model parameters. The Jacobian matrix of $\mathbf{F}(\mathbf{m}^k)$ is formulated as

$$\mathbf{J}\left(\mathbf{m}^k\right)=\begin{pmatrix}\frac{\partial f_1\left(\mathbf{m}^k\right)}{\partial m_1} & \cdots & \frac{\partial f_1\left(\mathbf{m}^k\right)}{\partial m_l}\\ \vdots & \ddots & \vdots\\ \frac{\partial f_n\left(\mathbf{m}^k\right)}{\partial m_1} & \cdots & \frac{\partial f_n\left(\mathbf{m}^k\right)}{\partial m_l}\end{pmatrix}, \tag{6.16}$$

where l is the number of relaxation-model parameters in the model, and n is the number of distinct frequencies at which the measurements were obtained. We implemented the first order central-difference formula to numerically approximate the derivative of $f_i(\mathbf{m}^k)$ with respect to j-th unknown relaxation-model parameter formulated as

$$\frac{\partial f_i\left(\mathbf{m}^k\right)}{\partial m_j}\approx\frac{f_i\left(m_j^k+\Delta mc_j^k\right)-f_i\left(m_j^k-\Delta mc_j^k\right)}{2\Delta m_j^k}, \tag{6.17}$$

where $\mathbf{m}^k = [m_1^k; m_2^k; \ldots; m_l^k]$, $\Delta mc_j^k = 0.001\, m_j^k$, $1 \le j \le l$, and $1 \le i \le n$. The higher order such as fourth order central-difference formula is applied in cases of the following chapters, which can be expressed as

$$\frac{\partial f_i(\mathbf{m}^k)}{\partial m_j} \approx \frac{-f_i\left(m_j^k + 2\Delta mc_j^k\right) + 8f_i\left(m_j^k + \Delta mc_j^k\right) - 8f_i\left(m_j^k - \Delta mc_j^k\right) + f_i\left(m_j^k - 2\Delta mc_j^k\right)}{12\Delta m_j^k}, \tag{6.18}$$

In this proposed scheme, we do not search for global minimum. Rather, we randomly initialize the inversion scheme to perform multiple inversions within predefined ranges for the unknown relaxation-model parameters. During the inversion, the estimates for the model parameters are selected as the solution of the inverse problem when: (a) the data misfit goes below a predefined threshold and (b) the frequency of occurrence of inversion-derived estimates are the highest among the multiple random initializations. As shown in Table 6.1, this selection criterion successfully identifies the estimates that are closest to the true values, which were defined a priori to generate the synthetic EM data. Data misfit error defined in this chapter is described in terms of the average absolute error or difference between modeled EM data based on the estimated relaxation-model parameters and the measured EM data. This average absolute data misfit error is expressed as

$$Error = \sqrt{\frac{1}{n} \cdot \sum_{i=1}^{n} \left\{ \left[\mathrm{real}\left(\mathbf{D}^{\mathrm{mod}}(\mathbf{m}^k) - \mathbf{D}^{\mathrm{meas}}\right)\right]_i^2 + \left[\mathrm{imag}\left(\mathbf{D}^{\mathrm{mod}}(\mathbf{m}^k) - \mathbf{D}^{\mathrm{meas}}\right)\right]_i^2 \right\}}, \tag{6.19}$$

where n is the number of distinct frequencies at which the measurements were obtained, subscript i denotes one of the n measurement frequencies, and k denotes the iteration number.

4.2 Damping parameter and iterative adjustment factor

Roy (1999) showed that an adaptive damping parameter made the inversion scheme more efficient and robust. The positive damping parameter λ is adjusted by the iterative adjustment factor v at each iteration to ensure fast convergence and accuracy of the estimated values of relaxation-model parameters. The damping parameter prevents singularities or near-singularities of the matrix $\left(\mathbf{W}_{\mathrm{d}}\mathbf{J}(\mathbf{m}^k)\right)^{\mathrm{T}} \cdot \left(\mathbf{W}_{\mathrm{d}}\mathbf{J}(\mathbf{m}^k)\right)$. The inversion formulation proposed in Eq. (6.14) simply moves down-gradient along the residual surface area for very large values of λ. Conversely, for very small values of λ, Eq. (6.14) reverts to GNA, which gives potentially fast but uncertain convergence (Aster et al., 2005). The general strategy is to use small values of λ but to switch to larger values of λ when the GNA-dominated inversion fails to make progress. That is to say, when the LMA leads to a reduction in $C(\mathbf{m})$, λ is reduced by dividing by

TABLE 6.1 The inversion-derived estimates of the parameters of Havriliak–Negami model (Eq. 6.3), the resistivity-specific formulation of Cole–Cole model (Eq. 6.7), and the resistivity-specific formulation of dual Cole–Cole model (Eq. 6.9) computed by inverting the synthetic multifrequency electromagnetic data.

Model	Symbol of model parameter	True value of model parameter	Histogram of initial guess of model parameter	Histogram of inversion-derived estimates	Estimate of model parameter
Havriliak–Negami Eq. (6.3)	ε_{r0}	100	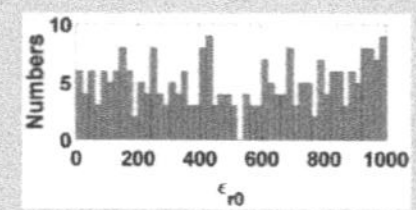	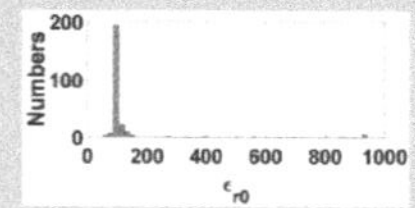	99.85
	$\varepsilon_{r\infty}$	25	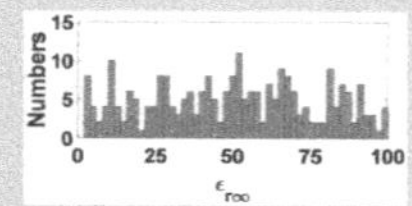	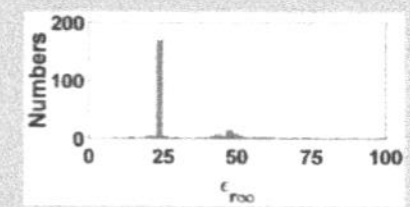	24.34
	τ (s)	1×10^{-3}	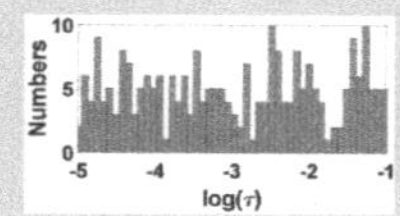	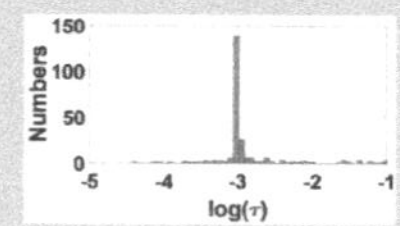	1×10^{-3}
	α	0.3	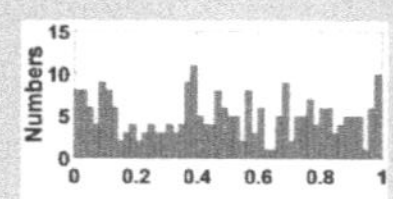	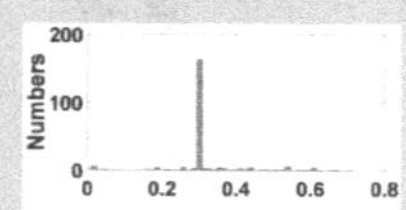	0.302

Continued

TABLE 6.1 The inversion-derived estimates of the parameters of Havriliak–Negami model (Eq. 6.3), the resistivity-specific formulation of Cole–Cole model (Eq. 6.7), and the resistivity-specific formulation of dual Cole–Cole model (Eq. 6.9) computed by inverting the synthetic multifrequency electromagnetic data.—cont'd

Model	Symbol of model parameter	True value of model parameter	Histogram of initial guess of model parameter	Histogram of inversion-derived estimates	Estimate of model parameter
	β	0.7	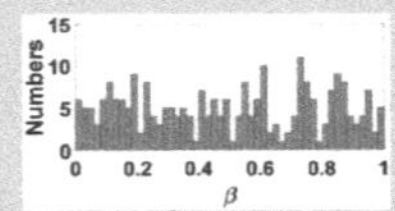	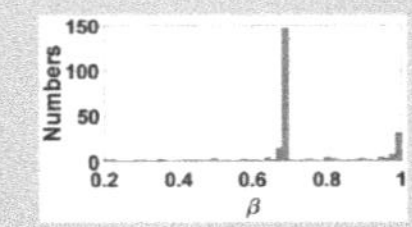	0.682
Resistivity-specific form of Cole–Cole model Eq. (6.7)	ρ_o (Ω)	50	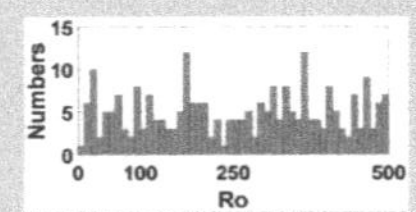	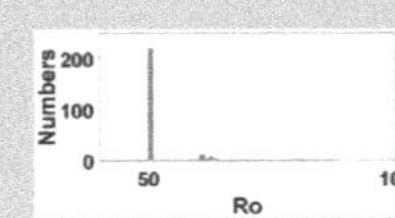	49.9
	m	0.5	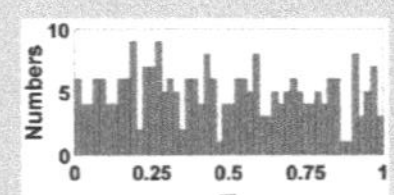	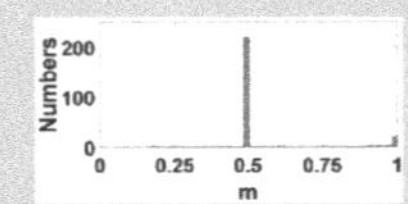	0.498
	τ (s)	1×10^{-3}	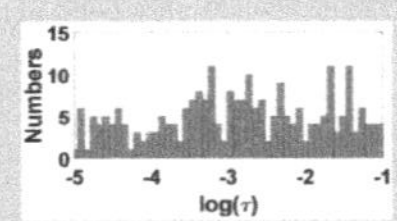	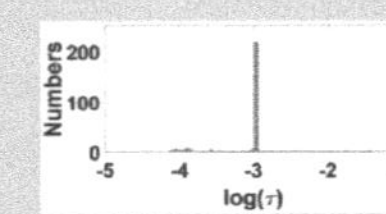	1×10^{-3}

	c	0.8	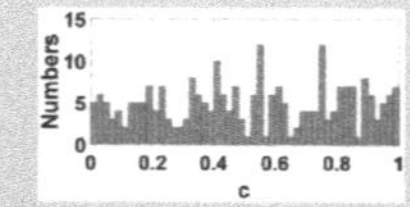	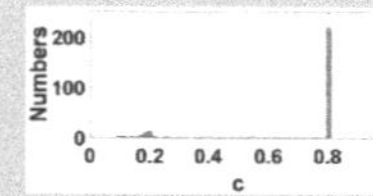	0.805
Resistivity-specific form of dual Cole–Cole model Eq. (6.9)	$\rho_o(\Omega)$	500	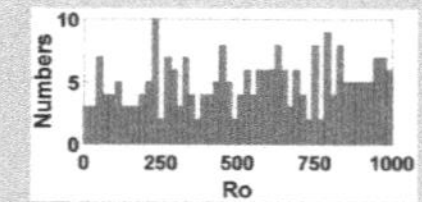	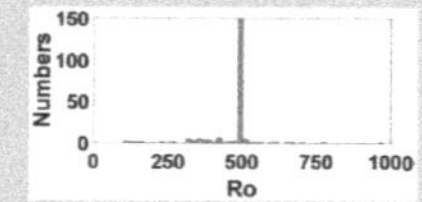	499.2
	m_1	0.4	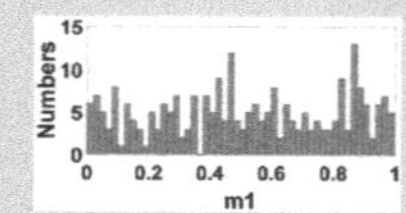	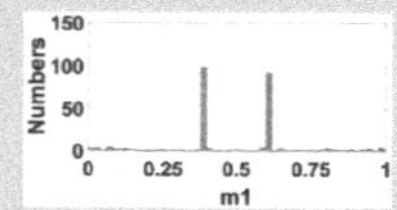	0.3974
	τ_1 (s)	1×10^{-1}	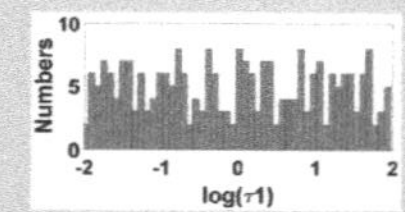	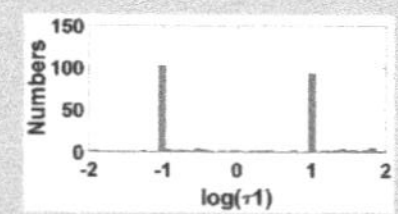	0.0989
	c_1	0.5	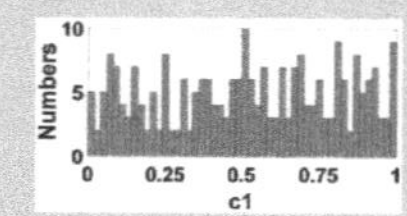	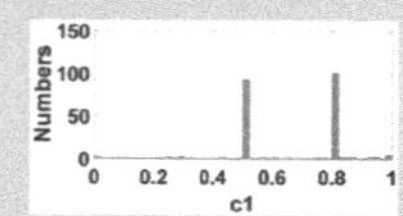	0.5097

Continued

TABLE 6.1 The inversion-derived estimates of the parameters of Havriliak–Negami model (Eq. 6.3), the resistivity-specific formulation of Cole–Cole model (Eq. 6.7), and the resistivity-specific formulation of dual Cole–Cole model (Eq. 6.9) computed by inverting the synthetic multifrequency electromagnetic data.—cont'd

Model	Symbol of model parameter	True value of model parameter	Histogram of initial guess of model parameter	Histogram of inversion-derived estimates	Estimate of model parameter
	m_2	0.6	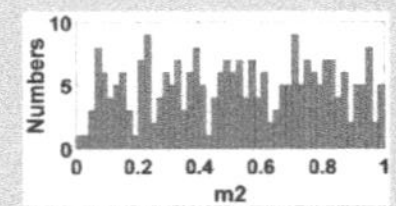	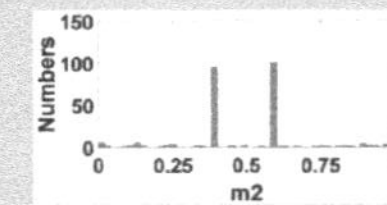	0.5976
	τ_2 (s)	1×10^1	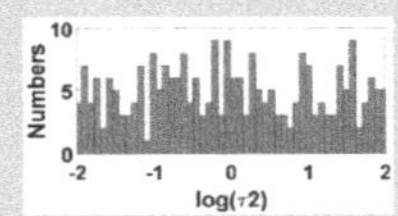	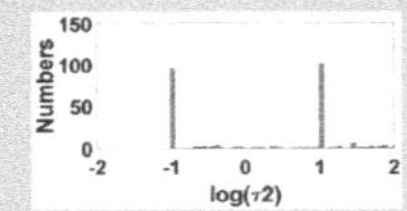	10.0885
	c_2	0.8	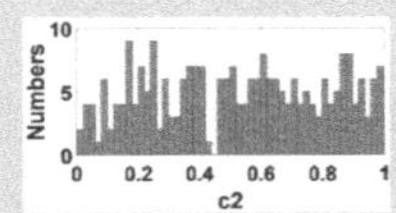	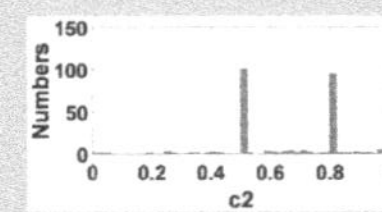	0.814

the factor v after each iteration. However, when the LMA does not lead to a reduction in $C(\mathbf{m})$, λ is increased by multiplying by the factor v after each iteration. One challenge associated with LMA is determining the value of λ and v. In Section 5.2, we present the sensitivity of inversion results to the initial values of λ and v. There are multiple combinations of initial values of λ and v that generate low data misfit. Such initial values of λ and v depend on the inverse problem and the available data. As shown in Section 5.2, the inversion of synthetic data and measured data exhibit good convergence and accuracy for specific combinations of initial values of λ and v. We chose the initial value of λ and v to be equal to 10. Another well-established heuristic of selecting suitable starting parameters was proposed by Newman and Alumbaugh (Newman and Alumbaugh, 1997) and was later used by Kemna (2005), as elaborated in page 67 and Eq. (3.58) of Kemna (2005). Such a choice led to proper scaling of the regularization terms.

4.3 Bounds of model parameters and jump-back-in step

Unlike incorporating the transformations functions of model parameters into cost function (Kim and Kim, 2011), we propose an alternative method to ensure the estimates are bounded during the inversion. For the estimation of each relaxation-model parameter, upper and lower bounds are set to ensure that the estimated values of the parameters are physically consistent, such that $l_j \leq m_j^k \leq u_j$, where l_j and u_j are the lower and upper bounds, respectively, of the j-th relaxation-model parameter to be estimated, m_j^k. The mid-point of the lower and upper bounds of the j-th relaxation-model parameter is M_j and m_j^{k-1} is the estimated value of the j-th parameter at the k-1-th iteration. At k-th iteration, when m_j^k exceeds the set bounds, a jump-back-in step is initiated using m_j^{k-1} and the upper and lower boundaries for the j-th model parameter. Fig. 6.2 illustrates the algorithm of the jump back-in step aimed at automating the step based on the distance between the estimated value of the j-th parameter m_j^{k-1} at the k-1-th iteration and the boundary. The j-th relaxation-model parameter after the jump back-in step is $m_j^{k'}$. Based on the algorithm illustrated in Fig. 6.2, Fig. 6.3 presents the four possible states before and after one of the two boundaries is exceeded by the model-parameter estimate.

In Fig. 6.2, there are four different scenarios when m_j^k exceeds the set bounds, algebraic manipulation will be applied to m_j^{k-1} that was within the bounds. After the algebraic manipulation, $m_j^{k'}$ will be obtained as the value of the j-th relaxation-model parameter at the k-th iteration that will be within the bounds. For the algebraic manipulation applied to m_j^{k-1}, we use an exponential function of the distance between m_j^{k-1} and one of the bounds, such that when the distance between the m_j^{k-1} and the bound, which m_j^k exceeds, is small the

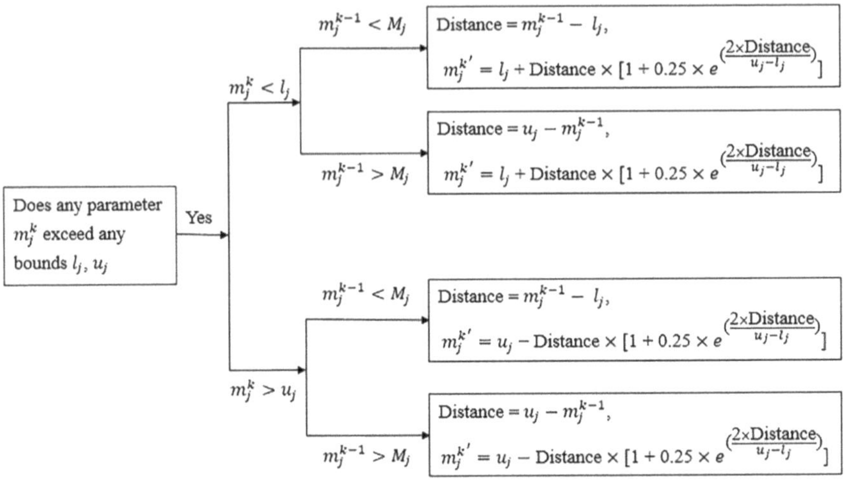

FIGURE 6.2 Flowchart of the jump back-in step to promote the exploration of the model-parameter space within the predefined bounds. One of the four possible scenarios is selected as the jump back-in step. *Adapted from Han Y, Misra S. A unified inversion scheme to process multifrequency measurements of various dispersive electromagnetic properties. J. Appl. Geophys. April, 2018;151:23–39.*

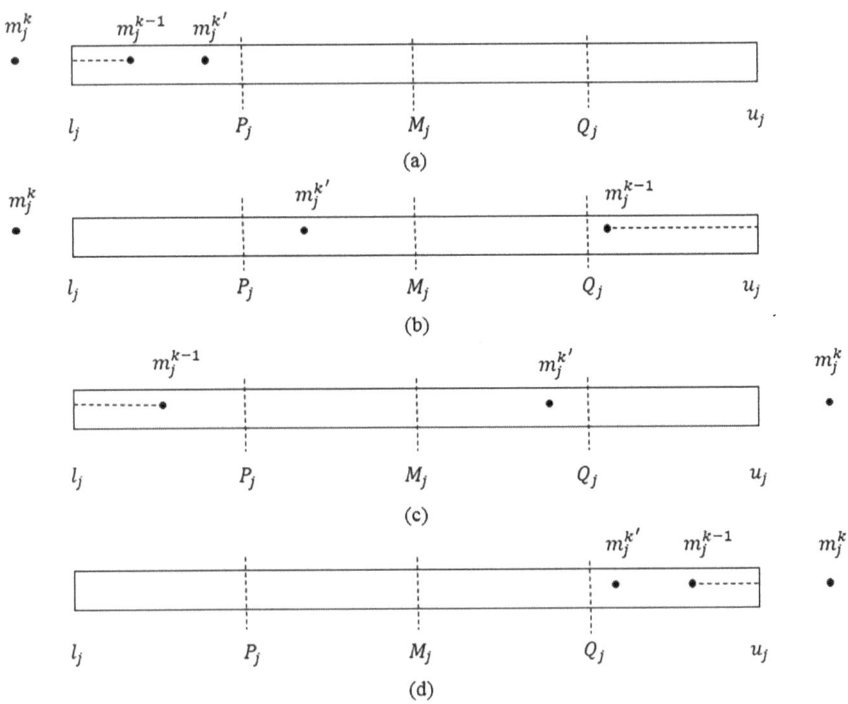

FIGURE 6.3 Illustration of the jump-back-in step for the four possible scenarios presented in the flowchart in Fig. 6.2. *Adapted from Han Y, Misra S. A unified inversion scheme to process multifrequency measurements of various dispersive electromagnetic properties. J. Appl. Geophys. April, 2018;151:23–39.*

$m_j^{k'}$ after jump back-in step will also close to that bound (similar to the situations shown in Fig. 6.3A and D). This is an important step when few estimates are close to the bounds (Han and Misra, 2018a). There is a factor "0.25" before the exponential function to guarantee that after the algebraic manipulation the estimates do not exceed the bounds. Upper and lower bounds of few relaxation models parameters, such as m, c, α, and β, are between 0 and 1. For few parameters, such as ρ_0 (low-frequency resistivity), the lower and upper bounds are generally 0 and 1.5 times of the maximum measured amplitude value, respectively. However, in our case studies, we use wider ranges for parameters, such as ρ_0, to demonstrate the convergence and accuracy of the unified inversion scheme. Fig. 6.3 shows four scenarios described in the flowchart in Fig. 6.2, when the j-th parameter at k-th iteration m_j^k exceeds the set bounds that is then transformed to $m_j^{k'}$ after the jump-back-in step to ensure the k-th iteration estimates are within bounds.

The jump-back-in step for processing or algebraically manipulating the estimates of model parameters when the estimates exceed the bounds is unique for the proposed bounded Levenberg–Marquardt algorithm. The jump-back-in step is also included in other deterministic inversion-based interpretation methods discussed in the following chapters. However, for proposed stochastic inversion-based interpretation method such as Markov-chain Monte Carlo (MCMC)-based inversion algorithm in Chapter 8, we apply a different method to keep the inverted model parameters within predefined bounds. When the inversion-derived estimates of the parameters exceed the lower or upper bounds, a new parameter vector is resampled from the proposal distribution until reaching a limit on number of resampling.

4.4 Jump-out of local minimum step

We implement a scheme to jump out of local minima when the rate of change in the difference/misfit between modeled data and measured data over 50 iterations is below a certain threshold. When a jump-out of local minima step is invoked, all the relaxation-model parameters are altered based on an automated algorithm. Fig. 6.4 illustrates the criteria of initiating jump out step. When the data misfit error is larger than a predefined threshold and the rate of change in error stabilizes over a predefined threshold for 50 iterations, the inversion-derived estimates for the parameters have reached a local minimum. This requires initiation of the jump out step to get out of local minimum and reach the global minimum. To be specific, the jump out is initiated when the decline rate r_1 for the misfit errors between adjacent iterations is lower than 0.01 and the decline rate r_2 between k-th and k-50-th iterations is lower than 0.1. For the j-th relaxation-model parameter, the midpoint of l_j and M_j, which is the midpoint of the lower and upper bounds, is P_j and the midpoint of M_j and u_j is Q_j. The jump out step includes reflection of inversion-derived

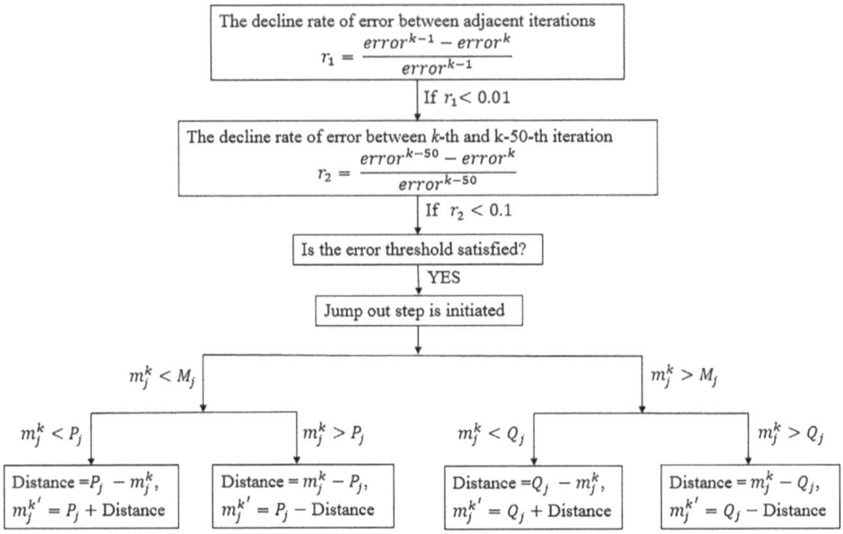

FIGURE 6.4 Flowchart of the jump-out step to escape three local minimum during the inversion process. This promotes the exploration of the model-parameter space. *Adapted from Han Y, Misra S. A unified inversion scheme to process multifrequency measurements of various dispersive electromagnetic properties. J. Appl. Geophys. April, 2018;151:23–39.*

estimates along a plane, which is an algebraic manipulation of the parameter about a reflection point. The jump out algorithm can accommodate moving out of three local minima during the inversion. This requires three reflections (algebraic manipulation) along P_j or Q_j, M_j, and boundary planes, respectively. Fig. 6.5A shows the process for the first reflection along the P_j plane. The j-th relaxation-model parameter after the jump out step is $m_j^{k'}$. The second reflection is similar to the first one but along the M_j plane. The third reflection uses either the lower boundary or upper boundary, i.e., m_j^k and $m_j^{k'}$ are equidistant from l_j and u_j, respectively, or vice versa. The three consecutive reflections are initiated only when the error remains higher than the predefined threshold. The algebraic manipulation of jump out is based on reflection points and reflection points are introduced to help the inversion algorithm automatically get out of the local minimum and converge to the global minimum. Fig. 6.5 better illustrates the process of three consecutive jumps required to avoid three local minima encountered during the inversion.

In the proposed unified inversion scheme, jump-back-in and jump-out of local minimum steps are critical to better explore the model-parameter space and find the most accurate estimates of the model parameter. Fig. 6.5 demonstrates the process of three consecutive jumps (reflections) required for jumping out of three local minima encountered during the inversion iteration. The first reflection is shown in Fig. 6.5A, which is based on P_j, the midpoint of

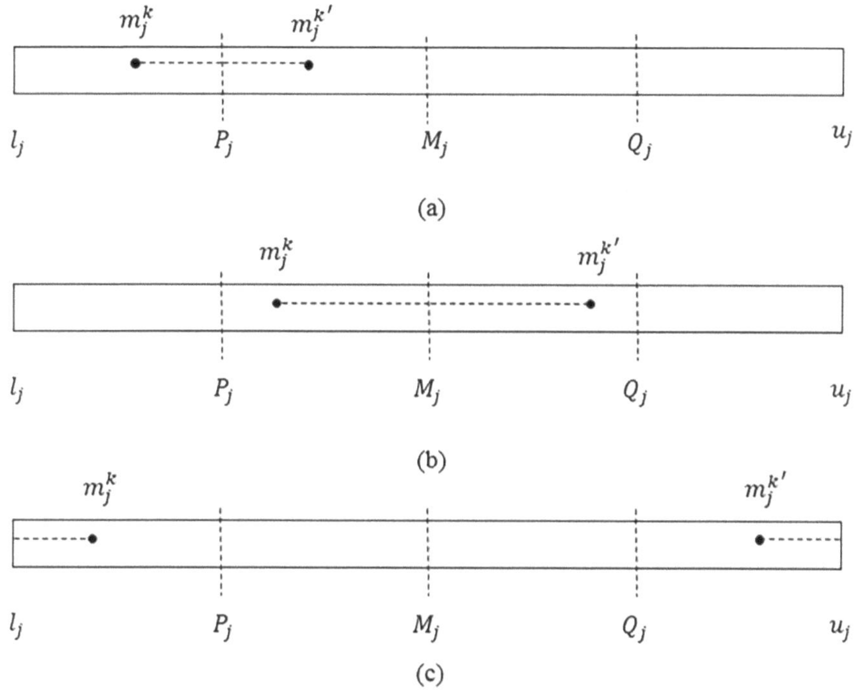

FIGURE 6.5 Illustration of three consecutive jump-out-of-minima steps required to avoid three local minima during the inversion for the three scenarios described in flowchart in Fig. 6.4. *Adapted from Han Y, Misra S. A unified inversion scheme to process multifrequency measurements of various dispersive electromagnetic properties. J. Appl. Geophys. April, 2018;151:23–39.*

l_j and M_j, which is the midpoint of the lower and upper bounds. The second reflection shown in Fig. 6.5B is based on M_j and the third reflection shown in Fig. 6.5C is based on lower and upper boundary. The jump-out of local minimum step is only included in the deterministic inversion scheme in this chapter, and not introduced in the other proposed deterministic or stochastic inversion-based interpretation methods and case studies in the following chapters.

5. Deterministic inversion of synthetic and laboratory electromagnetic measurements

5.1 Estimation of relaxation-model parameters by inverting synthetic electromagnetic data

Three sets of synthetic data were generated using Havriliak–Negami model (Eq. 6.3), resistivity-specific formulation of Cole–Cole model (Eq. 6.7), and resistivity-specific formulation of dual Cole–Cole model (Eq. 6.9) based on the parameter values listed in Column 3 of Table 6.1. About 2% Gaussian noise was added to the synthetic EM data. Following that, the synthetic data were inverted using the newly proposed inversion scheme coupled with a corresponding relaxation model. Column 4 of Table 6.1 lists the histogram for the 250 randomly chosen initial guesses for each of the model parameters. Column 5 lists the histogram of inversion-derived estimates that exhibit excellent convergence to the true parameter values. The performance of the newly proposed inversion algorithm with the jump back-in and jump-out steps is not sensitive to the initial guesses varying over three orders of magnitude. Table 6.2 focusses on the robustness of inversion of synthetic EM data in the presence of various amount of Gaussian noise ranging from 0% to 5% to estimate seven unknown model parameters of the resistivity-specific formulation of dual Cole—Cole model. In the presence of 5% noise, the estimation of model parameter τ_2 exhibited maximum relative error of 10%, whereas the estimates of the remaining parameters exhibited relative error of less than 1%.

5.2 Estimation of Havriliak—Negami model parameters by inverting laboratory electromagnetic data

This section presents deterministic inversion of the multifrequency complex permittivity measurements on poly (methyl acrylate), which is a petrochemical product, at 30°C, as reported by Havriliak and Negami (1966). The complex permittivity was measured using the General Radio Bridge (Type 1610-B) and Cell (Type 1690-A). The complex permittivity measurements were acquired at seven frequencies: 31, 100, 310 Hz, 1, 3.1, 10, and 150 kHz. Havriliak and Negami (1966) demonstrated that the α-relaxation process for poly (methyl acrylate) can be modeled using the Havriliak-Negami model, as described in Eq. (6.3). They graphically analyzed the plot of real part versus imaginary part of the complex permittivity measurements to determine the five parameters of Havriliak—Negami relaxation model that characterize the poly (methyl acrylate) sample under investigation. The EM data reported by Havriliak and Negami (1966) was processed using the newly proposed deterministic inversion. Our estimates of ε_{r0}, $\varepsilon_{r\infty}$, τ_0, α, and β using the proposed unified inversion scheme are close to those reported by the authors, as shown in Table 6.3. The Havriliak—Negami relaxation model predictions of real relative permittivity (ε'_r) and imaginary relative permittivity (ε''_r) based on the inversion-derived estimates of the five parameters match the measured complex relative permittivity measurements with an average absolute error (Eq. 6.19) less than 0.017, which is smaller than the error based on the reported parameter values by Havriliak and Negami (1966). Fig. 6.6 shows successful convergence of the inversion-derived estimates for five randomly chosen initial guesses of model parameter $\varepsilon_{r0}, \varepsilon_{r\infty}$, α, τ_0, and β in the ranges of 1 to 50, 1 to 25, 0.1 to 0.9, 10^{-5} s to 5×10^{-3} s, and 0.1 to 0.9, respectively. The bounds of the relaxation model parameters during inversion were: $0 \leq \varepsilon_{r0} \leq$ 100, $0 \leq$

TABLE 6.2 Inversion-derived estimates of resistivity-specific formulation of dual Cole–Cole model parameter computed by inverting the synthetic EM resistivity data containing various amounts of Gaussian noise. The true values of the model parameters are also presented in Table 6.1 in the section corresponding to the resistivity-specific formulation of dual Cole–Cole model.

Noise level	Estimated $R_o(\Omega)$	Estimated m_1	Estimated τ_1 (s)	Estimated c_1	Estimated m_2	Estimated τ_2 (s)	Estimated c_2
True values	500	0.4	0.1	0.5	0.6	10	0.8
Without noise	500	0.4	0.1	0.5	0.6	10	0.8
1% Gaussian noise	499.6	0.407	0.1	0.51	0.592	10.57	0.8
2% Gaussian noise	499.2	0.397	0.099	0.51	0.598	10.1	0.814
5% Gaussian noise	498	0.375	0.104	0.506	0.615	9	0.83

TABLE 6.3 Inversion-derived estimates for Havriliak–Negami model parameters computed by inverting the multifrequency complex permittivity measurements on methyl acrylate and their comparisons with estimates reported by Havriliak and Negami (1966).

Parameters	Inversion estimates	Reported estimates
ε_{r0}	6.4	6.48
$\varepsilon_{r\infty}$	3.67	3.705
α	0.32	0.345
τ_0 (s)	1.7×10^{-3}	2×10^{-3}
β	0.4	0.392
Average absolute error	0.017	0.07

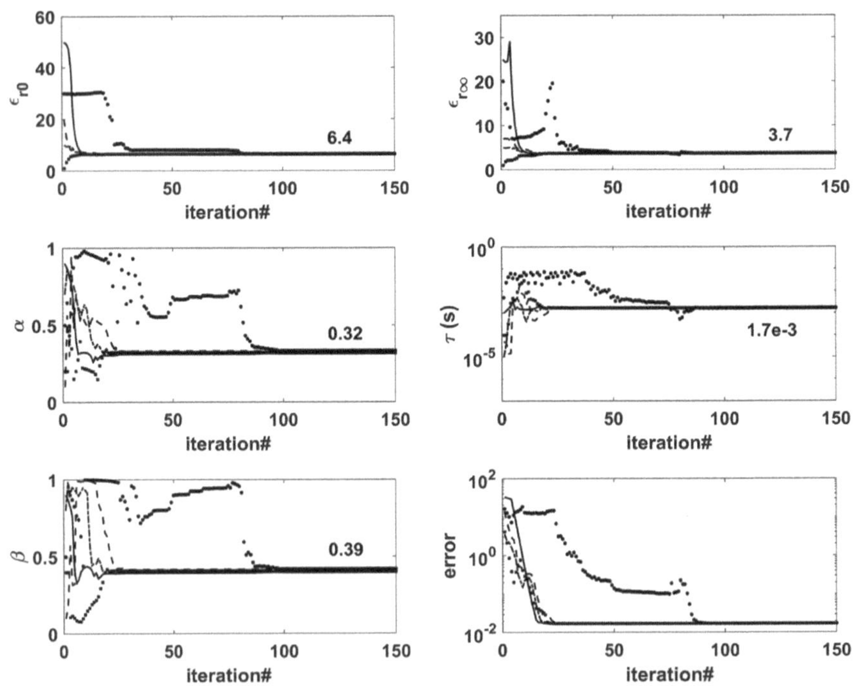

FIGURE 6.6 Convergence of inversion-derived estimates of the five relaxation-model parameters and data misfit error (bottom, right) as a function of iteration during the inversion of multifrequency measurement of complex relative permittivity on poly (methyl acrylate) solution, published by Havriliak and Negami (1966), assuming five different sets of initial guesses of the five relaxation-model parameters. *Modified from Han Y, Misra S. A unified inversion scheme to process multifrequency measurements of various dispersive electromagnetic properties. J. Appl. Geophys. April, 2018;151:23–39.*

$\varepsilon_{r\infty} \leq 100$, $0 \leq \alpha \leq 1$, $1 \times 10^{-5} < \tau_0 < 1 \times 10^{-1}$, and $0 \leq \beta \leq 1$ to ensure physical consistency. All estimates converge to the true value after 100 iterations irrespective of the choice of initial guesses. ε_{r0} and $\varepsilon_{r\infty}$ exhibit the fastest convergence among the five model parameters.

The inversion formulation uses damping factor λ and adjustment factor v that vary adaptively during the inversion. This improves the speed and accuracy of the estimation. Factor λ describes the dominance of Gauss-Newton convergence relative to gradient descent convergence. Factor v defines the rate of decrease in λ after each iteration. We conducted sensitivity study of λ and v, and observed that the inversion results of relaxation model parameters are sensitive to λ and v. Table 6.4 shows the inversion-derived estimates for nine cases that use different values of λ and v. Values of λ varied from 0.1 to 100 and that of v varied from 2 to 10. We demonstrate the efficacy of the values of λ and v by using the above multifrequency complex permittivity measurements on methyl acrylate solution requiring estimation of five relaxation-model parameters. We used the same initial guess for the nine cases (described in Table 6.4): $\varepsilon_{r0} = 20$, $\varepsilon_{r\infty} = 7$, $\alpha = 0.1$, $\tau_0 = 5 \times 10^{-5}$ s, and $\beta = 0.1$. Table 6.4 illustrates the estimated relaxation-model parameters are sensitive to the choice of initial values of λ and v. Notably, when comparing cases 7 and 8, change of v from two to three causes a large drop in the performance of the inversion scheme. A subsequent increase of v from 3 to 10 for cases 8 and 9 reduces the error back to small values.

To further present the sensitivity of inversion to initial values of λ and v, we plotted a two-dimensional surface plot, shown in Fig. 6.7, in which the logarithm of data misfit error is represented in grayscale. The same initial guess as above is used to generate this surface plot. The values of λ ranged from 0.01 to 100 and v ranged from 2 to 30 to generate the 2D surface plot. The logarithm of error due to data misfit varied from −2.7 to 5.4, which demonstrates that the newly developed modified bounded Levenberg–Marquardt inversion scheme is sensitive to the λ and v for purposes of estimating the relaxation-model parameters. Such sensitivity is dependent on the model, available data, and the desired estimation. Chapter 7 will discuss ways to mitigate such sensitivity of the inversion scheme. In the current chapter, we fix the initial values of λ and v to 10 for all the inversion cases because the inversion-derived estimates resulted in low data misfit error.

5.3 Estimation of Cole–Cole model parameters by inverting laboratory electromagnetic measurements

The proposed unified inversion scheme is applied to multifrequency measurement of complex conductivity of a mixture of Ottawa sand and 10% iron filings reported by Slater et al. (2006), as shown in Fig. 6.8. They quantified the induced polarization response of the mixture using the conductivity-

TABLE 6.4 Inversion-derived estimates and data misfit error for nine cases that use different initial values of damping factor λ and adjustment factor v. The performance of the inversion scheme is sensitive to the choice of initial values of λ and v.

Cases	λ	v	Inverted ε_{r0}	Inverted $\varepsilon_{r\infty}$	Inverted α	Inverted τ_0	Inverted β	error
Case 1	0.1	2	6.71	3.83	0.55	4.5×10^{-4}	0.94	7.2×10^{-3}
Case 2	0.1	3	6.4	3.66	0.31	1.7×10^{-3}	0.4	2.1×10^{-3}
Case 3	0.1	5	7.62	3.7	0.68	1.5×10^{-3}	0.97	5×10^{-2}
Case 4	10	2	6.69	3.82	0.53	5.1×10^{-4}	0.89	6.6×10^{-3}
Case 5	10	5	6.4	3.67	0.32	1.7×10^{-3}	0.4	2×10^{-3}
Case 6	10	10	6.4	3.67	0.32	1.7×10^{-3}	0.4	2×10^{-3}
Case 7	100	2	6.48	3.77	0.41	1×10^{-3}	0.58	3.7×10^{-3}
Case 8	100	3	5.25	4.89	0.27	2.3×10^{-4}	0.92	3.25
Case 9	100	10	6.4	3.67	0.32	1.7×10^{-3}	0.4	2×10^{-3}

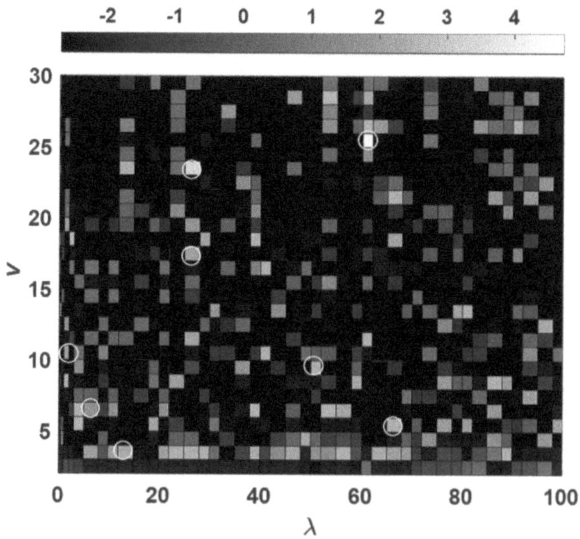

FIGURE 6.7 Grayscale surface plot of logarithm of data misfit error for various initial values of λ and v. The white regions denote the combinations of initial values of λ and v that result in very poor performance of the proposed inversion scheme. The black regions show the combinations that result in a robust performance of the inversion scheme. *Adapted from Han Y, Misra S. A unified inversion scheme to process multifrequency measurements of various dispersive electromagnetic properties. J. Appl. Geophys. April 2018;151:23–39.*

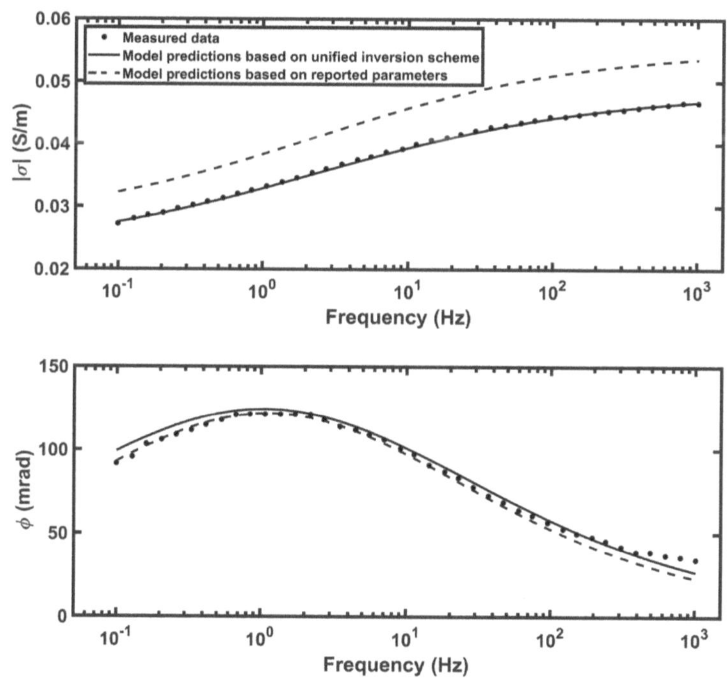

FIGURE 6.8 Comparisons of multifrequency measurements of conductivity magnitude $|\sigma|$ and phase ϕ of mixture of Ottawa sand and 10% iron filings as recorded by Slater et al. (2006) against the conductivity-specific formulation of Cole–Cole model predictions based on the inversion-derived estimates obtained using the newly proposed unified inversion scheme and those obtained by Slater et al. (2006). *Modified from Han Y, Misra S. A unified inversion scheme to process multifrequency measurements of various dispersive electromagnetic properties. J. Appl. Geophys. April, 2018;151:23–39.*

specific formulation of Cole–Cole model, as described in Eq. (6.8). They used the dynamic signal analyzer to measure conductivity magnitude $|\sigma| = \sqrt{\sigma'^2 + \sigma''^2}$, where σ' and σ'' are the real and imaginary parts of complex conductivity, and phase angle $\phi = \arctan\left(\frac{\sigma''}{\sigma'}\right)$ in the frequency range of 10^{-1} Hz and 10^3 Hz at 40 frequencies. They used the inversion algorithm developed by Kemna (2005), which applied a least-squares approach with Marquardt regularization to estimate the Cole–Cole model parameters. The estimates of the unknown model parameters reported by the authors are: $\sigma_0 = 0.0271$ S/m, $m = 0.51$, $\tau = 0.33$ s, and $c = 0.424$. In our inversion study, first we converted the magnitude and phase measured by Slater et al. (2006) to real and imaginary components of complex conductivity. Following that, inversion was performed on the complex conductivity dispersions. Our estimates of the four relaxation-model parameters are slightly different from those reported by the authors: $\sigma_0 = 0.0224$ S/m, $m = 0.54$, $\tau = 0.4$ s, and $c = 0.4$. The Cole–Cole relaxation model predictions based on our estimates of the unknown parameters match the complex conductivity measurements well, as shown in Fig. 6.8 and the absolute error is less than 2×10^{-4} S/m. The bounds of the unknown relaxation model parameters during inversion were: $0 \leq \sigma_0 \leq 1$, $0 \leq m \leq 1$, $1 \times 10^{-2} < \tau < 1 \times 10^2$, and $0 \leq c \leq 1$.

The proposed inversion scheme is then applied to the multifrequency complex impedance measurements on porous material at 19 frequencies in the range of 10^{-2}–10^4 Hz as reported by Jaggar and Fell (1988) and Xiang et al. (2001). They successfully modeled the measured induced polarization spectra using the impedance formulation of Cole–Cole model, as described in Eq. (6.6). Xiang et al. (2001) proposed a new direct inversion technique based on multifold least-square estimation, which solves for all the four unknown model parameters simultaneously without the need for assigning an initial guess of the unknown model parameters, thereby guarantying a unique solution. The solution is optimal in the least-squares sense and does not require iterative operator intervention. The estimates of the four unknown model parameters reported by the Xiang et al. (2001) were: $Z_0 = 21.9\ \Omega$, $m = 0.128$, $\tau = 0.0012$ s, and $c = 0.691$. Our estimates of the four relaxation-model parameters are close to those reported by the authors, as shown in Fig. 6.9. The Cole–Cole relaxation model predictions of impedance magnitude $|Z|$ and phase ϕ measurements based on our estimates of the four model parameters match the measured complex impedance measurements with an average absolute error less than 0.067 Ω, which is smaller than the error of 0.134 calculated based on the parameters reported by the authors. Fig. 6.9 shows that the inversion-derived estimates successfully converged for broad range of randomly chosen initial guesses for Z_0, m, τ, and c in the ranges of 1 Ω to 100 Ω, 0.1 to 0.9, 1×10^{-5} s to 1×10^{-2} s, and 0.1 to 0.9, respectively. The bounds of the unknown relaxation model parameters during inversion were: $0 \leq Z_0 \leq 100$, $0 \leq m \leq 1$, $1 \times 10^{-5} < \tau_0 < 1 \times 10^{-1}$, and $0 \leq c \leq 1$.

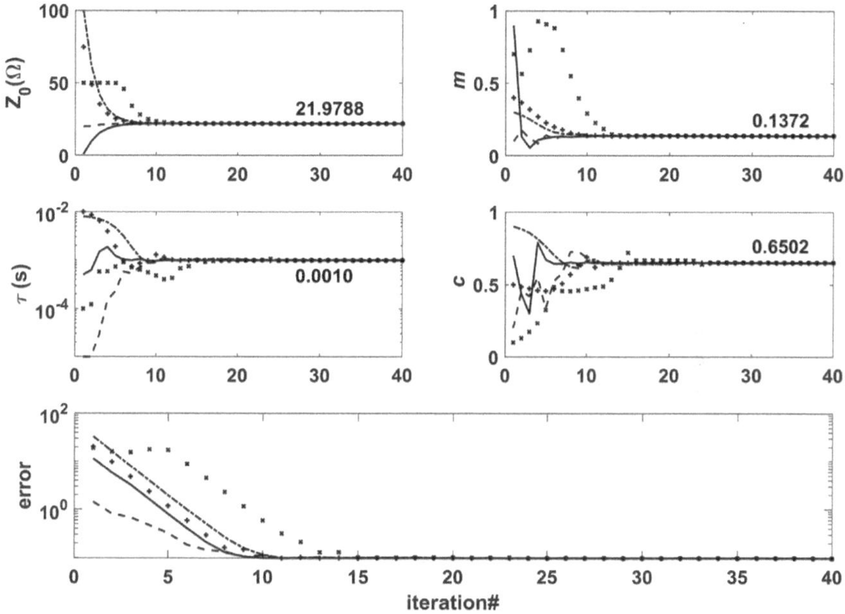

FIGURE 6.9 Convergence of inversion-derived estimates of the four relaxation-model parameters and data misfit error (bottom) as a function of iteration during the inversion of multifrequency measurement of complex impedance, published by Xiang et al. (2001), assuming five different sets of initial guesses of the four relaxation-model parameters. Convergence was achieved in 20 iterations irrespective of the initial guesses of the model parameters. *Modified from Han Y, Misra S. A unified inversion scheme to process multifrequency measurements of various dispersive electromagnetic properties. J. Appl. Geophys. April, 2018;151:23–39.*

5.4 Estimation of dual Cole–Cole model parameters by inverting laboratory electromagnetic measurements

The proposed inversion scheme is applied to the multifrequency resistivity measurements on unconsolidated silica-sand samples reported by Kemna et al. (2005), as shown in Fig. 6.10. We use Eq. (6.9) with $L = 2$ to model the measured data; in other words, we used a resistivity-specific formulation of dual Cole–Cole model. The samples have grain size of 125–250 μm and were saturated with 3×10^{-4} molar (M) KCl solution (Kemna et al., 2005). They measured complex resistivity at 71 frequencies from 1×10^{-3} Hz $- 1 \times 10^{5}$ Hz. The inversion-derived estimates successfully converged for broad ranges of randomly chosen initial guesses (Han and Misra, 2018a) for ρ_0, $\log(m_1)$, $\log(\tau_1)$, c_1, $\log(m_2)$, $\log(\tau_2)$, and c_2 in the ranges of 100–1000 Ωm, -4 to -1, -4 to -1 (τ_1 in s), 0.1 to 0.9, -4 to -1, -7.5 to -2 (τ_2 in s), 0.1 to 1, respectively. The logarithmic transformation is performed on some of the parameters to ensure that all the parameters are within one order of magnitude, which makes sure that the Jacobian matrix is nonsingular during the iterations.

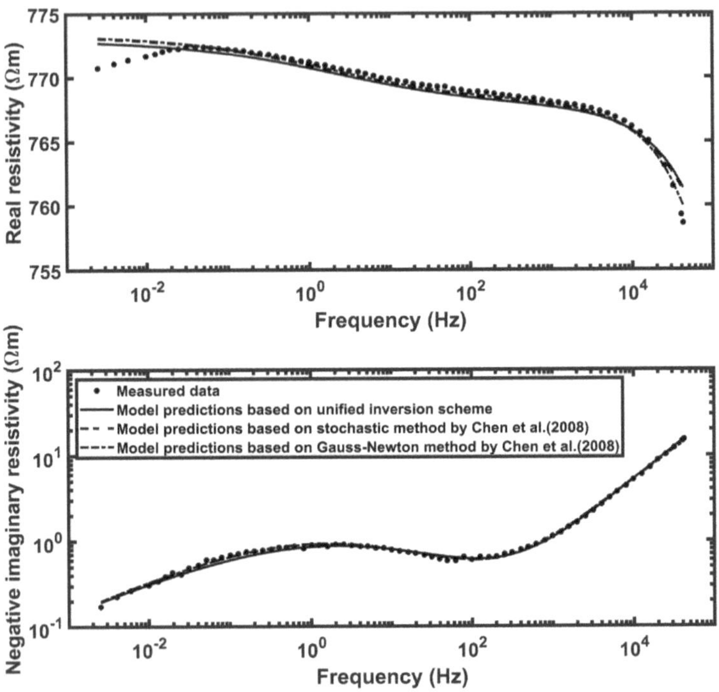

FIGURE 6.10 Comparison of the multifrequency measurements of real resistivity and negative imaginary resistivity of an unconsolidated silica-sand sample, reported by Kemna et al. (2005), against the dual Cole–Cole model predictions based on the inversion-derived estimates obtained using the newly proposed unified inversion scheme and those based on estimates computed by Chen et al. (2008). *Modified from Han Y, Misra S. A unified inversion scheme to process multifrequency measurements of various dispersive electromagnetic properties. J. Appl. Geophys. April, 2018;151:23–39.*

Our estimates for the seven unknown relaxation-model parameters are: $\rho_0 = 773\ \Omega\text{m}$, $m_1 = 0.0069$, $\log(\tau_1) = -1.1$ (τ_1 in s), $c_1 = 0.4$, $m_2 = 0.45$, $\log(\tau_2) = -7.1$ (τ_2 in s), and $c_2 = 0.78$, as listed in Table 6.5, which compares our inversion-derived estimates against those computed by Chen et al. (2008) using stochastic and deterministic inversion methods. The Cole–Cole relaxation model predictions based on our estimates of the seven parameters match the measured complex resistivity measurements with an absolute error of less than 1 Ωm, as shown in Fig. 6.10. The bounds of the unknown relaxation model parameters during inversion were: $1 \le \rho_0 \le 1000$, $-5 \le \log(m_1) \le 0$, $-10 \le \log(\tau_1) \le 0$, $0 \le c_1 \le 1$, $-5 \le \log(m_2) \le 0$, $-10 \le \log(\tau_2) \le 0$, $0 \le c_2 \le 1$.

The measurements reported by Kemna et al. (2005) are shown in Fig. 6.10. The real-part of the resistivity spectra can be interpreted as containing one peak because of low-frequency electrochemical polarization and one ill-defined

TABLE 6.5 **Inversion-derived estimates of the resistivity-specific formulation of dual Cole–Cole model computed by processing complex resistivity measurements on the silica-sand sample using the newly proposed unified inversion scheme against those obtained by Chen et al. (2008). Highest uncertainty in the estimation exists for m_2 followed by $\log(\tau_2)$.**

Symbol of dual Cole–Cole model parameter	Estimates based on the newly proposed inversion	Median of estimates computed by Chen et al. (2008) based on stochastic inversion	95% highest probability domain interval computed by Chen et al. (2008)	Estimates computed by Chen et al. (2008) based on Gauss-Newton inversion
$\rho_0(\Omega m)$	773	773.33	(772.38,774.37)	773.40
m_1	6.9e-3	6.9e-3	(6.7e-3,7e-3)	6.94e-3
$\log(\tau_1)$ (τ_1 in s)	−1.1	−0.972	(−1.005,−0.933)	−0.992
c_1	0.4	0.423	(0.413,0.433)	0.418
m_2	0.45	0.671	(0.305,1)	0.129
$\log(\tau_2)$ (τ_2 in s)	−7.1	−7.462	(−7.734,−6.948)	−6.406
c_2	0.78	0.736	(0.719,0.754)	0.765

peak because of high-frequency response such as Maxwell–Wagner polarization. In a strict sense, a dual Cole–Cole model with Maxwell–Wagner polarization model is required to fit the data. Debye decomposition can be used to generate unique characteristic parameters for such a spectra. Berube et al. (2017) discussed the limitations of Cole–Cole formulation when: (1) polarization peaks are partially defined, (2) when the *c*-exponent are lower than 0.1, and (3) in the presence of data uncertainty. The estimates of seven model parameters generated using the proposed inversion algorithm converged for a large range of initial guesses (Han and Misra, 2018c). Table 6.5 compares the estimates for the relaxation-model parameters corresponding to the unconsolidated silica-sand sample. The inversion-derived estimates are within the 95% highest probability domain computed by Chen et al. (2008) using stochastic-inversion. Although the stochastic or MCMC-based methods work well for a wide range of initial guesses, those methods are computationally expensive. However, compared to stochastic-inversion methods, the deterministic inversion methods are fast but do not quantify the uncertainties associated with the inversion-derived parameters.

6. Conclusions

A unified inversion scheme coupled with relaxation model was developed to process any multifrequency complex conductivity, complex resistivity, complex permittivity, and complex impedance measurements. The proposed inversion scheme was successfully coupled to various type of relaxation models, namely, Havriliak–Negami model, Pelton et al.'s complex impedance Cole–Cole model, Tarasov and Titov's complex conductivity Cole–Cole model, and Chen et al.'s complex resistivity model. The proposed inversion scheme is robust for various amounts of Gaussian noise ranging from 0% to 5%. The proposed inversion scheme is negligibly sensitive to the initial guesses of the relaxation-model parameters exhibiting convergence and accuracy up to 3-orders of magnitude variations in the initial guesses. The jump-back-in and jump-out of local minimum step proposed in this inversion scheme are important to get accurate and bounded parameter estimation. Jump-back-in and jump-out steps promote better exploration of the model-parameter space. The inversion scheme is sensitive to the choice of initial value of the damping factor and its iterative adjustment factor, which were fixed to a value of 10 in this chapter. The proposed unified inversion scheme will facilitate consistent and robust material characterization based on multifrequency measurements of various dispersive EM properties.

Nomenclature

Symbols and abbreviations

CEC Cation exchange capacity
CI Confidence interval
DSA Dynamic signal analyzer
EM Electromagnetic
GNA Gauss-Newton algorithm
HPD Highest probability domains
kHz KiloHertz
LMA Levenberg–Marquardt algorithm
MCMC Markov-chain Monte Carlo
mHz MilliHertz
SIP Spectral induced polarization

References

Aster, R.C., Borchers, B., Thurber, C.H., 2005. Parameter Estimation and Inverse Problems, first ed. Elsevier Academic Press, Amsterdam.

Bérubé, C.L., Chouteau, M., Shamsipour, P., Enkin, R.J., Olivo, G.R., August, 2017. Bayesian inference of spectral induced polarization parameters for laboratory complex resistivity measurements of rocks and soils. Comput. Geosci. 105, 51–64.

Binley, A., Slater, L.D., Fukes, M., Cassiani, G., December, 2005. Relationship between spectral induced polarization and hydraulic properties of saturated and unsaturated sandstone. Water Resour. Res. 41 (12), W12417.

Börner, F.D., Schön, J.H., October, 1991. A relation between the quadrature component of electrical conductivity and the specific surface area of sedimentary rocks. Log. Anal. 32 (5), 612–613.

Cao, Z., Chang, Y., Luo, Y., January, 2005. Inversion study of spectral induced polarization based on improved genetic algorithm. Progress in Electromagnetics Research Symposium (PIERS) Online. 1 (3), 266–270.

Chen, J., Kemna, A., Hubbard, S., November, 2008. A comparison between Gauss-Newton and Markov-chain Monte Carlo-based methods for inverting spectral induced-polarization data for Cole-Cole parameters. Geophysics 73 (6), F247–F259.

Cockett, R., Kang, S., Heagy, L.J., Pidlisecky, A., Oldenburg, D.W., September, 2015. SimPEG: an open source framework for simulation and gradient based parameter estimation in geophysical applications. Comput. Geosci. 85, 142–154.

Cole, K.S., Cole, R.H., April, 1941. Dispersion and absorption in dielectrics I. Alternating current field. J. Chem. Phys. 1, 341–351.

Davidson, D.W., Cole, R.H., December, 1951. Dielectric relaxation in glycerol, propylene glycol and n-propanol. J. Chem. Phys. 19, 1484–1490.

Dias, C.A., March, 2000. Developments in a model to describe low-frequency electrical polarization of rocks. Geophysics 65 (2), 437–451.

Fuller, B.D., Ward, S.H., January, 1970. Linear system description of the electrical parameters of rocks. IEEE Trans. Geosci. Electron. 8 (1), 7–18.

Gurin, G., Titov, K., Ilyin, Y., Tarasov, A., January, 2015. Induced polarization of disseminated electronically conductive minerals: a semi-empirical model. Geophys. J. Int. 200, 1555–1565.

Han, Y., Misra, S., September, 2017. Improved water-saturation estimates derived from inversion-based interpretation of broadband electromagnetic dispersion logs in organic-rich shale formations. SEG Ann. Int. Meet. 24–29.

Han, Y., Misra, S., May, 2018a. Joint petrophysical inversion of multifrequency conductivity and permittivity logs derived from subsurface galvanic, induction, propagation, and dielectric dispersion measurements. Geophysics 83 (3), D97–D112.

Han, Y., Misra, S., April, 2018b. Bakken Petroleum System characterization using dielectric-dispersion logs. Petrophysics 59 (2), 201–217.

Han, Y., Misra, S., April, 2018c. A unified inversion scheme to process multifrequency measurements of various dispersive electromagnetic properties. J. Appl. Geophys. 151, 23–39.

Han, Y., Misra, S., Simpson, G., 2017. Dielectric Dispersion Log Interpretation in Bakken Petroleum System: Presented at the SPWLA 58th Annual Logging Symposium.

Han, Y., Misra, S., Wang, H., Toumelin, E., February, 2019. Hydrocarbon saturation in a Lower-Paleozoic organic-rich shale gas formation based on Markov-chain Monte Carlo stochastic inversion of broadband electromagnetic dispersion logs. Fuel 243, 645–658.

Havriliak, S., Negami, S., 1966. A complex plane analysis of α-dispersions in some polymer systems. J. Polym. Sci. Part C: Polymer Symposia. 14 (1), 99–117.

Jaggar, S.R., Fell, P.A., June 1988. Forward and inverse Cole-Cole modeling in the analysis of frequency domain electrical impedance data. Explor. Geophys. 19 (3), 463–470.

Jin, Y., Misra, S., Homan, D., Rasmus, J., Revil, A., 2019. Mechanistic model of multi-frequency complex conductivity of porous media containing water-wet nonconductive and conductive particles at various water saturations. Adv. Water Resour. 130, 244–257.

Jin, Y., Misra, S., Homan, D., Rasmus, J., 2020. Effects of wettability of conductive particles on the multifrequency conductivity and permittivity of fluid-filled porous material. Fuel 268, 117411.

Kemna, A., 2005. Tomographic Inversion of Complex Resistivity: Theory and Application. Ph.D. thesis. Ruhr-University Bochum.

Kemna, A., Münch, H.M., Titov, K., Zimmermann, E., Vereecken, H., September, 2005. Relation of SIP relaxation time of sands to salinity, grain size and hydraulic conductivity. EAGE 11th Euro. Meet. Environ. Eng. Geophys. 4–7.

Kemna, A., Binley, A., Cassiani, G., Niederleithinger, E., Revil, A., Slater, L., Williams, K.H., Orozco, A.F., Haegel, F.H., Hördt, A., Kruschwitz, S., Leroux, V., Titov, K., Zimmermann, E., December, 2012. An overview of the spectral induced polarization method for near-surface applications. Near Surf. Geophys. 10 (6), 453–468.

Kemna, A., Räkers, E., Dresen, L., October 31–November 5, 1999. Field applications of complex resistivity tomography. SEG Ann. Int. Meet. SEG-1999-0331

Kim, H.J., Kim, Y., March, 2011. A unified transformation function for lower and upper bounding constraints on model parameters in electrical and electromagnetic inversion. J. Geophys. Eng. 8 (1), 21–26.

Kozhevnikov, N.O., Antonov, E.Y., July, 2012. Fast-decaying inductively induced polarization in frozen ground: a synthesis of results and models. J. Appl. Geophys. 82, 171–183.

Lazović, G., Vosika, Z., Lazarević, M., Simić-Krstić, J., Koruga, Đ., 2014. Modeling of bio-impedance for human skin based on fractional distributed-order modified cole model. FME Trans. 42 (1), 74–81.

Leroy, P., Revil, A., Kemna, A., Cosenza, P., Ghorbani, A., May, 2008. Complex conductivity of water-saturated packs of glass beads. J. Colloid Interface Sci. 321, 103–117.

Lesmes, D.P., Friedman, S.P., January, 2005. Relationships between the electrical and hydro-geological properties of rocks and soils. In: Rubin, Y., Hubbard, S.S. (Eds.), Hydrogeophysics. Springer, pp. 87–128 (Chapter 4).

Liu, M., Yang, J., Feng, J., Wang, T., Zhang, H., November, 2016. A discussion on the performance of seven existing models proposed to describe induced polarization. Geophysics 81 (6), E459–E469.

Loke, M.H., 2001. Tutorial: 2-D and 3-D Electrical Imaging Surveys. Course Notes for USGS Workshop 2-D and 3-D Inversion and Modeling of Surface and Borehole Resistivity Data.

Macnae, J., April, 2015. Comment on: Tarasov, A. & Titov, K., 2013, on the use of the cole–cole equations in spectral induced polarization, geophys. J. Int., 195, 352–356. Geophys. J. Int. 202, 529–532.

Marshall, D.J., Madden, T.R., October, 1959. Induced polarization, a study of its causes. Geophysics 24 (4), 790–816.

Mashimo, S., Umehara, T., Redlin, H., November, 1991. Structures of water and primary alcohol studied by microwave dielectric analyses. J. Chem. Phys. 95 (9), 6257–6260.

Misra, S., Han, Y., June 2016. Petrophysical interpretation of multi-frequency electromagnetic measurements in clay- and conductive-mineral-rich mudrock. SPWLA 57th Ann. Logg. Symp. 25–29.

Misra, S., Torres-Verdín, C., Homan, D., Rasmus, J., July 20, 2015. Complex electrical conductivity of mudrocks and source-rock formations containing disseminated pyrite. Unconvent. Res. Technol. Confer. 22.

Misra, S., Torres-Verdín, C., Revil, A., Rasmus, J., Homan, D., March, 2016. Interfacial polarization of disseminated conductive minerals in absence of redox-active species—Part 2: effective electrical conductivity and dielectric permittivity. Geophysics 81 (2), E159–E176.

Newman, G.A., Alumbaugh, D.L., February, 1997. Three-dimensional massively parallel electromagnetic inversion—I. Theory. Geophys. J. Int. 128 (2), 345–354.

Nordsiek, S., Weller, A., November, 2008. A new approach to fitting induced-polarization spectra. Geophysics 73 (6), F235–F245.

Pelton, W.H., Ward, S.H., Hallof, P.G., Sill, W.R., Nelson, P.H., April, 1978. Mineral discrimination and removal of inductive coupling with multifrequency IP. Geophysics 43 (3), 588–609.

Revil, A., Florsch, N., May, 2010. Determination of permeability from spectral induced polarization in granular media. Geophys. J. Int. 181 (3), 1480–1498.

Revil, A., February, 2012. Spectral induced polarization of shaly sands: influence of the electrical double layer. Water Resour. Res. 48 (2), W02517.

Revil, A., Eppehimer, J.D., Skold, M., Karaoulis, M., Godinez, L., Prasad, M., May, 2013. Low-frequency complex conductivity of sandy and clayey materials. J. Colloid Interface Sci. 398, 193–209.

Revil, A., Florsch, N., Mao, D., September, 2015. Induced polarization response of porous media with metallic particles—Part 1: a theory for disseminated semiconductors. Geophysics 80 (5), D525–D538.

Revil, A., Murugesu, M., Prasad, M., Le Breton, M., July, 2017. Alteration of volcanic rocks: a new non-intrusive indicator based on induced polarization measurements. J. Volcanol. Geoth. Res. 341, 351–362.

Roy, I.G., January, 1999. An efficient non-linear least-squares 1D inversion scheme for resistivity and IP sounding data. Geophys. Prospect. 47 (4), 527–550.

Schurr, J.M., September, 1964. On the theory of the dielectric dispersion of spherical colloidal particles in electrolyte solution. J. Phys. Chem. 68 (9), 2407–2413.

Seigel, H.O., July, 1959. Mathematical formulation and type curves for induced polarization. Geophysics 24 (3), 547–565.

Seleznev, N., Hou, C.Y., Freed, D., Habashy, T., Feng, L., Fellah, K., Xu, G., June, 2017. Coherent interpretation of wideband electromagnetic measurements in the Millihertz to Gigahertz frequency range. SPWLA 58th Ann. Logg. Symp. 17–21.

Slater, L.D., Ntarlagiannis, D., Wishart, D., March, 2006. On the relationship between induced polarization and surface area in metal-sand and clay-sand mixtures. Geophysics 71 (2), A1–A5.

Šumi, F., December, 1965. Prospecting for non-metallic minerals by induced polarization. Geophys. Prospect. 13 (4), 603–616.

Tarasov, A., Titov, K., July, 2013. On the use of the Cole-Cole equations in spectral induced polarization. Geophys. J. Int. 195 (1), 352–356.

Tathed, P., Han, Y., Misra, S., December, 2018a. Hydrocarbon saturation in Bakken Petroleum System based on joint inversion of resistivity and dielectric dispersion logs. Fuel 233, 45–55.

Tathed, P., Han, Y., Misra, S., May, 2018b. Hydrocarbon saturation in upper Wolfcamp shale formation. Fuel 219, 375–388.

Torres-Verdin, C., Druskin, V.L., Fang, S., Knizhnerman, L.A., Malinverno, A., November, 2000. A dual-grid nonlinear inversion technique with applications to the interpretation of dc resistivity data. Geophysics 65 (6), 1733–1745.

Weigand, M., Kemna, A., January, 2016. Debye decomposition of time-lapse spectral induced polarisation data. Comput. Geosci. 86, 34–45.

Wong, J., July, 1979. An electrochemical model of the induced-polarization phenomenon in disseminated sulfide ores. Geophysics 44 (7), 1245–1265.

Xiang, J., Jones, N.B., Cheng, D., Schlindwein, F.S., September, 2001. Direct inversion of the apparent complex-resistivity spectrum. Geophysics 66 (5), 1399–1404.

Zisser, N., Kemna, A., Nover, G., May, 2010. Relationship between low-frequency electrical properties and hydraulic permeability of low-permeability sandstones. Geophysics 75 (3), E131–E141.

Chapter 7

Deterministic inversion of galvanic resistivity, induction resistivity, propagation resistivity, and dielectric dispersion logs

Yifu Han[1], Siddharth Misra[2,3]
[1]*Schlumberger Technology Corporation, Beijing, China;* [2]*Harold Vance Department of Petroleum Engineering, College of Engineering, Texas A&M University, TX, United States;* [3]*Department of Geology and Geophysics, Texas A&M University, TX, United States*

1. Introduction of multifrequency electromagnetic logs joint interpretation

Borehole-based subsurface electromagnetic (EM) logs, namely galvanic resistivity (laterolog), induction resistivity, propagation resistivity, and dielectric dispersion logs, are commonly used for water saturation estimation in hydrocarbon-bearing formations. EM logs exhibit frequency dependence primarily due to the interfacial polarization effects arising from clay grain surfaces, conductive minerals, and charge blockage in pore throats. Interfacial polarization effects adversely affect the log-derived water saturation estimates, especially when there is low porosity, high formation water salinity, presence of conductive mineral disseminations such as pyrite and high clay minerals concentration (Han and Misra, 2018a; Tathed et al., 2018a,b). Conventional EM log interpretation methods estimate water saturation and other subsurface properties by separately interpreting the galvanic, induction, propagation, and dielectric dispersion logs using various empirical models or mixing laws. This approach leads to significant variations and uncertainties in petrophysical estimations from each logging tool or measurement technique.

To limit operational expenses in near-wellbore subsurface characterization, operators commonly employ only one of the four EM logging tools operating at a specific frequency. Also, it is not a common practice to measure

Multifrequency Electromagnetic Data Interpretation for Subsurface Characterization
https://doi.org/10.1016/B978-0-12-821439-8.00006-9

permittivity along with resistivity when using the laterolog and induction logging tools. Deployment of all the four EM tools in a single well followed by a joint interpretation of the multifrequency electrical conductivity and relative permittivity acquired over a broadband frequency range for water saturation estimation is also not a common practice. Wang and Poppitt (2013) reported first-of-its-kind continuous broadband multifrequency electrical conductivity and relative permittivity logs derived from EM induction, logging while drilling (LWD) propagation, and dielectric dispersion logs acquired in an organic-shale formation at seven log-acquisition frequencies.

The interfacial polarization effects arising from clay and conductive mineral grains influence charge carrier migration, accumulation/depletion, and diffusion processes in brine saturated porous geological formations; thereby altering the electrical and dielectric properties of the porous media (Misra et al., 2017; Misra and Han, 2016a,b). Dielectric relaxation models, such as Cole–Cole model, are frequency-dependent phenomenological model that can predict the electrical conductivity and permittivity spectra arising due to the interfacial polarization effects. However, being phenomenological model, the parameters of dielectric relaxation models do not represent petrophysical properties (Han and Misra, 2018b). We propose a deterministic inversion-based joint petrophysical interpretation of multifrequency electrical conductivity and relative dielectric permittivity logs derived from various combinations of the four aforementioned downhole EM logs acquired in clay- and conductive mineral-rich geological formations.

The proposed joint deterministic interpretation method uses the mechanistic PPIP-SCAIP model or PS model (explained in earlier chapters) that accounts for the interfacial polarization effects arising from clay and conductive mineral grains; thereby generating physically consistent water saturation estimates. The proposed joint deterministic interpretation can be coupled with any other mechanical model of interfacial polarization effects other than the PPIP-SCAIP model or PS model. The proposed deterministic inversion-based interpretation also generates estimates of formation water conductivity or equivalent salinity, surface conductance of clay, and average radius of clays and conductive mineral grains. The proposed method is applied to one field case and three synthetic geological formations, with varying clay type, conductive mineral properties, and water saturation. Furthermore, the sensitivity of inversion-derived estimates to the presence of various types of noise in the EM logs is investigated. The joint petrophysical deterministic inversion algorithm is applied to field broadband dispersion EM data acquired in the organic-rich shale formation (Wang and Poppitt, 2013). Water saturation, formation brine conductivity, surface conductance of clay, and radius of clay were almost consistently estimated (except for one case combination) in this shale formation using various combinations of available EM logs.

The joint inversion of relative permittivity and electrical conductivity logs derived from a combination of the four aforementioned EM logs is challenging because of the limited data points corresponding to the discrete log-acquisition frequencies, which ranges from four to nine frequencies depending on the available EM log combinations. Such a joint inversion is more challenging than that of laboratory EM measurements of complex conductivity/permittivity, which can be measured at 10 to 100 discrete frequencies in the same broadband frequency range. Added to that, in most cases, relative permittivity logs are not derived from galvanic resistivity and induction logs; consequently, the proposed joint interpretation will most likely lack permittivity data at laterolog and induction log frequencies. Limited and missing data will make it harder to solve the nonuniqueness issues in the proposed inversion. Besides that, there exist four to eight unknown/assumed model parameters in the PS model that adversely affect the nonuniqueness issues of the joint inversion.

2. Previous studies on deterministic interpretation of multifrequency permittivity and resistivity/conductivity measurements

A joint inversion of galvanic resistivity (laterolog) and EM induction log was proposed by Mezzatesta et al. (1994) for improved evaluation of residual and movable hydrocarbon saturation. Glinskikh et al. (2014) implemented the Nelder-Mead simplex (direct search) algorithm coupled with dispersed clay mixing models (de Lima and Sharma, 1990) for petrophysical inversion of high-frequency induction logs in the range between 0.875 and 14 MHz acquired from a shaly sandstone formation in oilfields of West and East Siberia to estimate water saturation, porosity, and clay fraction in the formation. LWD propagation deep resistivity tool operated at 125 kHz, 500 kHz, and 2 MHz was deployed and implemented in a high-angle well in a carbonate reservoir in onshore Abu Dhabi (Al-Ameri et al., 2015). They implemented an inversion technique to estimate the horizontal resistivity, vertical resistivity, and dip angle by minimizing a cost function consisting of the sum of data misfit and sum of constraints of formations models. In the inversion scheme, Levenberg–Marquardt algorithm (LMA) was preferred over the steepest-descent method to increase the speed of inversion. Hizem et al. (2008) used SMD model to jointly invert the four-frequency dielectric dispersion logs in the range of 20 MHz to 1 GHz to estimate the water saturation, formation water salinity, and rock textural parameters.

Inversion of multifrequency data is common for laboratory EM measurements. For example, Jougnot et al. (2010) developed a mechanistic model which accounts for the double layer polarization of grains, stern layer polarization of grains, and Maxwell–Wagner polarization. They coupled a simplex inversion algorithm with the model to interpret spectral induced polarization complex conductivity data of partially saturated clay-rich rock samples in the frequency range from 10 mHz to 45 kHz. Critical water saturation and stern layer contribution to surface conductivity of clay for each core sample were estimated using this method.

Various inversion algorithms have been developed to process the conductivity/resistivity and permittivity measurements. The least square method is widely used and L_2 norm objective function includes the data-misfit term (Lin et al., 1984), regularization term (Al Marzooq et al., 2014), and constraints on formation models (Al-Ameri et al., 2015). The Levenberg–Marquardt type algorithms have been implemented as the least square method in EM induction log inversion (Lin et al., 1984), LWD propagation resistivity log inversion (Al-Ameri et al., 2015), and multifrequency complex conductivity laboratory measured data inversion (Revil et al., 2015). Other inversion algorithms used to process the conductivity/resistivity and permittivity data are Nelder-Mead simplex (direct search) algorithm (de Lima and Sharma, 1990) and a combination of modified Gauss–Newton method and Monte Carlo algorithm (Zhou et al., 2016). Thiel et al. (2012) developed a robust Gauss–Newton inversion algorithm by introducing Huber error function to handle outliers in LWD propagation log measurements. The Huber function behaves like the L_2 norm for small error values and approaches the L_1 norm for large error values. In this chapter, we propose a modified LMA-based deterministic inversion method coupled with PS model to simultaneously process multifrequency dielectric permittivity and electrical conductivity logs acquired in the broadband frequency range of 1 kHz to 1 GHz. The proposed inversion algorithm has good convergence for wide range of initial guesses, and the algorithm can be generalized because we identified a set of fixed parameters required for the robust and reliable joint inversion.

3. Proposed modified bounded Levenberg–Marquardt nonlinear inversion scheme

We proposed a deterministic inversion scheme (Han and Misra, 2018c) for inverting synthetic multifrequency EM logs. We then improved and revised the proposed joint deterministic inversion algorithm for inverting field data or broadband electrical conductivity and relative permittivity dispersion logs. To avoid repetition in describing the proposed inversion scheme, in this chapter, we only present the revised joint deterministic inversion algorithm for inverting field data or broadband EM logs. In Chapter 8, we propose a Bayesian-based interpretation method which can quantify the uncertainty of the inversion estimates to process multifrequency EM logs (Han et al., 2019).

The nonlinear deterministic inverse problem discussed in this chapter involves the estimation of 3–5 PS model parameters that generate least misfit between the measured or synthetic logs and the PS model predictions of multifrequency effective permittivity and electrical conductivity of clay- and conductive mineral-bearing formations. To demonstrate the efficacy of the proposed deterministic inversion scheme, synthetic logs with 5% Gaussian noise are simulated using the PS model at the log-acquisition frequencies of various combinations of the four aforementioned EM tools for conductive mineral- and clay-bearing formations.

3.1 Methodology

In the proposed deterministic inversion scheme, n-sized measured data vectors $\mathbf{Perm}^{\text{meas}}$ and $\mathbf{Cond}^{\text{meas}}$ represent the log-derived or synthetically generated effective permittivity and electrical conductivity values of a conductive mineral- and clay-bearing formation acquired at n distinct EM log-acquisition angular frequencies, namely, $\omega_1, \omega_2, \ldots, \omega_n$. $\mathbf{Perm}^{\text{meas}}$ contains the measurements of relative permittivity, and $\mathbf{Cond}^{\text{meas}}$ contains the measurements of conductivity at various log-acquisition frequencies. The goal is to estimate l-sized PS model parameter vector $\mathbf{m}^{\mathbf{prop}}$ that best represents the petrophysical properties, such as water saturation, formation water conductivity or salinity, clay surface conductance, and average size of IP-causing grains, of the subsurface formation under investigation. To that end, PS model is used to generate the n-sized modeled data vectors $\mathbf{Perm}^{\text{mod}}(\mathbf{m})$ and $\mathbf{Cond}^{\text{mod}}(\mathbf{m})$, comprising PS model predictions at n distinct log-acquisition frequencies for a specific l-sized PS model parameter vector $\mathbf{m}$. The goal of deterministic inversion is to modify the PS model predictions of $\mathbf{Perm}^{\text{mod}}(\mathbf{m})$ and $\mathbf{Cond}^{\text{mod}}(\mathbf{m})$ by modifying the model parameter vector $\mathbf{m}$ till there is a good fit/agreement of the modeled $\mathbf{Perm}^{\text{mod}}(\mathbf{m})$ and $\mathbf{Cond}^{\text{mod}}(\mathbf{m})$ with the measured $\mathbf{Perm}^{\text{meas}}$ and $\mathbf{Cond}^{\text{meas}}$, respectively. The model parameter vector $\mathbf{m}$ for which the best fit is obtained indicates the best estimate of subsurface properties (i.e., the PS model parameters) denoted by $\mathbf{m}^{\mathbf{prop}}$.

Cost function to be minimized during the fitting process accounts for the large difference in the standard deviations of relative permittivity and electrical conductivity logs at distinct log-acquisition frequencies. In doing so, the inversion scheme requires minimizing quadratic cost function $C(\mathbf{m})$ formulated as

$$C(\mathbf{m}) = \left\| \frac{\mathbf{D}^{\text{mod}}(\mathbf{m}^{\text{est}}) - \mathbf{D}^{\text{meas}}}{\boldsymbol{\sigma}} \right\|_2^2, \tag{7.1}$$

where $\mathbf{D}^{\text{mod}}(\mathbf{m}^{\text{est}})$ represents modeled data vector, comprising modeled relative permittivity and conductivity, based on the estimates of model parameters at a specific iteration, $\mathbf{D}^{\text{meas}}$ represents the measured data vectors, and $\boldsymbol{\sigma}$ is the standard deviation vector for the broadband relative permittivity and electrical conductivity logs. We defined a $2n$-sized data misfit vector $\mathbf{F}(\mathbf{m})$ for the purposes of minimization. The data misfit vector $\mathbf{F}(\mathbf{m})$ is thus expressed as a combination of seven electrical conductivity and seven relative permittivity measurements in the broadband frequency as follows

$$\mathbf{F}(\mathbf{m}^k) = \left[f_1(\mathbf{m}^k), f_2(\mathbf{m}^k), \ldots f_{14}(\mathbf{m}^k)\right]^T, \tag{7.2}$$

where $f_i(\mathbf{m}^k)$ is expressed as

$$f_i(\mathbf{m}^k) = \frac{\mathbf{D}^{\text{mod}}(\mathbf{m}^{\text{est}})_i - \mathbf{D}_i^{\text{meas}}}{\boldsymbol{\sigma}_i}, \tag{7.3}$$

where $i = 1, 2, \ldots, 14$, in which the first seven elements ($i = 1, 2, \ldots, 7$) represent the relative permittivity measurements and the remaining seven elements ($i = 8, 9, \ldots, 14$) represent the electrical conductivity measurements for the seven frequencies of measurement. The elements in data misfit vector $\mathbf{F}(\mathbf{m})$ are in similar order of magnitude after dividing the broadband relative permittivity and electrical conductivity logs with the standard deviation vector, as shown in the cost function (Eqs. 7.1 and 7.2). Therefore, the weight matrix and weight factor are not needed for the field case study (as described by Han and Misra, 2018c). To ensure the generality of the inversion algorithm, we did not implement a reference vector $\mathbf{m}_R$, which represents the prior knowledge of parameter values and mitigates the nonuniqueness issues in the modified bounded LMA. For the field case study, the modified Levenberg–Marquardt inversion scheme is expressed as

$$\left[\left(\mathbf{J}\left(\mathbf{m}^k\right)\mathbf{W}_s\right)^{\mathrm{T}} \cdot \left(\mathbf{J}\left(\mathbf{m}^k\right)\mathbf{W}_s\right) + \lambda\mathbf{I}\right] \cdot \mathbf{W}_s^{-1} \cdot \Delta\mathbf{m}^k = -\left(\mathbf{J}\left(\mathbf{m}^k\right)\mathbf{W}_s\right)^{\mathrm{T}} \cdot \left(\mathbf{F}\left(\mathbf{m}^k\right)\right), \tag{7.4}$$

where $\mathbf{J}(\mathbf{m}^k)$ is the $n \times l$ Jacobian matrix derived of $\mathbf{F}(\mathbf{m}^k)$, which is already discussed and described in Chapter 6; $\mathbf{I}$ is $l \times l$ identity matrix, superscript T is matrix transpose operator, $\Delta\mathbf{m}^k$ is the l-sized correction vector generated at the k-th iteration that determines the direction toward the convergence, and λ is the damping parameter, which is adaptively adjusted by a constant damping factor v during the course of the iteration. $\mathbf{W}_s$ is $l \times l$ scaling diagonal matrix that ensures that the Jacobian matrix remains nonsingular during the inversion iterations. $\mathbf{W}_s$ is expressed as

$$\mathbf{W}_s = \begin{pmatrix} 10^{\alpha_1} & \cdots & 0 \\ \vdots & \ddots & \vdots \\ 0 & \cdots & 10^{\alpha_l} \end{pmatrix}, \tag{7.5}$$

where the parameters (α_1, α_2, ..., α_l) are the exponents in the diagonal denote the order of magnitude difference between the expected estimates for the l unknown PS-model parameters. The values of α_1, α_2, ..., α_l correspond to the unknown PS-model parameters, namely water saturation, formation water conductivity, surface conductance of clay, clay grain size, and conductive mineral grain size, which was assumed to be -1, 0, -6, -6, and -4, respectively, for the cases investigated in this chapter. Without the first-of-its-kind scaling matrix $\mathbf{W}_s$, the Jacobian matrix tends to become singular or badly scaled during the inversion iteration because the corresponding elements in the Jacobian matrix tend to become much larger than the other elements. This scaling matrix is not always mandatory if we perform the logarithmic transformation of the unknown parameters, which can also keep the unknown parameters are in the same order of magnitude. In Chapter 8, we performed the logarithmic transformation of the unknown PS-model parameters rather than introducing the scaling matrix.

The inverted unknown parameters at each iteration must be within the predefined lower and upper limits. The bounds of the unknown PS model petrophysical parameters during inversion are:

$0 \leq$ water saturation $(s_w) \leq 1$; $0 \leq$ bulk conductivity of brine $(C_w) \leq 25$ S/m; $1 \times 10^{-8} \leq$ surface conductance of clay $(\lambda_c) \leq 1 \times 10^{-5}$ S; $0 \leq$ radius of spherical clay grains $(r_c) \leq 10$ μm; and $1 \leq$ radius of conductive mineral grains $(r_i) \leq 1000$ μm.

We incorporated the jump-back-in step which is proposed and discussed in Chapter 6 into the inversion scheme to ensure that all the inverted unknown parameters at each iteration are within the predefined bounds. We define error as the difference between modeled EM data based on the inversion-derived PS-model parameter estimates at k-th iteration and the measured or synthetic EM data. The relative error at k-th iteration is expressed as

$$Rel.\ Err = \sqrt{\frac{1}{n} \cdot \sum_{i=1}^{n} \left(\frac{\mathbf{Perm}^{\mathrm{mod}}(\mathbf{m}^k) - \mathbf{Perm}^{\mathrm{meas}}}{\mathbf{Perm}^{\mathrm{meas}}} \right)_i^2 + \left(\frac{\mathbf{Cond}^{\mathrm{mod}}(\mathbf{m}^k) - \mathbf{Cond}^{\mathrm{meas}}}{\mathbf{Cond}^{\mathrm{meas}}} \right)_i^2} \tag{7.6}$$

4. Sensitivity of PPIP-SCAIP model on synthetic layers

Hydrocarbon-bearing source-rock formations and mudrock formations generally contain clays and electrically conductive mineral inclusions, such as pyrite, magnetite, and graphitic-precursor carbon. The frequency-dependence of electrical conductivity and effective dielectric permittivity of geological formations containing such inclusions are due to the electric-field-induced interfacial polarization effects and associated charge relaxation and accumulation around the host-inclusion interfaces. The proposed mechanistic electrochemical PPIP-SCAIP model or PS model in Chapter 4 evaluates the dependence of effective complex conductivity of geological mixtures on the conductivity of host and inclusion material, size and shape of inclusions, and the measurement frequency. Unlike the discussed sensitivity of PS model predictions to various model parameters in Chapter 4, the sensitivity of the PS model predictions is investigated for three specific synthetic formations in this chapter. Table 7.1 displays the assumed petrophysical properties of the three synthetic formations. Synthetic Layer 1 is an oil-filled formation containing conductive mineral and smectite-type conductive clay; Synthetic Layer 2 is a water-filled formation containing conductive mineral and kaolinite-type poorly conductive clay; and Synthetic Layer 3 is a brine-filled formation containing smectite-type conductive clay and low volume fraction of conductive mineral.

TABLE 7.1 Assumed petrophysical properties of the three synthetic layers used for this chapter.

Parameters	Unit	Layer 1	Layer 2	Layer 3
Volume fraction of conductive mineral, V_i	%	5	3	1
Bulk conductivity of conductive mineral, S_i	S/m	1000	5000	1000
Relative permittivity of conductive mineral, ε_{ri}		30	30	30
Diffusion coefficient of conductive mineral, D_i	m^2/s	10^{-6}	5×10^{-6}	10^{-6}
Radius of conductive mineral grains, r_i	**μm**	**30**	**100**	**50**
Volume fraction of clay, V_c	%	30	60	50
Relative permittivity of clay, ε_{rc}		5	5	5
Surface conductance of clay, λ_c	**S**	**5×10^{-6}**	**10^{-6}**	**5×10^{-6}**
Radius of spherical clay grains, r_c	**μm**	**0.3**	**1**	**0.5**
Volume fraction of sand, V_s	%	45	20	29
Surface conductance of sand, λ_s	S	10^{-9}	10^{-9}	10^{-9}
Radius of sand grains, r_s	μm	500	500	500
Total porosity of rock, ϕ_T	%	20	17	19
Bulk conductivity of brine, C_w	**S/m**	**3**	**1**	**10**
Relative permittivity of brine, ε_{rb}		80	80	80
Diffusion coefficient of brine, D_b	m^2/s	10^{-9}	10^{-9}	2×10^{-9}
Relative permittivity of hydrocarbon, ε_{rhy}		3	3	3
Water saturation, s_w	%	**20**	**90**	**80**

The unknown parameters that will be estimated using the proposed inversion scheme are shown in bold. Porosity of the three synthetic layers is kept around at 20%.

Kaolinite is the most prominent member of the kaolin group of clays and is classified as 1:1 layer minerals because unit cells are composed of one tetrahedral unit joined to one octahedral unit. Kaolinite-type clay generally occurs as discrete particles with diameter as large as 20 microns but the median diameter is generally about 1 micron. Smectite (montmorillonite) is the dominant member of the montmorillonite group of clay minerals and is classified as 2:1 layer minerals because the unit cells are composed of two tetrahedral units enclosing a single octahedral unit. The bonding between unit cells is weak by comparison with kaolinite and is the reason for the small diameter of discrete smectite particles, generally under 0.5 micron. The PS model assumes that the shape of clay grains is spherical. The radius of spherical clay grains is 1 micron in Synthetic Layer 2 containing kaolinite-type clay, and the radius of spherical clay grains is 0.3 micron and 0.5 micron in Synthetic Layers 1 and 3, respectively, containing smectite-type clay.

Leroy and Revil (2004) reported surface conductance of two sandstones, in one the primary clay type is smectite and in the other it is kaolinite. They measured the electrical conductivity of sandstones and used an electrical conductivity model to determine the normalized surface conductivity of smectite and kaolinite. They reported the normalized surface conductivity of kaolinite is much lower than that of smectite at pore fluid conductivity of 1 S/m. So, for the Synthetic Layer 2 (kaolinite-type clay), the surface conductance of clay is assumed to be 1×10^{-6} S and that in Synthetic Layers 1 and 3 is 5×10^{-6} S (conductive clay).

The proposed mechanistic electrochemical PPIP-SCAIP (PS) model is calibrated at low frequency (1 Hz) using Archie's equation with default values of Archie's equation parameters, and calibrated at high frequency (1 GHz) using CRIM for synthetic deterministic inversion applications discussed in this chapter. The PS model can also be calibrated at low frequency using Simandoux model for the synthetic and field cases of Bayesian-based inversion applications discussed in Chapter 8.

Figs. 7.1–7.5 show the sensitivities of PS-model predictions to the variations in five model parameters, namely s_w, C_w, λ_c, r_c, and r_i, for the three Synthetic Layers 1, 2, and 3 presented in plots A and D, B and E, and C and F, respectively. Our study indicates that effective permittivity and electrical

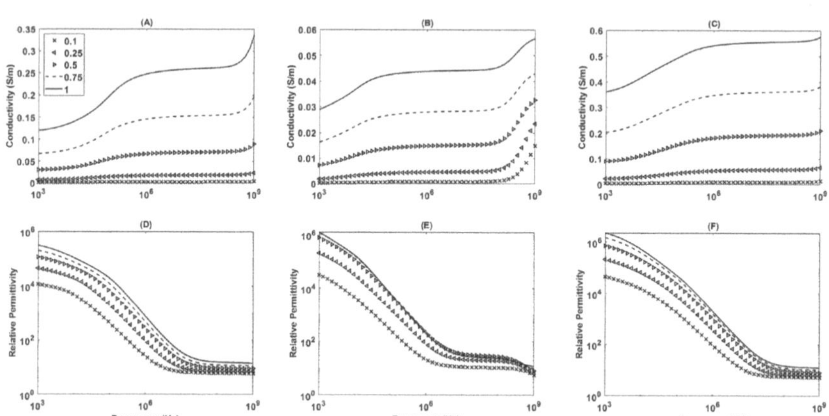

FIGURE 7.1 Sensitivities of PS model predictions of electrical conductivity and relative permittivity to variation in water saturation in saturation unit (fraction). The PS model predictions of electrical conductivity and relative permittivity for Synthetic Layers 1, 2, and 3 are presented in subplots A and D, B and E, and C and F, respectively. *Source: Authors.*

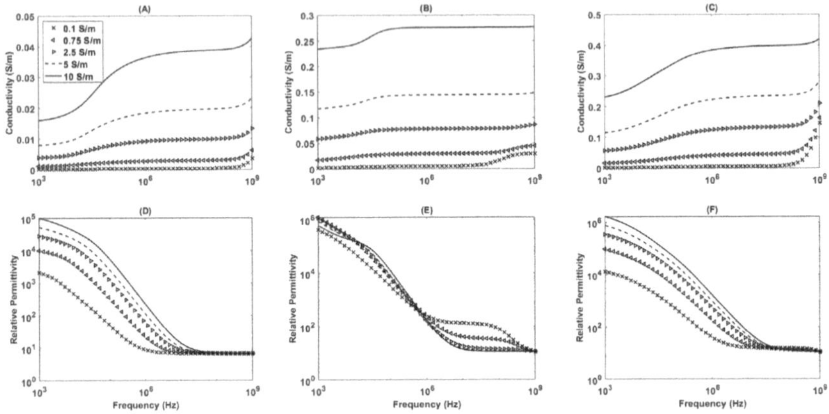

FIGURE 7.2 Sensitivities of PS model predictions of electrical conductivity and relative permittivity to variation in brine conductivity in S/m. The PS model predictions of electrical conductivity and relative permittivity for Synthetic Layers 1, 2, and 3 are presented in subplots A and D, B and E, and C and F, respectively. *Source: Authors.*

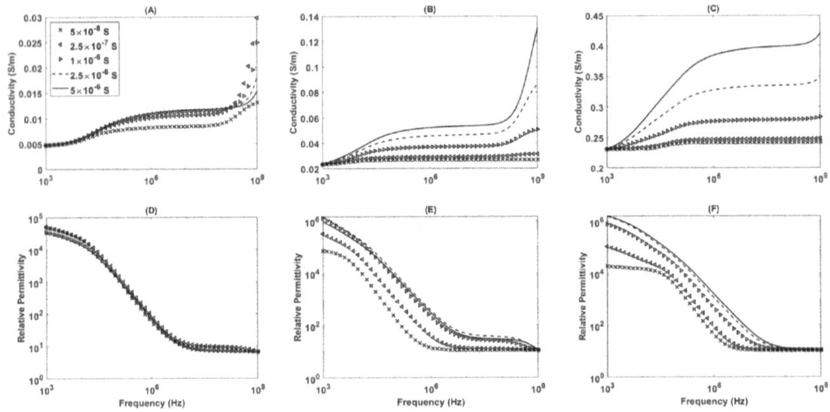

FIGURE 7.3 Sensitivities of PS model predictions of electrical conductivity and relative permittivity to variation in surface conductance of clay in S. The PS model predictions of electrical conductivity and relative permittivity for Synthetic Layers 1, 2, and 3 are presented in subplots A and D, B and E, and C and F, respectively. *Source: Authors.*

conductivity predictions in the entire frequency range of 1 kHz to 1 GHz (50 logarithmic-spaced frequencies) are sensitive to the first four parameters, whereas only those at low frequencies (less than around 1 MHz) are sensitive to the variation in radius of conductive minerals (similar to observations made by Misra et al. (2016)). The three synthetic layers exhibit substantially different dispersion spectra. Synthetic Layer 3 (Plots C and F) exhibits the largest magnitude and dispersion of electrical conductivity and effective permittivity due to the presence of brine that contributes charge carriers that

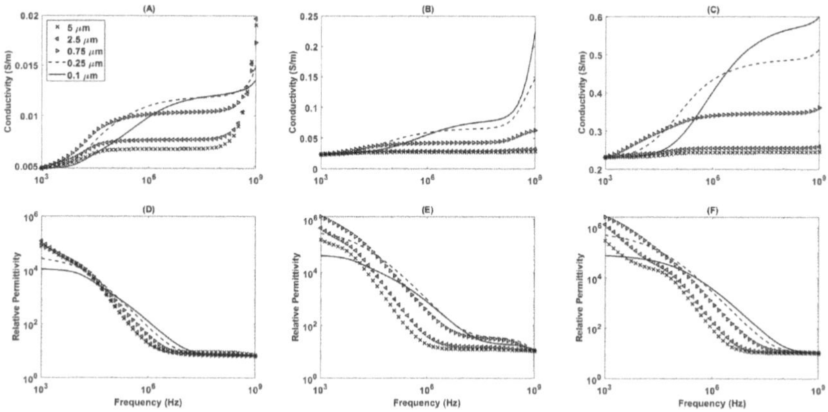

FIGURE 7.4 Sensitivities of PS model predictions of electrical conductivity and relative permittivity to variation in radius of spherical clay grains in μm. The electrical conductivity and effective permittivity estimates for Synthetic Layers 1, 2, and 3 are (A) and (D), (B) and (E), and (C) and (F), respectively. *Source: Authors.*

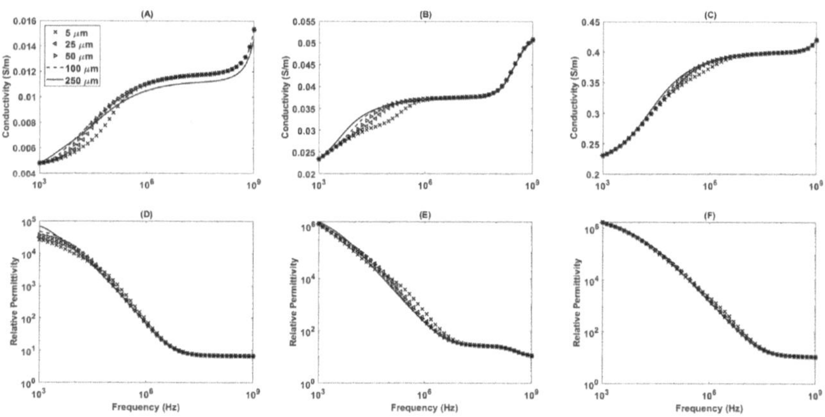

FIGURE 7.5 Sensitivities of PS model predictions of electrical conductivity and relative permittivity to variation in radius of spherical conductive mineral grains in μm. The electrical conductivity and effective permittivity estimates for Synthetic Layers 1, 2, and 3 are (A) and (D), (B) and (E), and (C) and (F), respectively. *Source: Authors.*

cause large interfacial polarization. Compared with electrical conductivity dispersions, effective permittivity dispersions are generally more substantial for frequencies below 2 MHz.

4.1 Sensitivity of PPIP-SCAIP model to water saturation

The PS model predictions of relative permittivity and electrical conductivity in the frequency range of 1 kHz to 1 GHz are highly sensitive to s_w in the range of 0.1 to 1 saturation unit. Conductivity and permittivity of all the three synthetic layers exhibit frequency dispersions. For the smectite-type conductive clay-rich Synthetic Layer 3, the electrical conductivity increases from 0.36 to 0.58 S/m when s_w is 100%, whereas the electrical conductivity increases from 0.023 to 0.068 S/m when s_w is 25% in the frequency range from 1 kHz to 1 GHz. The relative changes with increasing frequency of electrical conductivity increase with the values of s_w for the three synthetic layers, as shown in Fig. 7.1 (A—C). Revil (2013) also reported increase in both relative permittivity and electrical conductivity for higher water content in clay-rich samples. On the log-log plot, the relative permittivity spectra of Synthetic Layers 2 and 3 are very similar irrespective of the water saturation. Unlike other synthetic layers, conductive mineral-rich and kaolinite-type nonconductive clay-rich Synthetic Layer 2 (subplot B and E in Fig. 7.1) exhibits sharp frequency dispersion in electrical conductivity at all five water saturation levels for frequencies greater than 100 MHz.

4.2 Sensitivity of PPIP-SCAIP model to salinity

Fig. 7.2 reports the effect of change in brine conductivity from 0.1 to 10 S/m on the PS model predictions of relative permittivity and electrical conductivity. The relative change in electrical conductivity as a function of frequency drastically increases with increase in C_w from 0.1 to 10 S/m for the three synthetic layers. Interestingly, electrical conductivity dispersions of Synthetic Layers 1 and 3 are larger compared to that of Synthetic Layer 2 (subplot B and E), which is a poorly conductive kaolinite bearing layer. For the Synthetic Layer 1 (subplot A and D), the relative permittivity decreases from 9.8×10^4 to 6.7 at high-brine conductivity of 10 S/m, whereas the relative permittivity drops from 1×10^4 to 6.7 at low-brine conductivity of 0.75 S/m for the frequency range from kHz to GHz. At frequencies lower than 1 MHz, brine-filled Synthetic Layer 3 (subplot C and F) exhibits one-order of magnitude higher relative permittivity compared to oil-filled Synthetic Layer 1 (subplot A and D). Kaolinite-bearing Synthetic Layer 2 exhibits negligible variation in permittivity with brine conductivity for frequencies lower than 1 MHz. The modeled relative permittivity for Synthetic Layer 2 at frequency higher than 1 MHz decreases with increasing water conductivity or salinity unlike the other two layers.

4.3 Sensitivity of PPIP-SCAIP model to surface conductance of clay

Figs. 7.3 and 7.4 demonstrate the influence of two clay properties, namely surface conductance and size of clay grains, on the relative permittivity and electrical conductivity. Both these parameters have significant effects on the relative permittivity dispersion. The clay surface conductance lower than 10^{-8} S generates negligible electrical conductivity dispersion yet appreciable relative permittivity dispersion. The relative permittivity dispersion at low-clay conductance is governed by the volume fraction of clays and conductive minerals. The magnitude and dispersion of electrical conductivity tend to increase with increasing surface conductance of clay except for Synthetic Layer 1. For clay surface conductance values higher than 1×10^{-6} S, substantial dispersion in electrical conductivity is observed for Synthetic Layers 1 and 2 for frequencies higher than 100 MHz, whereas that for Synthetic Layer 3 (subplot C and F) is more visible in the lower frequencies (lower than 1 MHz). All synthetic layers, specifically Synthetic Layer 1 (subplot A and D), show a complicated trend for the variation in relative permittivity spectra with change in the surface conductance of clays, wherein maximum dispersion in permittivity is observed for clay surface conductance value between 2.5×10^{-7} S to 5×10^{-6} S. For example, the highest magnitude of relative permittivity dispersion is observed for the Synthetic Layer 1 (subplot A and D) with clay surface conductance of 2.5×10^{-7} S and that for Synthetic Layer 2 (subplot B and E) with clay surface conductance of 5×10^{-6} S.

4.4 Sensitivity of PPIP-SCAIP model to radius of spherical clay grain

Unlike other formation properties, variation in the size of the spherical clay grains results in dispersion of electrical conductivity and relative permittivity that are distinct and uncorrelated for the three synthetic layers, as shown in Fig. 7.4. The magnitude and dispersion of electrical conductivity are higher for smaller clay sizes of the clay-rich water-bearing Synthetic Layers 2 and 3 due to the increase in the surface area available for the interfacial polarization. The electrical conductivity and relative permittivity of hydrocarbon-bearing Synthetic Layer 1 are relatively less sensitive to variations in clay size, as compared to those of Synthetic Layers 2 and 3.

4.5 Sensitivity of PPIP-SCAIP model to radius of spherical conductive mineral grain

Fig. 7.5 shows the effects of variations of radius of conductive minerals for all synthetic layers. The electrical conductivity seems to insensitive to these variations for frequencies higher than around 1 MHz, whereas relative permittivity seems insensitive in the entire frequency range from 1 kHz to 1 GHz. Below 1 MHz, electrical conductivity increases with the radius of conductive minerals; this trend is not obvious for Synthetic Layer 1 (subplots A and D). Consequently, we do not present the inversion results for conductive minerals estimate for some of the case studies in this chapter because of the low sensitivity of electrical conductivity and permittivity to the size of conductive mineral.

5. Joint deterministic interpretation of multifrequency electromagnetic measurements

5.1 Inversion of synthetic logs generated at induction and dielectric dispersion frequencies

As mentioned earlier, a subsurface log acquisition plan generally does not involve all the four types of EM logging tools. There have been several oil-field log-acquisition efforts involving induction and dielectric dispersion logs (Tathed et al., 2018a,b; Misra et al., 2016; Pirrone et al., 2011), induction and propagations logs (Altman et al., 2008), and propagation and dielectric dispersion logs (Little et al., 2010). Notably, most of the existing induction log interpretation or inversion techniques only generate resistivity estimates and assume the permittivity of the formation to be one that is unrealistic in the presence of large volume fractions of clays and conductive minerals (Misra and Han, 2016a). Application of the proposed inversion technique to estimate petrophysical properties of the formation is challenging due to the following reasons: (1) the EM logs are acquired in a well at limited number of frequencies, generally between two and five frequencies (Han and Misra, 2018a); (2) permittivity logs are not generated during the laterolog and induction log interpretation, and (3) different volumes of a formation are investigated at different log-acquisition frequencies. Therefore, in this section, we apply the modified bounded LMA-based inversion method to synthetic electrical conductivity and relative dielectric permittivity data generated under the constraints of measurements obtained at less than six discrete frequencies and in absence of permittivity data at induction or laterolog frequencies, wherein the effective relative permittivity is assumed to be 1. The inversion algorithm is tested on synthetic data containing 5% Gaussian noise.

Synthetic electrical conductivity and relative permittivity data (Fig. 7.6) are generated using the PS model at one induction resistivity (26 kHz) and four dielectric dispersion (20 MHz, 100 MHz, 400 MHz, and 1 GHz) log-acquisition frequencies. The proposed deterministic inversion algorithm is applied to the five-frequency synthetic data to estimate three model parameters, namely, water saturation, bulk conductivity of brine, and surface conductance of clay. About 5% Gaussian noises are added to the synthetic electrical conductivity and effective permittivity data, as shown in Fig. 7.6. Inversion-derived estimates of the three PS model parameters are: $s_w = 0.79$, $C_w = 10.6$ S/m, and $\lambda_c = 4.7 \times 10^{-6}$ S, which are reported in Tables 7.3–7.5, and are close to the original parameter values, listed in Table 7.1, assumed for Synthetic Layer 3. The PS model predictions based on inversion-derived estimates of the three parameters match the synthetic electrical conductivity and effective permittivity (with 5% Gaussian noise) with relative error of less than 3×10^{-2}. Inversion-derived estimates for Synthetic Layer 1 are consistently more accurate compared to those for Synthetic Layer 3, which is discussed in

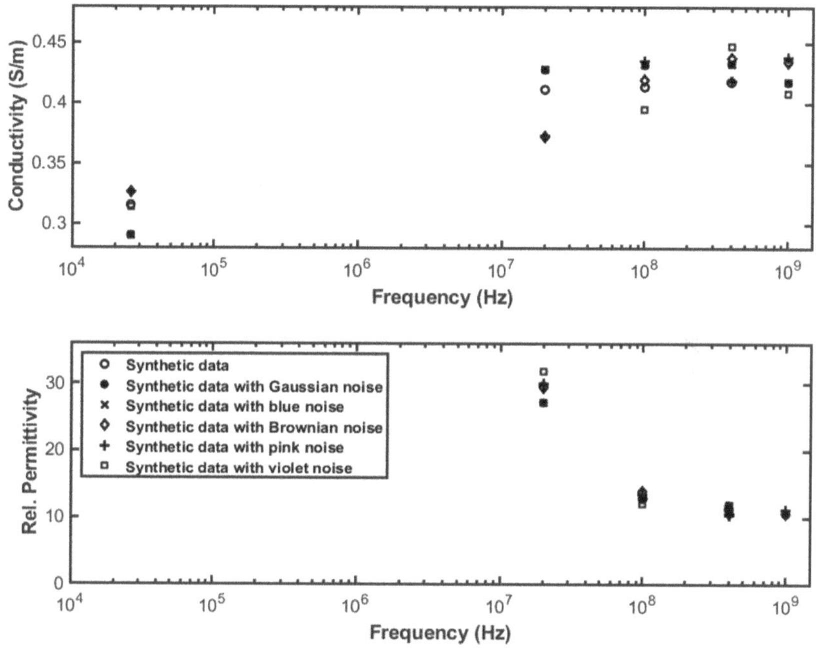

FIGURE 7.6 Synthetic electromagnetic data without noise and with 5% Gaussian noise, blue noise, Brownian noise, pink noise, and violet noise at one induction (~kHz) and four dielectric dispersion (~MHz to 1 GHz) frequencies generated for Synthetic Layer 3. The synthetic data at induction frequency do not have a corresponding relative permittivity value, in accordance with common practices of log acquisition. *Source: Authors.*

Section 6.2, because the EM data for Synthetic Layer 1 were acquired at 9 frequencies compared to 5 frequencies for Synthetic Layer 3.

The statistical assumptions on the noise model are always critical. Various types of noises were added to the synthetic data to evaluate the noise-agnostic properties of the proposed inversion algorithm. Fig. 7.6 shows the synthetic data predicted by PS model for the Synthetic Layer 3 without and with different types of these noise types. Section 6.3 discusses the inversion-derived estimates of s_w, C_w, and λ_c obtained by inverting synthetic data with these noise types. Inversion-derived estimates are most reliable in the presence of 5% Brownian noise and least reliable in the presence of Blue noise.

5.2 Inversion of synthetic logs generated at propagation and dielectric dispersion frequencies

In various oil-field log-acquisition effort, EM propagation tool is deployed during LWD followed by dielectric dispersion tool during the wireline logging run.

TABLE 7.2 Test cases based on the different initial values of λ and v.

Cases	λ	v	Inverted s_w	Inverted C_w (S/m)	Inverted λ_c (S)	Inverted r_c (μm)	Inverted r_i (μm)	Relative error
Case 1	0.1	2	0.205	2.87	7.3×10^{-6}	0.287	33.9	0.0233
Case 2	0.1	3	0.204	2.89	7.1×10^{-6}	0.287	34.1	0.0215
Case 3	0.1	5	0.202	2.96	5.6×10^{-6}	0.285	35.3	0.0214
Case 4	10	2	0.206	2.86	7.3×10^{-6}	0.287	33.9	0.0233
Case 5	10	5	0.201	2.96	5.5×10^{-6}	0.285	35.2	0.0216
Case 6	10	10	0.201	2.97	5.3×10^{-6}	0.285	35.4	0.0215
Case 7	100	2	0.206	2.84	7.3×10^{-6}	0.289	33.1	0.0231
Case 8	100	3	0.204	2.89	7.2×10^{-6}	0.286	34.4	0.0215
Case 9	100	10	0.201	2.97	5.4×10^{-6}	0.285	35.4	0.0230

TABLE 7.3 Relative errors in water saturation estimates obtained using the joint inversion of multifrequency electrical conductivity and effective permittivity data containing 5% Gaussian noise for three synthetic layers.

Synthetic layer	Estimated s_w	True s_w	Rel. Error in s_w (%)
1	0.201	0.2	0.65
2	0.89	0.9	1.1
3	0.79	0.8	1.25

TABLE 7.4 Relative errors in brine conductivity estimates obtained using the joint inversion of multifrequency electrical conductivity and effective permittivity data containing 5% Gaussian noise for three synthetic layers.

Synthetic layer	Estimated C_w (S/m)	True C_w (S/m)	Rel. Error in C_w (%)
1	2.97	3	1
2	1.03	1	3
3	10.6	10	6

TABLE 7.5 Relative errors in clay surface conductance estimates obtained using the joint inversion of multifrequency electrical conductivity and effective permittivity data containing 5% Gaussian noise for three synthetic layers.

Synthetic layer	Estimated λ_c (S)	True λ_c (S)	Rel. Error in λ_c (%)
1	5.2×10^{-6}	5×10^{-6}	4
2	1.05×10^{-6}	1×10^{-6}	5
3	4.7×10^{-6}	5×10^{-6}	6

Such tool deployment in a well will typically provide electrical conductivity and relative permittivity data at six frequencies in the range of 1 MHz to 1 GHz. To simulate this scenario, PS model is used to generate synthetic effective permittivity and electrical conductivity data at propagation (400 kHz, 2 MHz), and dielectric dispersion (20 MHz, 100 MHz, 400 MHz, 1 GHz) log-operation frequencies acquired in a conductive mineral- and clay-bearing synthetic formations. Parameters mentioned in Table 7.1 for the Synthetic Layer 2 are used as default

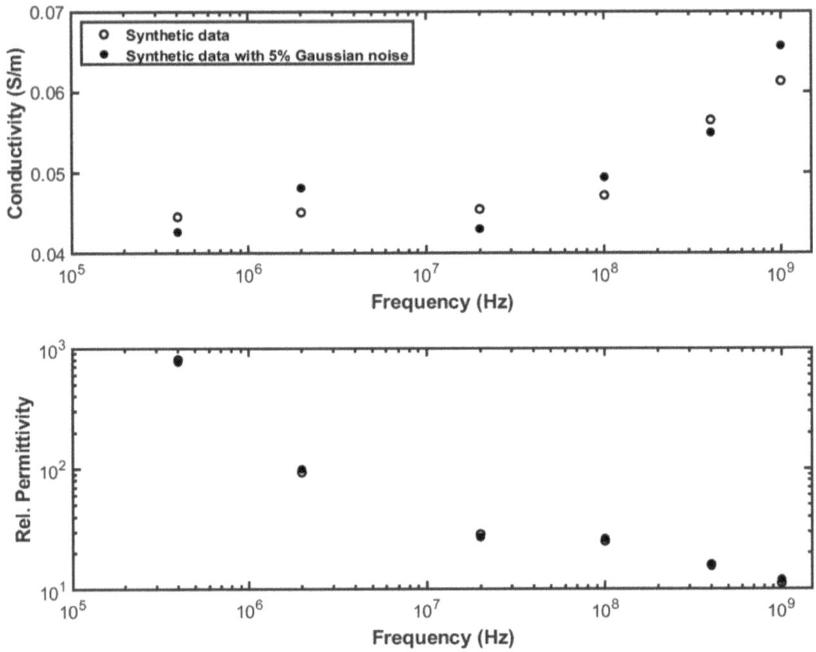

FIGURE 7.7 Synthetic electromagnetic data with and without 5% Gaussian noise at two propagation (~MHz) and four dielectric dispersion (~MHz to 1 GHz) frequencies generated for Synthetic Layer 2. *Source: Authors.*

parameters and the objective of inversion of the six-frequency synthetic data is to accurately estimate the s_w, C_w, λ_c, and r_c. Fig. 7.7 shows the synthetic effective permittivity and electrical conductivity data with Gaussian noise predicted by PS model for Synthetic Layer 2 at propagation and dielectric dispersion log-acquisition frequencies.

Inversion-derived estimates for the four PS model parameters are: $s_w = 0.89$, $C_w = 1.03$ S/m, $\lambda_c = 1.05 \times 10^{-6}$ S, and $r_c = 1.03$ μm, which are reported in Tables 7.3–7.6, and are close to the original parameter values, listed in Table 7.1, assumed for Synthetic Layer 2 that were used to generate the synthetic data shown in Fig. 7.7. The PS model predictions based on our estimates of the four parameters match the synthetic effective permittivity and electrical conductivity with relative error of less than 5×10^{-2}. Inversion-derived estimates for Synthetic Layer 1 are consistently more accurate compared to those for Synthetic Layer 2 except for the radius of clay estimates because the EM data for Synthetic Layer 1 were acquired at 9 frequencies compared to 6 frequencies for Synthetic Layer 2.

6. Sensitivity analysis and accuracy of the joint deterministic interpretation method

6.1 Sensitivity analysis of inversion to factors λ and v implemented in the modified bounded Levenberg–Marquardt algorithm

The inversion formulation uses λ and v (shown in Eq. 7.4) that vary adaptively during the inversion iteration to improve the speed and accuracy of the estimation. Factor λ describes the dominance of Gauss–Newton convergence relative to gradient descent convergence. Factor v defines the rate of decrease in λ after each iteration. We conducted sensitivity study of λ and v, and observed that the inversion-derived estimates of subsurface formation properties are not sensitive to λ and v.

We demonstrate the efficacy and sensitivity analysis of the values of λ and v through the synthetic broadband frequency effective permittivity and electrical conductivity data inversion case that involves a five-parameter PS-model inversion. The synthetic EM data with and without 5% Gaussian noise was generated for a conductive mineral- and clay-bearing formation, similar to Synthetic Layer 1, by using PS model at laterolog (1 kHz), induction (26 kHz, 52 kHz), propagation (400 kHz, 2 MHz), and dielectric dispersion (20 MHz, 100 MHz, 400 MHz, 1 GHz) log-acquisition frequencies. This presents an ideal and uncommon situation when electrical conductivity and effective permittivity of a given depth in a conductive mineral- and clay-bearing geological formation are measured at nine frequencies using the four EM downhole tools available in the industry. Fig. 7.8 shows the synthetic effective permittivity and electrical conductivity generated using the PS model at four aforementioned EM logging frequency ranges. When plotting the relative permittivity as a function of frequency, logarithmic scale is used in y-axis that compresses the effect of noise in permittivity.

The inversion results validate the accuracy of the deterministic interpretation method in estimating s_w, C_w, λ_c, r_c, and r_i. Contour plots in Fig. 7.9A and B depict LMA-based inversion method for synthetic EM data without Gaussian noise case in terms of error minimization for a specific set of initial guesses of the model parameters, namely $s_w = 0.5$, $C_w = 7$ S/m, $\lambda_c = 5\times10^{-7}$ S, $r_c = 0.2$ μm, and $r_i = 50$ μm. The contours present relative error of data misfit in logarithmic scale; for example, contour of -1 indicates a relative error of 0.1. Fig. 7.9A illustrates the changes in brine conductivity and water saturation estimates, and Fig. 7.9B illustrates the changes in the estimates of clay properties: surface conductance of clay and radius of spherical clay grains during the first 15 iterations that resulted in the convergence of inversion-derived estimates. To avoid repetition in reporting the observed trends, we do not present the changes in the estimates of radius of

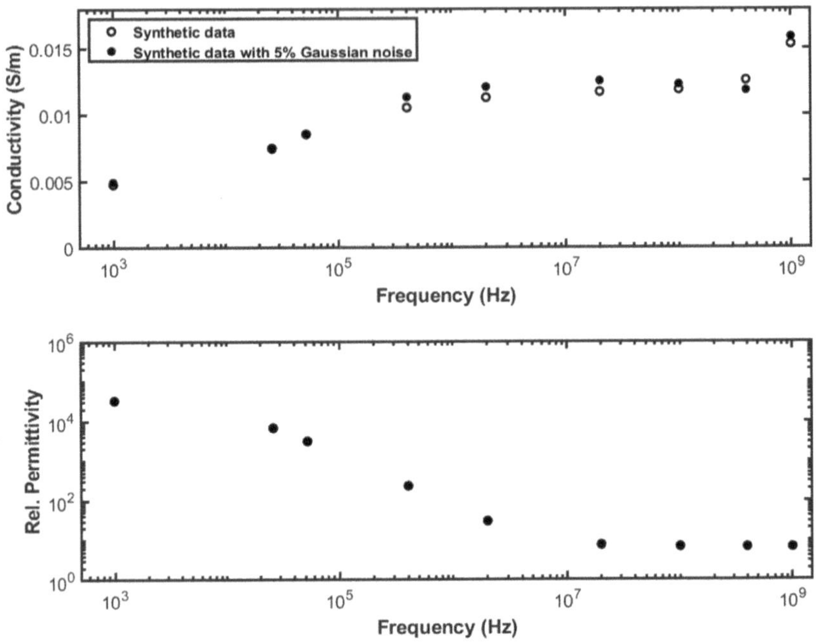

FIGURE 7.8 Synthetic electromagnetic data with and without 5% Gaussian noise at one laterolog (~Hz), two induction (~kHz), two propagation (~MHz), and four dielectric dispersion frequencies (~MHz to 1 GHz) generated assuming parameters for Synthetic Layer 1. *Source: Authors.*

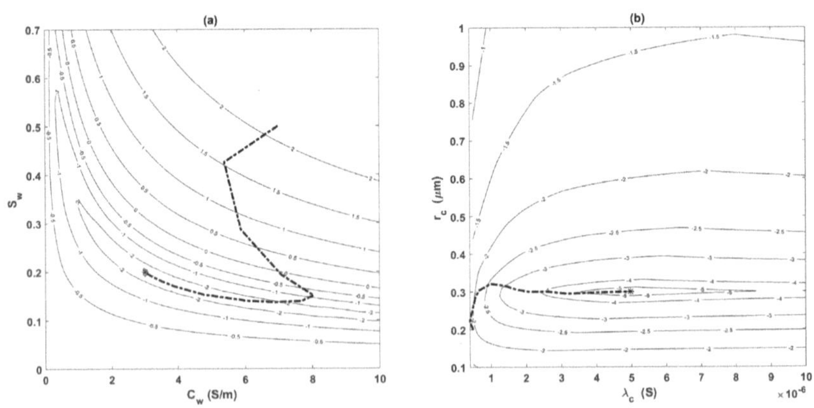

FIGURE 7.9 Changes in (A) brine conductivity and water saturation estimates and those in (B) surface conductance of clay and radius of spherical clay with associated relative error of data misfit obtained during the modified bounded LMA-based inversion of synthetic EM data without Gaussian noise shown in Fig. 7.8 (representing the electrical conductivity and effective permittivity response of Synthetic Layer 1). The dash line identifies the estimates generated by the modified bounded LMA after each iteration. The start point for both plots is: $S_w = 0.5$, $C_w = 7$ S/m, $\lambda_c = 5 \times 10^{-7}$ S, $r_c = 0.2$ μm, and $r_i = 50$ μm. The end or converged points are denoted by asterisk. *Source: Authors.*

TABLE 7.6 Relative errors in clay radius estimates obtained using the joint inversion of multifrequency electrical conductivity and effective permittivity data containing 5% Gaussian noise for two synthetic layers.

Synthetic layer	Estimated r_c (μm)	True r_c (μm)	Rel. Error in r_c (%)
1	0.285	0.3	5
2	1.03	1	3

TABLE 7.7 Relative errors in conductive mineral radius estimates obtained using the joint inversion of multifrequency electrical conductivity and effective permittivity data containing 5% Gaussian noise for one synthetic layer.

Synthetic layer	Estimated r_i (μm)	True r_i (μm)	Rel. Error in r_i (%)
1	35.4	30	18

conductive mineral grains. Tables 7.3–7.7 list the inversion-derived estimates obtained by processing the synthetic EM response of Synthetic Layer 1 containing 5% Gaussian noise. Water saturation estimates followed by brine conductivity estimates are generated at higher accuracy compared to the λ_c, r_c, and r_i estimates. The modified bounded LMA-based estimates for the five unknown PS model parameters are (also, listed in Tables 7.3–7.7) $s_w = 0.201$, $C_w = 2.97$ S/m, $\lambda_c = 5.2 \times 10^{-6}$ S, $r_c = 0.285$ μm, and $r_i = 35.4$ μm. These estimates are close to the original parameter values, listed in Table 7.1, assumed for Synthetic Layer 1 that were used to generate the synthetic broadband frequency EM data with 5% Gaussian noise, as shown in Fig. 7.8. The PS model predictions based on our estimates of the five parameters match the noise contaminated synthetic effective permittivity and electrical conductivity with relative error of close to 1×10^{-2}.

Table 7.2 shows the inversion results of 5% Gaussian noise contaminated EM responses for nine cases that use different values of λ and v. Values of λ varied from 0.1 to 100 and that of v varied from 2 to 10. We used the same initial guess for the nine cases (described in Table 7.2): $s_w = 0.6$, $C_w = 0.5$ S/m, $\lambda_c = 7 \times 10^{-6}$ S, $r_c = 0.1$ μm, and $r_i = 10$ μm. Table 7.2 illustrates the estimated subsurface properties of formation, except for the surface conductance of clay, are not sensitive to the λ and v. To conduct more sensitivity studies of λ and v, we plotted a two-dimensional (2D) colored surface plot (Fig. 7.10). The initial guess used to generate this surface plot is same as above. Fig. 7.10 shows 2D colored surface plot in which the relative error is

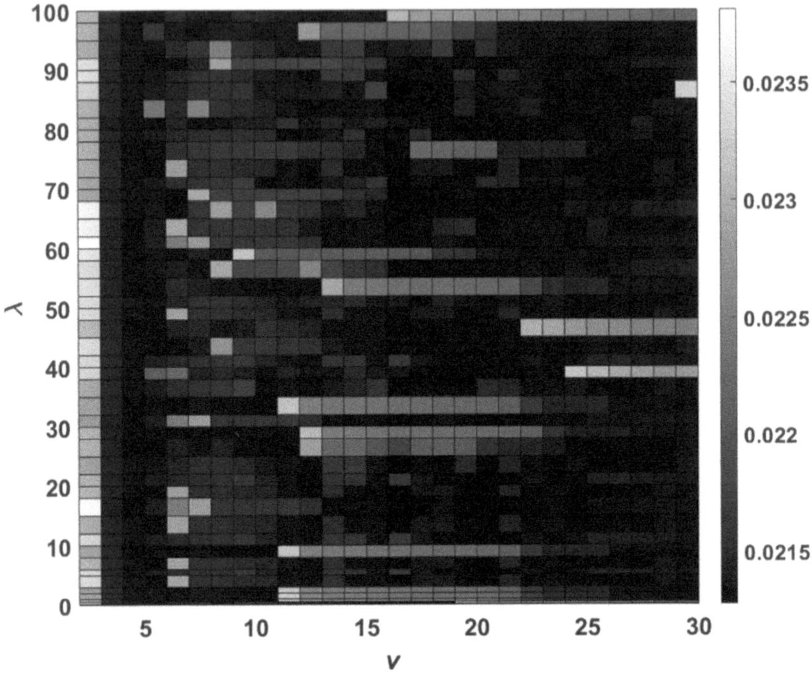

FIGURE 7.10 The two-dimensional colored surface plot of logarithm of data misfit error for various initial values of λ and v. *Source: Authors.*

represented in grayscale. Total of 47 values of λ in the range of 0.01–100 were selected, while v varied from 2 to 30 at one interval to get this 2D surface plot. The value of v equaling to two gives the relative largest error. Overall, the relative error varied from 0.0215 to 0.0235 in almost all cases, which demonstrates that the newly developed modified bounded Levenberg–Marquardt algorithm is not sensitive to λ and v. In this chapter, we set the values of λ and v to 10 in all the synthetic inversion cases.

6.2 Errors in inversion-derived estimates for various synthetic layers

The synthetic multifrequency electrical conductivity and effective permittivity data were computed using the PS model. Synthetic EM data corresponding to Synthetic Layer 1 was computed at nine frequencies, including laterolog, induction, propagation, and dielectric dispersion tool operation frequencies. Synthetic EM data corresponding to Synthetic Layer 2 was computed at six frequencies, including two propagation and dielectric dispersion tool operation frequencies. Synthetic EM data corresponding to Synthetic Layer 3 was computed at five frequencies, including induction and dielectric dispersion tool operation frequencies. Inversion derived estimates for Synthetic Layer 1 have least errors compared to those for Synthetic Layer 2 and 3 except for the radius of clay estimates due to more modeled data at acquisition frequencies of EM tools.

6.3 Inversion-derived estimates when using various noise models

Gaussian noise is a statistical noise having normally distributed probability density function. It is the most widely used noise added to synthetic data to test the model sensitivity to noise. Pink noise is a signal having its power spectral density inversely proportional to frequency. Brownian noise, also called random walk noise, has spectral density inversely proportional to frequency. Brownian noise has large energy at low frequencies that damps rapidly with increase in the frequency. Violet noise and Blue noise have a power density that increases 6 and 3 dB per octave, respectively, with increasing frequency in a limited frequency range. Pink noise has a frequency spectrum that is linear in logarithmic scale having equal power in bands that are proportionally wide and every octave contains same amount of energy. We added 5% Gaussian noise, Pink noise, Brownian noise, Violet noise, and Blue noise to the EM responses in Synthetic Layer 3 inversion case. Table 7.8 shows the inversion-derived estimates obtained by processing the synthetic EM data containing the various types of noise. The inversion algorithm is most resistant to Brownian noise and least resistant to Blue noise.

7. Application of the joint deterministic interpretation of multifrequency electromagnetic measurements in an organic-rich shale formation

Broadband EM measurements in the frequency range of 1 Hz to 1 GHz are generally performed on geological core samples in laboratory environment. However, there is only one example of subsurface broadband EM dispersion data acquisition reported by Wang and Poppitt (2013), which was processed later by Han and Misra (2017). Wireline induction logging tool operating at

TABLE 7.8 Inversion-derived water saturation, brine conductivity, and surface conductance of clay estimates obtained using the joint inversion of induction and dielectric dispersion synthetic log data containing 5% Gauss noise, Pink noise, Brownian noise, Violet noise, and Blue noise.

Type of noise	Est. s_w	Est. C_w (S/m)	Est. λ_c (S)
Gaussian noise	0.79	10.6	4.7×10^{-6}
Pink noise	0.785	10.2	5.01×10^{-6}
Brownian noise	0.792	10.1	4.84×10^{-6}
Violet noise	0.77	10.8	4.7×10^{-6}
Blue noise	0.826	9.3	4.8×10^{-6}

26 kHz, LWD propagation logging tool operating at 1 and 2 MHz, and wireline dielectric-dispersion logging tool operating at 20 MHz, 100 MHz, 260 MHz, and 1 GHz were deployed in an organic-rich shale formation in Eastern Europe for acquiring broadband electrical conductivity- and relative permittivity-dispersion logs. Conductive pyrite minerals are present in the formation. In this field case, contrary to conventional acquisition practices, multifrequency permittivity logs were acquired at the induction-log acquisition frequency.

We selected two depths in the organic-rich shale formation under investigation. XX09 m is a pyrite- and clay-rich zone and XX52 m is clay rich zone. The joint deterministic petrophysical inversion algorithm processed the broadband electrical conductivity and relative permittivity dispersion logs acquired at these two discrete depths. In Chapter 8, we use a Markov-chain Monte Carlo-based Bayesian interpretation method which can quantify the uncertainty of the inversion estimates to process continuous EM dispersion logs in homogeneous and isotropic intervals. Table 7.9 displays the s_w, C_w, λ_c, r_c, and r_i estimates and other petrophysical properties for these two discrete depths. Based on the estimations, depth XX09 m is low salinity, oil-bearing zone, whereas depth XX52 m is low-salinity, water-bearing zone. For those two discrete depths, the radius of clay is around 0.1 μm and surface conductance of clay is low, which is lower than 1×10^{-7} S. The radius of conductive pyrite mineral in depth XX09 m is around 170 μm. Fig. 7.11 present the comparison of modeled and measured broadband relative permittivity and electrical conductivity data for depths XX09 m and XX52 m, respectively.

Inversion-derived formation property estimates obtained using different combinations of EM logs were compared to find out the best combination of EM logging tools for the desired near-wellbore characterization. Tables 7.10 and 7.11 display the s_w, C_w, λ_c, and r_c estimates for the depths XX52 and XX09 m, respectively, for the various combination of EM logs. The water saturation and brine conductivity estimates are relatively similar for all the log combinations. For the depth XX52 m, the combination of induction and dielectric dispersion logs and the combination of induction, propagation, and dielectric dispersion logs led to the largest data misfit. The combination of induction and dielectric dispersion logs generated 1-order of magnitude higher surface conductance and radius of clay estimates compared to those obtained from other combinations. For the depth XX09 m, water saturation estimates are slightly different for the various combinations, such that the largest difference is 0.14 saturation unit. Dielectric dispersion log interpretation generates the highest water saturation of 0.53, and the combination of LWD propagation and dielectric dispersion logging tool generates the lowest water saturation of 0.39, as listed in Table 7.11. Brine conductivity, surface conductance of clay, and radius of clay estimates are relatively similar when

TABLE 7.9 Inversion-derived and assumed petrophysical properties of two depths in the organic-rich formation.

Parameters	Unit	Depth XX52	Depth XX09
Volume fraction of pyrite grains, V_i	%	0	2.3
Bulk conductivity of pyrite, S_i	S/m	–	1000
Relative permittivity of pyrite, ε_{ri}		–	30
Diffusion coefficient of pyrite, D_i	m^2/s	–	1×10^{-5}
Radius of pyrite grains, r_i	**μm**	–	**172**
Volume fraction of clay, V_c	%	60	47
Relative permittivity of clay, ε_{rc}		6	6
Surface conductance of clay, λ_c	**S**	**6.6×10^{-8}**	**2.5×10^{-8}**
Radius of spherical clay grains, r_c	**μm**	**0.13**	**0.1**
Volume fraction of sand, V_s	%	31	43
Surface conductance of sand, λ_s	S	10^{-9}	10^{-9}
Radius of sand grains, r_s	μm	500	500
Total porosity of rock, ϕ_T	%	9	7.7
Bulk conductivity of brine, C_w	**S/m**	**0.48**	**0.38**
Relative permittivity of brine, ε_{rb}		80	80
Diffusion coefficient of brine, D_b	m^2/s	1.5×10^{-9}	1.5×10^{-9}
Relative permittivity of hydrocarbon, ε_{rhy}		3	3
Water saturation, s_w	%	**100**	**46**

Joint deterministic inversion of induction, propagation, and dielectric dispersion logs was performed at the two discrete depths. The five inversion-derived estimates are shown in bold.

using different combinations of EM logs for the depth XX09 m. Radius of pyrite estimates vary with log combinations ranging from 172 to 352 μm. The combination of induction, propagation, and dielectric dispersion logs and the combination of propagation and dielectric dispersion logs exhibit high data misfit for the second field cases. Estimates derived from only dielectric dispersion logs exhibit smallest data misfit in both the cases; however, the application dielectric dispersion logs are suitable in low-salinity formation and in wells with low-borehole rugosity. Combining various EM logs accounts for the effects due to polarization mechanisms in various frequency ranges; thereby facilitating improved water saturation estimates.

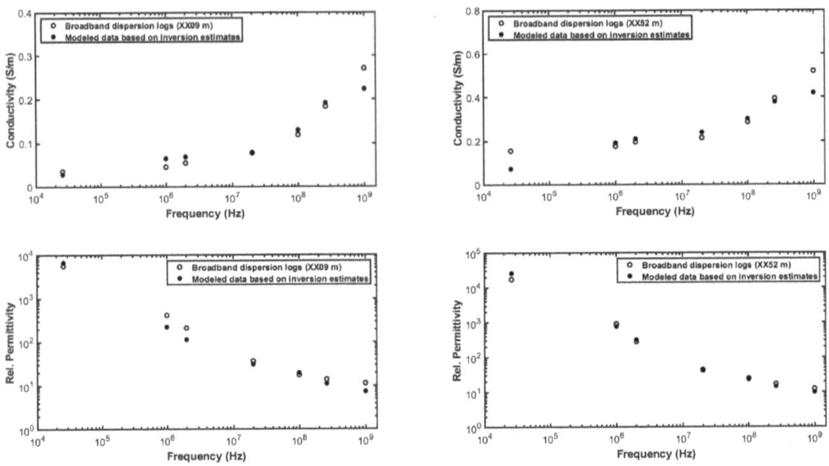

FIGURE 7.11 Comparison of broadband electrical conductivity- and relative permittivity-dispersion measurements against those modeled using the inversion-derived estimates for depth XX09 and XX52 m in the organic-rich shale formation. *Source: Authors.*

TABLE 7.10 Estimated petrophysical properties and data misfit in depth XX52 m in an organic-rich formation when performing petrophysical inversion of various combinations of logging tools.

Logging tool combination	Est.s_w (fraction)	Est. C_w (S/m)	Est.λ_c (S)	Est.r_c (μm)	Data misfit (fraction)
Induction + propagation + dielectric dispersion	1	0.48	6.6×10^{-8}	0.13	0.2316
Dielectric dispersion	1	0.58	7.2×10^{-8}	0.13	0.1272
Induction + dielectric dispersion	1	0.35	5.9×10^{-7}	1.31	0.2587
Propagation + dielectric dispersion	1	0.49	7.2×10^{-8}	0.14	0.1327

TABLE 7.11 Estimated petrophysical properties and data misfit in depth XX09 m in an organic-rich formation when performing petrophysical inversion of various combinations of logging tools.

Logging tool combination	Est.s_w (fraction)	Est.C_w (S/m)	Est.λ_c (S)	Est.r_c (μm)	Est.r_i (μm)	Data misfit (fraction)
Induction + propagation + dielectric dispersion	0.46	0.38	2.5×10^{-8}	0.1	172	0.2275
Dielectric dispersion	0.53	0.43	2.6×10^{-8}	0.1	392	0.1367
Induction + dielectric dispersion	0.43	0.42	2.5×10^{-8}	0.1	194	0.1386
Propagation + dielectric dispersion	0.39	0.54	2.8×10^{-8}	0.1	352	0.2385

Nomenclature

Symbols and abbreviations

EM Electromagnetic
kHz KiloHertz
MHz MegaHertz
GHz GigaHertz
S_w Water Saturation
C_w Connate water Conductivity(S/m)
ϕ_T Total porosity of the formation
ε_{rhy} Hydrocarbon relative permittivity
ω Angular frequency
ε_{rb} Relative permittivity of brine
V_i Volumetric fraction of pyrite grains
S_i Bulk conductivity of pyrite
ε_{ri} Relative permittivity of pyrite
D_i Diffusion coefficient of pyrite
r_i Radius of pyrite grains
V_c Volumetric fraction of clay
ε_{rc} Relative permittivity of clay
λ_c Surface conductance of clay
r_c Radius of spherical clay grains
V_s Volumetric fraction of nonconductive grains
λ_s Surface conductance of nonconductive grains
r_s Radius of nonconductive grains
D_b Diffusion coefficient of brine

References

Al Marzooq, A., Al Ghamdi, T., Salim, A., Sassi, K.H., Badri, M., Saldungaray, P., May 2014. An integrated workflow using spectroscopy, dielectric dispersion, and NMR for the characterization of shale plays—application to a Silurian shale. In: SPWLA 55th Annual Logging Symposium, pp. 18–22.

Al-Ameri, A., Bandyopadhyay, I., Balbaheeth, A.H., Bittar, M., Aki, A., Essam, R., July 2015. Improved formation evaluation with inversion techniques using logging while drilling azimuthal deep resistivity sensor – a case study. In: SPWLA 56th Annual Logging Symposium, pp. 18–22.

Altman, R., Anderson, B., Rasmus, J., Lüling, M.G., May 2008. Dielectric effects and resistivity dispersion on induction and propagation-resistivity logs in complex volcanic lithologies: a case study. In: SPWLA 49th Annual Logging Symposium, pp. 25–28.

de Lima, O.A.L., Sharma, M.M., October 1990. A grain conductivity approach to shaly sandstones. Geophysics 55 (10), 1347–1356.

Glinskikh, V.N., Nesterova, G.V., Epov, M.I., May–June 2014. Forward modeling and inversion of induction logs from shaly sand reservoirs using petrophysical conductivity models. Russ. Geol. Geophys. 55 (5–6), 793–799.

Han, Y., Misra, S., September 2017. Improved water-saturation estimates derived from inversion-based interpretation of broadband electromagnetic dispersion logs in organic-rich shale formations. In: SEG Annual International Meeting, pp. 24–29.

Han, Y., Misra, S., April 2018a. Bakken Petroleum System characterization using dielectric-dispersion logs. Petrophysics 59 (2), 201–217.

Han, Y., Misra, S., April 2018b. A unified inversion scheme to process multifrequency measurements of various dispersive electromagnetic properties. J. Appl. Geophys. 151, 23–39.

Han, Y., Misra, S., May 2018c. Joint petrophysical inversion of multifrequency conductivity and permittivity logs derived from subsurface galvanic, induction, propagation, and dielectric dispersion measurements. Geophysics 83 (3), D97–D112.

Han, Y., Misra, S., Wang, H., Toumelin, E., February 2019. Hydrocarbon saturation in a Lower-Paleozoic organic-rich shale gas formation based on Markov-chain Monte Carlo stochastic inversion of broadband electromagnetic dispersion logs. Fuel 243, 645–658.

Hizem, M., Budan, H., Deville, B., Faivre, O., Mosse, L., Simon, M., September 2008. Dielectric dispersion: a new wireline petrophysical measurement. In: SPE Annual Technical Conference and Exhibition, pp. 21–24.

Jougnot, D., Ghorbani, A., Revil, A., Leroy, P., Cosenza, P., January 2010. Spectral induced polarization of partially saturated clay-rocks: a mechanistic approach. Geophys. J. Int. 180 (1), 210–224.

Leroy, P., Revil, A., March 2004. A triple-layer model of the surface electrochemical properties of clay minerals. J. Colloid Interface Sci. 270 (2), 371–380.

Lin, Y., Gianzero, S., Strickland, R.W., June 1984. Inversion of induction logging data using the least squares technique. In: SPWLA 25th Annual Logging Symposium, pp. 10–13.

Little, J.D., Julander, D.R., Knauer, L.C., Aultman, J.T., Hemingway, J.L., June 2010. Dielectric dispersion measurements in California heavy oil reservoirs. In: SPWLA 51st Annual Logging Symposium, pp. 19–23.

Mezzatesta, A.G., Payton, C.C., Strack, K.M., Tabarovsky, L.A., April 1994. Simultaneous inversion of galvanic and induction logging measurements to improve resolution. In: Latin American and Caribbean Petroleum Engineering Conference, pp. 27–29.

Misra, S., Han, Y., August 2016a. Petrophysical interpretation of multi-frequency electromagnetic measurements in clay- and conductive-mineral-rich mudrocks. In: Unconventional Resources Technology Conference, pp. 1–3.

Misra, S., Han, Y., September 2016. Improved oil-in-place estimates in clay-and pyrite-bearing shales based on inversion of multi-frequency electromagnetic measurements. In: AAPG Eastern Section Meeting, pp. 25–27.

Misra, S., Lüling, M.G., Rasmus, J., Homan, D.M., Barber, T.D., June 2016. Dielectric effects in pyrite-rich clays on multifrequency induction logs and equivalent laboratory core measurements. In: SPWLA 57th Annual Logging Symposium, pp. 25–29.

Misra, S., Torres-Verdín, C., Homan, D., Rasmus, J., July 2017. Dispersive and directional effective electrical conductivity and dielectric permittivity of conductive-mineral-bearing samples derived from multi-frequency tensor electromagnetic induction measurements. Geophysics 82 (4), D211–D223.

Pirrone, M., Bona, N., Galli, M.T., Pampuri, F., Faivre, O., Han, M., Hizem, M., Mosse, L., March 2011. An innovative dielectric dispersion measurement for better evaluation of thin layered reservoirs applied in a south Italy well. In: Offshore Mediterranean Conference and Exhibition, pp. 23–25.

Revil, A., January 2013. Effective conductivity and permittivity of unsaturated porous materials in the frequency range 1mHz-1GHz. Water Resour. Res. 49 (1), 306–327.

Revil, A., Aal, G.Z.A., Atekwana, E.A., Mao, D., Florsch, N., September 2015. Induced polarization response of porous media with metallic particles - part 2: comparison with a broad database of experimental data. Geophysics 80 (5), D539–D552.

Tathed, P., Han, Y., Misra, S., December 2018a. Hydrocarbon saturation in Bakken Petroleum System based on joint inversion of resistivity and dielectric dispersion logs. Fuel 233, 45–55.

Tathed, P., Han, Y., Misra, S., May 2018b. Hydrocarbon saturation in upper Wolfcamp shale formation. Fuel 219, 375–388.

Thiel, M., Omeragic, D., Habashy, T., November 2012. Robust nonlinear parametric Gauss-Newton inversion for well-logging applications. In: SEG Annual International Meeting, pp. 4–9.

Wang, H., Poppitt, A., June 2013. The broadband electromagnetic dispersion logging data in a gas shale formation. In: SPWLA 54th Annual Logging Symposium, pp. 22–26.

Zhou, J., Li, H., Rabinovich, M., D'Arcy, B., June 2016. Interpretation of azimuthal propagation resistivity measurements: modeling, inversion, application and discussion. In: SPWLA 57th Annual Logging Symposium, pp. 25–29.

Section 4

Applications of multifrequency electromagnetic data and interpretation

Chapter 8

Stochastic inversion based interpretation of multifrequency electromagnetic logs from a European organic-rich shale formation

Yifu Han[1], Siddharth Misra[2,3]
[1]*Schlumberger Technology Corporation, Beijing, China;* [2]*Harold Vance Department of Petroleum Engineering, College of Engineering, Texas A&M University, TX, United States;* [3]*Department of Geology and Geophysics, Texas A&M University, TX, United States*

1. Introduction

Interpretation of multifrequency electromagnetic (EM) responses in organic-rich shale formation is challenging because of high-clay content, low porosity, high tortuosity of pores, and the presence of conductive minerals, such as pyrite. In the presence of such petrophysical conditions, estimates of hydrocarbon saturation obtained from multimineral analysis solver and various subsurface EM logs, such as induction resistivity log, dielectric permittivity log, or dielectric dispersion logs, are not consistent. Conventional interpretation of EM logs using empirical models tends to break down in organic-rich shale formation because of the unaccounted interfacial polarization effects of minerals and the unaccounted dispersive behavior of the electrical conductivity and dielectric permittivity. Several of the above-mentioned issues can be addressed by processing multifrequency EM logs using the mechanistic models to accurately estimate hydrocarbon saturation and other petrophysical properties of organic-rich shale formations.

In this chapter, we discuss the simultaneous deployment and joint-interpretation of EM induction resistivity, logging while drilling (LWD) propagation resistivity, and dielectric dispersion logging tools in a well drilled in an organic-rich shale formation in Poland Lubelskie (Lublin

Multifrequency Electromagnetic Data Interpretation for Subsurface Characterization
https://doi.org/10.1016/B978-0-12-821439-8.00002-1

Basin). Multifrequency EM measurements (also referred as the broadband EM dispersion measurements) were acquired at seven EM-log-acquisition frequencies continuously across a depth interval of 1500 m. This was the first ever acquisition of broadband EM dispersion logs, comprising relative dielectric permittivity, and electrical conductivity, measured at seven EM-log-acquisition frequencies in the range of 1 kHz to 1 GHz. The interpretation of the multi-frequency EM logs was done using Markov-chain Monte-Carlo (MCMC)-based stochastic inversion scheme coupled with the mechanistic clay-pyrite interfacial polarization model (referred as the PPIP-SCAIP model). In doing so, we simultaneously estimated the hydrocarbon saturation, connate water conductivity, and surface conductance of clay in the shale formation. PPIP-SCAIP model is described in detail in Chapters 3 and 4.

The proposed Markov-chain Monte Carlo-based stochastic inversion scheme is robust to noise and exhibits fast convergence. The estimated hydrocarbon saturation is also compared against those derived from resistivity log, dielectric dispersion logs interpretation, and multimineral analysis solver. The estimated hydrocarbon saturation, connate water conductivity, and surface conductance of clay in two specific homogeneous and isotropic intervals of the organic-rich shale formation are in the ranges of 0.2–0.9, 0.1 S/m to 0.25 S/m, and 1×10^{-7} to 1×10^{-6} S, respectively.

1.1 The European organic-rich shale formation

The lower Paleozoic basins in Eastern Europe, Poland, are considered to have prolific unconventional oil and gas resources (Karcz et al., 2013). Lublin Basin, which is located in the transition area between the East European Craton and the Variscan orogenic belt, has the most complicated tectonic structure compared with other Paleozoic basins in Poland and contains fine clastic rocks that have hydrocarbon-bearing potential. The burial and thermal maturity in Lublin Basin is also different from other sedimentary basins in eastern Poland because of the complex geological structure. The thermal maturity or vitrinite reflectance in Lublin Basin ranges from 0.6% Ro to 3.0% Ro showing an increasing trend from northeast to southwest region. TOC varies from 0.5 to 4.5 wt%, and the thickness of the stratigraphic intervals varies from 10–50 to 60–180 m (Karcz et al., 2013). The mineral constituents primarily include varying amount of quartz, illite, chlorite, kaolinite, kerogen, pyrite, calcite, and phosphate. The lower Wenlock shale and Llandovery interval which displays much higher gas saturation in shallower zone are key interest formations in Lublin Basin (Papiernik et al., 2017). Organic-rich unconventional shale gas formations in Lublin Basin are fine-grained shales, claystones, and mudstones (Botor et al., 2019), exhibiting high-clay content, high TOC, low porosity, high tortuosity, and the presence of conductive minerals such as pyrite (Wang and Poppitt, 2013). In the presence of such petrophysical conditions, hydrocarbon saturation estimates obtained from the interpretation of various subsurface near-wellbore EM logs, such as induction resistivity log and dielectric dispersion log, and multimineral analysis solver are not consistent (Han and Misra, 2018a; Tathed et al., 2018a). In addition, hydrocarbon saturation estimates derived from interpretation of EM logs by using conventional saturation models tend to break down in organic-rich shale formation because of the unaccounted interfacial-polarization effects and the unaccounted frequency-dependence (dispersion) of conductivity and permittivity (Han and Misra, 2018a; Tathed et al., 2018a, b).

1.2 Deployment of multiple electromagnetic logging tools

Chapter 1 presents the devices and tools commonly used for EM measurements in the subsurface. Generally, one type of EM logging tool with very narrow log-acquisition frequency range is run in a wellbore for purposes of reservoir characterization (Han and Misra, 2018b). This chapter presents the first ever acquisition of broadband EM dispersion logs, comprising relative dielectric permittivity and electrical conductivity, measured at seven EM-log-acquisition frequencies in the range of 1 kHz to 1 GHz. The induction resistivity tool is typically operated at frequency between 10 and 80 kHz. In this study, the induction resistivity was operated at 26 kHz with depth of investigation around 36 inches (Wang and Poppitt, 2013). In this study, the LWD propagation tool was operated at two discrete frequencies, 1 and 2 MHz, with depth of investigation around 15 and 10 inches, respectively (Wang and Poppitt, 2013). The dielectric tool operates at four distinct frequencies ranging from 10 MHz to 1 GHz. In this study, it was operated at 20, 100, 260 MHz, and 1 GHz. The LWD propagation and new-generation dielectric tools measure the attenuation, phase shift, and travel time for EM waves though the formation between two receivers located in the tool, which are then transformed to conductivity and relative permittivity of the formation at multiple frequencies (Han and Misra, 2018b).

1.3 Limitations of conventional electromagnetic-log-interpretation models

Archie equation is an empirical model for interpreting the subsurface resistivity measurements acquired in clean sandstone formations at low EM frequencies. Archie equation does not account for the effects of conductive clay minerals and the interfacial-polarization effects around the clays and conductive minerals. Surface-charge-bearing clay minerals in shaly sand formations can significantly change the bulk electrical resistivity; therefore, the water saturation estimates in shaly sand formations derived using Archie equation are unreliable. Simandoux model for shaly sands considers both the brine conductivity and the surface conductivity of clay minerals. Simandoux model (Eq. 8.1) was developed based on laboratory investigations on synthetic homogeneous mixtures of sand, conductive clay minerals, and water.

$$\frac{1}{R_t} = \frac{\phi_t^m}{aR_w} S_w^n + \frac{V_{sh}}{R_{sh}} \tag{8.1}$$

However, Simandoux equation, like Archie equation, does not account for the frequency dependence of conductivity and ignores permittivity (i.e., effects of charge storage/accumulation). Formation resistivity acquired at different frequencies when interpreted using models, like Archie's equation and Simandoux Equation, will lead to distinct estimates of water saturation because of the frequency dispersion of resistivity/conductivity. Furthermore, such estimates will be unreliable because of the models ignore charge accumulation and storage that gives rise to permittivity. More importantly, Simandoux equation and Archie equation ignore the effects of conductive minerals, such as graphite and pyrite. In this chapter, water saturation is estimated by processing induction resistivity using Simandoux equation with the typical values of the cementation exponent ($m = 1.33$) and tortuosity factor ($a = 1.65$) (Asquith, 1980). The estimates of water saturation are then compared with those computed using the stochastic inversion of multifrequency EM logs (Wang and Poppitt, 2013).

Complex refractive index (CRI) model is widely used to process the dielectric permittivity log acquired around 1 GHz. CRI model accounts for orientation polarization of ions and dipoles, while assuming negligible interfacial polarization at grain surface. The tortuosity, rock fabric, pore-network, and minerals spatial distributions of geomaterials are not considered in CRI model (Brovelli and Cassiani, 2008). CRI model has a more generalized form known as the Lichtenecker-Rother (LR) model, which introduces a geometrical parameter to indicate the angle between the principal axis of the polarizing material (i.e., inclusions, layers, or pores) and the applied EM field (Zakri et al., 1998). All these models discussed so far are frequency independent that fail to account the effect of polarization phenomena leading to dispersions in conductivity and permittivity as a function of the operating frequency. Among the frequency-dependent models, Stroud-De-Milton (SMD) model is a mixing law that can calculate frequency-dependent complex dielectric permittivity of brine-saturated clean sandstone as a function of total porosity, water saturation, water salinity, relative permittivity of conductive and nonconductive phases, and rock textural parameter describing cementation/tortuosity (Tathed et al., 2018a, b). The mechanistic model of complex permittivity of brine water is well defined in the previous Chapter 2. SMD model has been reliable for simple, porous sandstones. However, SMD model is not applicable for clay-rich and pyrite-rich formations because SMD model, like most of the conventional EM-log-interpretation models, ignores the interfacial-polarization effects of clay and conductive minerals (Han and Misra, 2018a). Nonetheless, we apply the joint SMD and LR model to process dielectric dispersion logs (Han and Misra, 2018a; Tathed et al., 2018a, b) to obtain water saturation estimates.

1.4 Estimation of hydrocarbon saturation based on stochastic inversion of multifrequency electromagnetic logs

Formation evaluation in conventional reservoirs requires the estimation of water saturation, which can be derived from the interpretation of EM logs, such as induction resistivity log or dielectric dispersion log. However, accurate water saturation estimation is more challenging in unconventional formations because of the complex rock fabric, high-clay content, low porosity, high tortuosity, and the presence of conductive minerals. Conventional EM log-interpretation models trend to break down under such petrophysical conditions. The interpretation of various EM logs measured by different EM logging tools gives inconsistent water saturation estimates. Han and Misra (2018a) reported that the water-saturation estimates derived from various EM logs using various interpretation models can vary more than 0.4 saturation units for a single depth, which is also illustrated in Chapter 9. Moreover, the core-scale laboratory measurements of water saturation are also not consistent with well-log-derived saturation estimates (Han and Misra, 2018a).

The assessment of hydrocarbon saturation in unconventional reservoirs can be improved by jointly processing multifrequency EM logs (broadband EM dispersion logs) using the mechanistic clay-pyrite dispersion model, also known as the PPIP-SCAIP model. In this chapter, multifrequency EM logs acquired by EM induction resistivity tool, LWD propagation tool, and dielectric dispersion tool are simultaneously processed to estimate water saturation, connate water conductivity, and surface conductance of clay in certain depth intervals of organic-rich unconventional shale formation in Lublin Basin. PPIP-SCAIP model is described in detail in Chapters 3 and 4.

For the desired interpretation of multifrequency EM logs, we couple the stochastic inversion scheme with the PPIP-SCAIP model, which is a mechanistic model of the effects of interfacial polarization of clay and conductive particles/inclusions on the conductivity and permittivity of fluid-filled porous material containing both clays and conductive minerals. PPIP-SCAIP can accurately quantify the frequency-dependent relative permittivity and electrical conductivity dispersion effects by considering the interfacial-polarization effects of clay and conductive mineral. Although the size distribution and shape of inclusions can be also modeled using the PPIP-SCAIP model, we assume that both the clay and conductive minerals are spherical in geometry and of constant size across the formation. The petrophysical parameters assumed in the PPIP-SCAIP model are listed in Table 8.1. The PPIP-SCAIP model is calibrated to the Simandoux equation at low frequency (1 Hz) and to the CRIM at high frequency (1 GHz). The application of PPIP-SCAIP model in the inversion-based EM log interpretation assumes clay and pyrite grains are uniformly distributed in the

TABLE 8.1 Petrophysical properties assumed for the Synthetic Layer 1 and Synthetic Layer 2.

Parameters/Properties	Unit	Layer 1	Layer 2
Volumetric fraction of pyrite grains, V_i	%	0	3
Bulk conductivity of pyrite, S_i	S/m		1000
Relative permittivity of pyrite, ε_{ri}			30
Diffusion coefficient of pyrite, D_i	m^2/s		5×10^{-5}
Radius of pyrite grains, r_i	μm		10
Volumetric fraction of clay, V_c	%	50	55
Relative permittivity of clay, ε_{rc}		5.8	5.8
Surface conductance of clay, λ_c	S	3×10^{-7}	2×10^{-7}
Radius of spherical clay grains, r_c	μm	0.6	0.25
Low frequency conductivity in pure-clay formation, C_{sh}	S/m	0.4	0.35
Volumetric fraction of sand (nonconductive grains), V_s	%	42	33
Relative permittivity of sand, ε_{rs}		4.65	4.65
Surface conductance of sand, λ_s	S	1×10^{-9}	1×10^{-9}
Radius of sand grains, r_s	μm	100	100
Total porosity of rock, ϕ_T	%	8	9
Bulk conductivity of water, C_w	S/m	3.5	1
Relative permittivity of water, ε_{rw}		56	57
Diffusion coefficient of water, D_w	m^2/s	1.5×10^{-9}	1.5×10^{-9}
Relative permittivity of hydrocarbon, ε_{rhy}		2	2
Water saturation, s_w	%	90	50
Tortuosity factor, a		1.65	1.65
Cementation exponent, m		1.33	1.33
Saturation exponent, n		2	2
Formation temperature, T	°F	200	200

Five parameters are to be estimated for the Layer 1 and 6 parameters are to be estimated for the Layer 2.

formation, which is homogeneous and isotropic. However, Maxwell–Wagner polarization effects (Misra et al., 2016a), and the effects of effective porosity, pore-network tortuosity, isolated pores, and the connectivity of clay and pyrite grains on the conductivity and permittivity of the formation are not taken into account in the PPIP-SCAIP model.

2. Stochastic inversion of multifrequency electromagnetic logs

2.1 Bayesian formulation

2.1.1 Bayesian framework

We propose a Bayesian framework to estimate the unknown petrophysical parameters of the PPIP-SCAIP model representing a sedimentary geological formation made of quartz, clay, and conductive minerals. In the Bayesian framework, the broadband EM dispersion logs (i.e., multifrequency EM logs) acquired at seven log-acquisition frequencies are treated as the input data vector and the estimates for three to four unknown petrophysical model parameters along with corresponding probability distributions are computed using a stochastic inversion approach.

MCMC sampling-based stochastic method provides a global approach for estimating the unknown model parameters and the uncertainty in the estimations (Chen et al., 2008). Chapter 10 provides more details about the stochastic inversion and the MCMC sampling-based inversion technique. In our formulation, $\boldsymbol{d}$ denotes the data vector comprising the broadband EM dispersion logs, $\boldsymbol{m}$ represents the unknown petrophysical model parameter vector, and the prior knowledge of the unknown petrophysical parameter vector $\boldsymbol{m}$ is denoted as $\boldsymbol{I}$, also referred as the hypothesis or assumed model. The Bayesian formulation for the posterior probability density function of model parameter $\boldsymbol{m}$ given data $\boldsymbol{d}$ and prior information $\boldsymbol{I}$ is expressed as:

$$p(\boldsymbol{m}|\boldsymbol{d},\boldsymbol{I}) = \frac{p(\boldsymbol{m}|\boldsymbol{I})p(\boldsymbol{d}|\boldsymbol{m},\boldsymbol{I})}{p(\boldsymbol{d}|\boldsymbol{I})} \tag{8.2}$$

where $p(\boldsymbol{m}|\boldsymbol{d},\boldsymbol{I})$ is the posterior distribution of model parameter $\boldsymbol{m}$ given data $\boldsymbol{d}$ and prior information $\boldsymbol{I}$ that needs to be estimated in the stochastic inversion task. Other components of Eq. (8.2) are: $p(\boldsymbol{m}|\boldsymbol{I})$ is the prior distribution of model parameter vector $\boldsymbol{m}$ given the prior information $\boldsymbol{I}$, $p(\boldsymbol{d}|\boldsymbol{m},\boldsymbol{I})$ is the likelihood of acquiring data $\boldsymbol{d}$ given the prior information $\boldsymbol{I}$ and the model parameter vector $\boldsymbol{m}$, and $p(\boldsymbol{d}|\boldsymbol{I})$ is a normalizing constant, such that $p(\boldsymbol{d}|\boldsymbol{I}) = \int p(\boldsymbol{m}|\boldsymbol{I})p(\boldsymbol{d}|\boldsymbol{m},\boldsymbol{I})d\boldsymbol{m}$. In this framework, posterior describes the model parameters of interest ($\boldsymbol{m}$) that are computed after using the data $\boldsymbol{d}$, whereas the prior describes model parameters of interest ($\boldsymbol{m}$) that are assumed or hypothesized before observing or using the data $\boldsymbol{d}$. The posterior $p(\boldsymbol{m}|\boldsymbol{d},\boldsymbol{I})$ is the joint probability distribution of model parameters of interest that we want to infer based on the measured data $\boldsymbol{d}$ and an assumed model/hypothesis (i.e., the prior information about the model parameters, $\boldsymbol{I}$). The goal of Bayesian inference is to compute the posterior distribution $p(\boldsymbol{m}|\boldsymbol{d},\boldsymbol{I})$ of model parameter $\boldsymbol{m}$ by assuming its proportionality to the product of likelihood function and the prior distribution. Because of the consideration of proportionality, there is a drastic reduction in the computation cost required to calculate the denominator in Eq. (8.2), i.e., the integral operation to compute the normalizing constant $p(\boldsymbol{d}|\boldsymbol{I})$ over the entire model parameter space. The use of MCMC sampling within Bayesian framework makes it possible to obtain the posterior distribution of model parameter $\boldsymbol{m}$ without calculating the normalizing constant.

2.1.2 *Prior distribution of model parameters* p(m|I)

The prior distribution $p(\boldsymbol{m}|\boldsymbol{I})$ of the unknown petrophysical parameters of the PPIP-SCAIP model can be obtained from prior knowledge of the formation under investigation. For example, the prior distribution of water saturation can be determined either from core-scale Dean–Stark and retort core measurements or log-scale EM logs, such as induction resistivity or dielectric dispersion logs interpretation (Han and Misra, 2018a; Tathed et al., 2018a, b) or reservoir-scale seismic interpretation (Wu et al., 2005). The thermal neutron capture cross-section (sigma) subsurface measurement helps in determining the prior distribution of water conductivity or salinity (Donadille et al., 2016). The prior distribution of surface conductance of clay can be related to the cation exchange capacity of clay and the prior distribution of radius of clay and pyrite can be determined from scanning electron microscope or energy dispersive spectroscopy images of rocks obtained from the certain depths of the formation. However, the prior distribution of petrophysical parameters obtained from prior knowledge can be subjective and site-specific. In this study, we assume each petrophysical parameter is independent of others and the prior distributions are noninformative uniform distribution over the prior parameter ranges.

2.1.3 *Likelihood function* p(d|m, I)

The likelihood function can be expressed as a curve fitting function in the form of a normal distribution (Minsley, 2011) expressed as:

$$p(\boldsymbol{d}|\boldsymbol{m},\boldsymbol{I}) = \left[(2\pi)^N|\boldsymbol{C}_d|\right]^{-1/2} e^{\left[-\frac{1}{2}\left(\frac{d^{meas}-\boldsymbol{d}(\boldsymbol{m})}{\sigma}\right)^T C_d^{-1}\left(\frac{d^{meas}-\boldsymbol{d}(\boldsymbol{m})}{\sigma}\right)\right]} \tag{8.3}$$

where $\boldsymbol{d}^{\text{meas}}$ is the measured data vector that comprises the broadband EM dispersion logs, $\boldsymbol{d}(\boldsymbol{m})$ is the modeled data vector computed using the PPIP-SCAIP model based on the estimate of the model parameter vector $\boldsymbol{m}$, $\boldsymbol{C}_d$ is the data covariance matrix which is a diagonal matrix with entries equal to the estimated data error variance for each measurement, $\boldsymbol{\sigma}$ is the standard deviation vector for the broadband permittivity and conductivity logs, and N is the number of measurements/elements in the data vector, which is 14 in the study. The normal distribution likelihood function assumes that the data error is Gaussian without any outlier. The likelihood function should be defined in the form of Laplace distribution with a broader tail if the data errors are expected to be non-Gaussian with some outliers (Chen et al., 2008). For purposes of our study, $\boldsymbol{d}^{\text{meas}}$ is the measured data vector comprising relative dielectric permittivity and electrical conductivity measured at seven EM log-acquisition frequencies; hence N is 14. The PPIP-SCAIP model is used as the forward model to generate the modeled data vectors $\boldsymbol{Perm}(\boldsymbol{m})$ and $\boldsymbol{Cond}(\boldsymbol{m})$, which are the predictions of dispersive relative dielectric permittivity and dispersive electrical conductivity, respectively, arising from the interfacial polarization effects because of clay and conductive minerals at the seven distinct log-acquisition frequencies, namely, ω_1, ω_2, ..., ω_7, for a specific model parameter vector $\boldsymbol{m}$. The modeled data vectors can be expressed as

$$\boldsymbol{Perm}(\boldsymbol{m}) = [Perm(\boldsymbol{m}, \omega_1), Perm(\boldsymbol{m}, \omega_2), \ldots Perm(\boldsymbol{m}, \omega_7)], \tag{8.4}$$

and

$$\boldsymbol{Cond}(\boldsymbol{m}) = [Cond(\boldsymbol{m}, \omega_1), Cond(\boldsymbol{m}, \omega_2), \ldots Cond(\boldsymbol{m}, \omega_7)], \tag{8.5}$$

where $Perm(\boldsymbol{m}, \omega_i)$ and $Cond(\boldsymbol{m}, \omega_i)$ represent the PPIP-SCAIP based model predictions of relative dielectric permittivity and electrical conductivity calculated at specific log-acquisition angular frequency ω_i for a specific model vector $\boldsymbol{m}$. The 14-dimensional modeled data vector $\boldsymbol{d}(\boldsymbol{m})$ is thus expressed as a combination of the seven modeled relative dielectric permittivity and seven modeled electrical conductivity measurements as

$$\boldsymbol{d}(\boldsymbol{m}) = [\boldsymbol{Perm}(\boldsymbol{m}), \boldsymbol{Cond}(\boldsymbol{m})]^T, \tag{8.6}$$

2.2 Metropolis–Hastings sampling algorithms

More information about the MCMC sampling based on the Metropolis–Hastings algorithm is provided in Chapter 10. For purposes of stochastic inversion, MCMC sampling is used to compute the posterior distribution of model parameters of interest that is used to quantify the reliable estimates of the model parameters along with their uncertainties. MCMC generates credible range of model parameters allowed by the data. Various MCMC sampling algorithms have been used in Bayesian inference problems (Mosegaard and Tarantola, 1995; Brookst, 1998). One popular sampling algorithm is Metropolis–Hastings (MH) algorithm that can draw samples from any probability distribution. As more samples are drawn, MH algorithm produces distribution of values that approximate the desired probability distribution. In our study, the MH sampling algorithm is used to select model parameters $\boldsymbol{m}$ from posterior distribution $p(\boldsymbol{m}|\boldsymbol{d}, \boldsymbol{I})$ without explicitly calculating the posterior distribution, such that the next sample value only depends on the current sample value (i.e., a Markov chain). This reduces the computational effort required to calculate the normalizing constant, which involves the computationally expensive integral operation over the entire model parameter space.

We apply the two-step Metropolis–Hastings sampling algorithm (Metropolis et al., 1953; Hastings, 1970) to obtain a sequence of random samples to generate a Markov chain to perform the desired Monte Carlo approximation. Step 1 generates proposal distribution that proposes the next step, and Step 2 determines whether to accept the proposed next step based on an acceptance probability. First, a new model parameter vector $\boldsymbol{m}'$ is drawn from a proposal distribution, $q(\boldsymbol{m}'|\boldsymbol{m})$, which is a function of the current model parameter vector $\boldsymbol{m}$ in the Markov chain. Following that, the proposed model parameter $\boldsymbol{m}'$ is either accepted or rejected according to an acceptance probability (Minsley, 2011). The sequence of model parameters thus obtained can then be used to approximate the posterior distribution. Removing samples from the burn-in period reduces the dependence on starting value by ensuring the stabilization of Markov chain. Autocorrelations because of Markov chain can be reduced through thinning.

2.2.1 Proposal distribution for sampling the unknown model parameters, q(m′|m)

The efficiency of MCMC sampling algorithms strongly depends on the choice of proposal distribution $q(\boldsymbol{m}'|\boldsymbol{m})$. Computational costs increase with poor selection of the proposal distribution. When the proposed model parameters are too exploratory, most of the proposed samples for the next step are rejected for falling in the low-probability regions. However, limited exploration cannot escape local minima because several accepted proposed samples are not far from the previous high-probability regions (Minsley, 2011). Typically, a new model parameter $\boldsymbol{m}'$ is proposed based on multivariate Gaussian distribution that has the current model parameter $\boldsymbol{m}$ as its mean. The proposal distribution (Minsley, 2011) is then expressed as

$$q(\boldsymbol{m}'|\boldsymbol{m}) = \left[(2\pi)^k|\boldsymbol{C}_k|\right]^{-1/2} e^{\left[-\frac{1}{2}\boldsymbol{m}^T\boldsymbol{C}_k^{-1}\boldsymbol{m}\right]} \tag{8.7}$$

where k is the number of unknown model parameter space. In our study, k is three for field application cases and five or six for synthetic cases. $\boldsymbol{C}_k$ is the proposal covariance matrix, defined as the linearized estimation of the posterior model covariance (Malinverno, 2002; Aster et al., 2005), formulated as

$$\boldsymbol{C}_k \approx \left[\boldsymbol{J}^T\boldsymbol{C}_d^{-1}\boldsymbol{J}\right]^{-1} \tag{8.8}$$

where $\boldsymbol{J}$ is the Jacobian matrix of data misfit vector $[\boldsymbol{d}^{\text{meas}} - \boldsymbol{d}(\boldsymbol{m})]$. In doing so, it is ensured that the unknown petrophysical parameters with high sensitivity have smaller search ranges and unknown petrophysical parameters with low sensitivity have a wider search ranges over the entire model parameter space (Minsley, 2011). This formulation of the covariance matrix is critical for the successful exploration and exploitation of the model parameter space.

2.2.2 Acceptance probability $\alpha(m|m')$

The current model parameter $\boldsymbol{m}$ is updated at each iteration to build the Markov chain, where the current state depends only on the previous state. At each step, a new candidate sample $\boldsymbol{m}'$ is drawn from the proposal distribution, referred as $q(\boldsymbol{m}'|\boldsymbol{m})$. The candidate sample is either accepted or rejected according to model acceptance probability (Bérubé, 2017). The model acceptance probability (Chib and Greenberg, 1995) can be defined as

$$\alpha(\boldsymbol{m}|\boldsymbol{m}') = \min\left[1, \frac{p(\boldsymbol{m}'|\boldsymbol{d},\boldsymbol{I})}{p(\boldsymbol{m}|\boldsymbol{d},\boldsymbol{I})}\frac{q(\boldsymbol{m}|\ \boldsymbol{m}')}{q(\boldsymbol{m}'|\ \boldsymbol{m})}\right] \tag{8.9}$$

$p(\boldsymbol{m}|\boldsymbol{d},\boldsymbol{I})$ is the target posterior distribution from which samples need to be drawn, but it is computationally expensive to quantify this distribution. However, ratio $\frac{p(\boldsymbol{m}'|\boldsymbol{d},\boldsymbol{I})}{p(\boldsymbol{m}|\boldsymbol{d},\boldsymbol{I})}$ can be computed because the computationally expensive normalization constant is common between the two probabilities and cancels out. The formulation of acceptance probability α as described in Eq. (8.9) selects the samples of the Markov chain from the higher probability regions of the model space while escaping local minima. The proposed model has higher chance to be accepted if the proposed model has higher posterior probability than current model (Minsley, 2011). The Markov chain always moves from $\boldsymbol{m}$ to the drawn candidate $\boldsymbol{m}'$ if the model acceptance probability is greater than 1. When the model acceptance probability $\alpha(\boldsymbol{m}|\boldsymbol{m}')$ is smaller than 1, a random sample u is drawn from a uniform density between 0 and 1 and the drawn sample $\boldsymbol{m}'$ is accepted if $\alpha(\boldsymbol{m}|\boldsymbol{m}')$ is greater than u; or else it is rejected. In this work, we assume proposal distribution is multivariate Gaussian distribution, which is symmetric $q(\boldsymbol{m}|\boldsymbol{m}') = q(\boldsymbol{m}'|\boldsymbol{m})$. In doing so, the acceptance probability reduces to the ratio of likelihood functions for the proposed model and current model (Bérubé et al., 2017).

2.3 Convergence monitoring

We run three Markov chains by starting from different sets of initial values covering the default range of each unknown petrophysical parameter and the total number of iterations is T. The drawn samples in the first half of iterations ($0.5T$) depend on initial values, so the first half of iterations for each Markov chain are thrown away as they are in the burn-in period (Chen et al., 2008). The scale reduction score is calculated based on the variance within each Markov chain sequence and the variance between the three independent Markov chain sequence. The scale reduction score is used to monitor the convergence of three independent Markov chains to the approximation of the target posterior distribution (Gelman and Rubin, 1992). Markov chains are considered to be converged if the scale reduction score for each estimated parameter is less than 1.2. The default maximum number of iterations in the proposed application of stochastic inversion-based log interpretation method for the European organic-rich shale formation under investigation is 50,000. The maximum number of iterations or length of Markov chains will be adjusted if the Markov chains do not converge or the scale reduction score for any estimated parameter is larger than 1.2 for certain depths/cases.

3. Results and discussions

3.1 Application of the inversion-based interpretation method to synthetic layers

For purposes of testing the newly developed MCMC-based stochastic inversion scheme coupled with the PPIP-SCAIP model, we process synthetic multifrequency EM logs generated for two distinct synthetic layers very similar to the predominant clay- and pyrite-rich zones in the European organic-rich shale formation under investigation. Synthetic Layer 1 represents a clay rich ($V_c = 0.5$), low porosity ($\phi_T = 0.08$), and water bearing ($s_w = 0.9$) formation. Synthetic Layer 2 identifies a clay rich ($V_c = 0.55$), pyrite rich ($V_i = 0.03$), low porosity ($\phi_T = 0.09$), and hydrocarbon bearing ($s_w = 0.5$) formation. Table 8.1 lists the petrophysical properties assumed for the two synthetic clay- and pyrite-rich layers. The synthetic multifrequency EM logs are computed using the PPIP-SCAIP model based on the assumed petrophysical properties listed in Table 8.1. 5% Gaussian noise is added to each of the synthetic multifrequency EM logs.

Table 8.2 shows the prior ranges and initial values of petrophysical parameters, which are inverted in the two synthetic cases used in our study. The prior ranges of petrophysical parameters provide the range or boundary of

TABLE 8.2 The default prior ranges and initial values of unknown petrophysical properties/parameters assumed for the three Markov chains of the proposed stochastic inversion methodology.

Model parameters	Prior ranges	Initial-1	Initial-2	Initial-3
s_W	(0.01,1)	0.4	0.6	0.8
C_w (S/m)	(0.01,20)	0.75	0.25	2.5
r_c (μm)	(0.1,10)	2	0.5	1.5
λ_c (S)	$(1\times10^{-9}, 1\times10^{-6})$	2×10^{-7}	1×10^{-8}	5×10^{-8}
C_{sh} (S/m)	(0.25,1)	0.25	0.5	0.75
r_i (μm)	(1,100)	25	1	5

each parameter in the model-parameter space. In Synthetic Layer 1, five petrophysical parameters are estimated using the proposed inversion scheme, whereas six petrophysical parameters are estimated for the Synthetic Layer 2. Figs. 8.1 and 8.3 show the petrophysical parameters estimated for Synthetic Layers 1 and 2, respectively, at each iteration of the three Markov chains when inverting the synthetic multifrequency EM measurements containing 5% Gaussian noise. The scale reduction score method applied to monitor convergence requires at least two Markov chains to converge. We ran three Markov chains for inverting the EM data. The three Markov chains from very different sets of starting points can help to detect possible local convergence (Chen et al., 2008). The inversion process converges after 15,000 iterations for the Synthetic Layer 1, while it only takes 5000 iterations for the Synthetic Layer 2. Figs. 8.2 and 8.4 show the distributions of estimated petrophysical parameters for the Synthetic Layers 1 and 2, respectively. All the accepted samples in the second half of each Markov chain are used to generate the distributions of estimated petrophysical parameters in Figs. 8.1 and 8.3 for Synthetic Layers 1 and 2, respectively. Table 8.3 shows the real values and estimated 95% highest probability domains (HPDs) for each petrophysical parameter. Within the 95% HPD, the certainty of the inversion-derived estimate is more than 95%. The estimated 95% HPD for all parameters are close

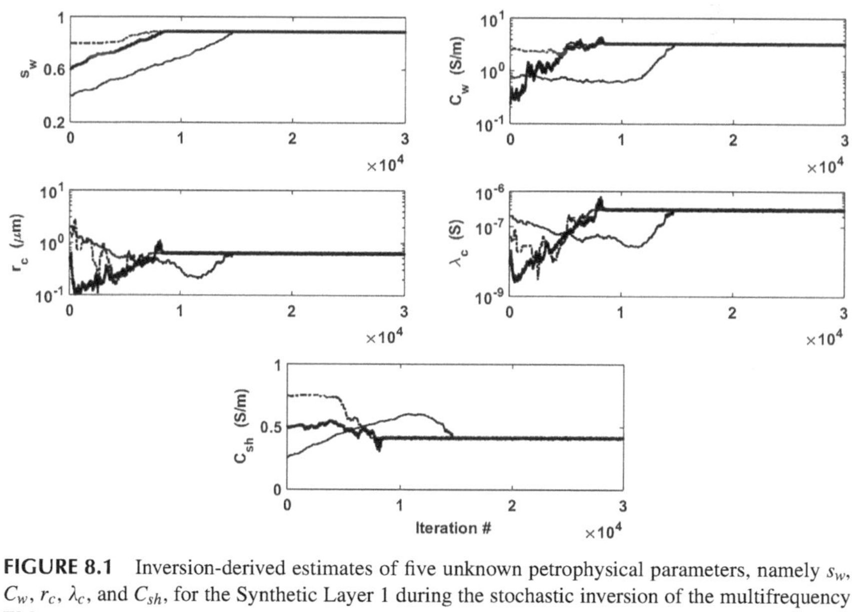

FIGURE 8.1 Inversion-derived estimates of five unknown petrophysical parameters, namely s_w, C_w, r_c, λ_c, and C_{sh}, for the Synthetic Layer 1 during the stochastic inversion of the multifrequency EM measurements containing 5% Gaussian noise for the 3 Markov chains using the Metropolis–Hastings sampling method. The inversion-derived estimates of all parameters converge to stable values after 15,000 iterations (Han et al., 2019). *Credit: Adapted from Han et al. (2019)*

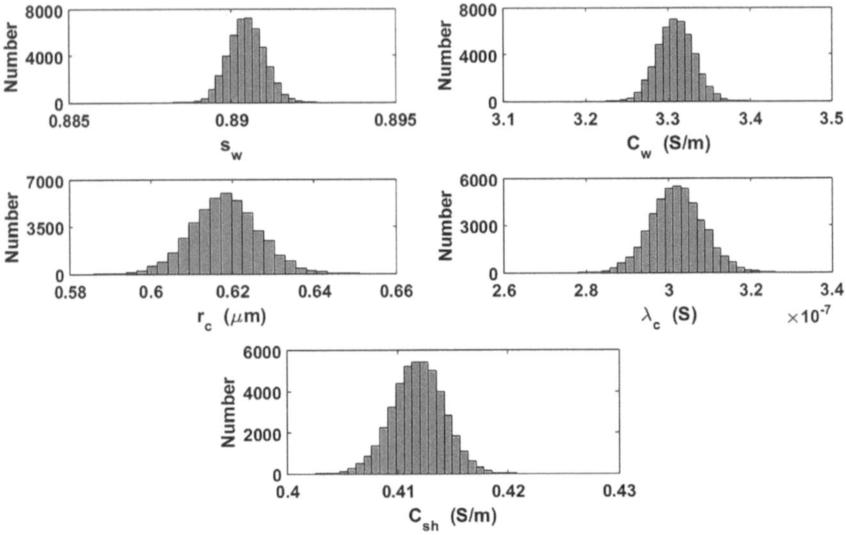

FIGURE 8.2 Histograms of the five estimated petrophysical parameters, namely s_w, C_w, r_c, λ_c, and C_{sh}, for the Synthetic Layer 1. All the accepted samples in the second half of each Markov chain are used to generate the histograms (Han et al., 2019). *Credit: Adapted from Han et al. (2019)*

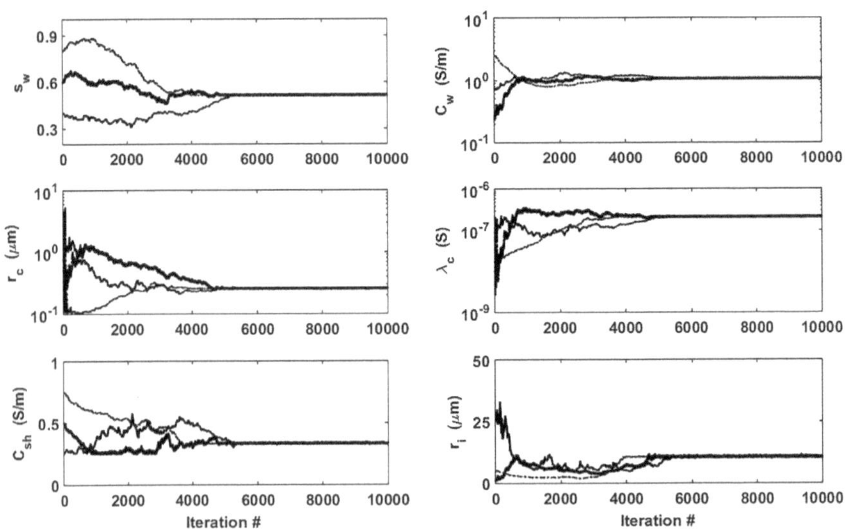

FIGURE 8.3 Inversion-derived estimates of six unknown petrophysical parameters, namely s_w, C_w, r_c, λ_c, C_{sh}, and r_i, for Synthetic Layer 2 during the stochastic inversion of multifrequency EM measurements containing 5% Gaussian noise for 3 Markov chains using the Metropolis–Hastings sampling method. The inversion-derived estimates of all parameters converge to stable values after 5000 iterations (Han et al., 2019). *Credit: Adapted from Han et al. (2019)*

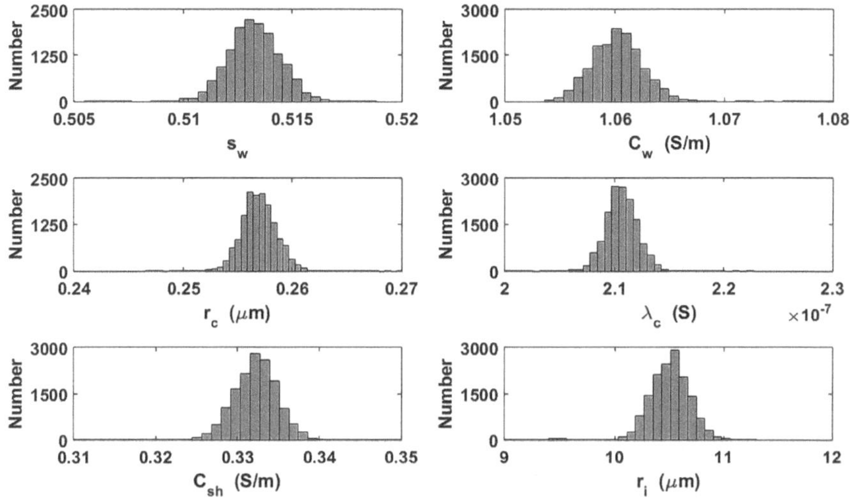

FIGURE 8.4 Histograms of six estimated petrophysical parameters, namely s_w, C_w, r_c, λ_c, C_{sh}, and r_i, for the Synthetic Layer 2. All the accepted samples in the second half of each Markov chain are used to generate the histograms (Han et al., 2019). *Credit: Adapted from Han et al. (2019)*

TABLE 8.3 Estimated 95% highest probability domains for the petrophysical parameters of Synthetic Layers 1 and 2 computed using the stochastic-inversion methodology.

Model parameters	Layer 1 true values	95% HPD	Layer 2 true values	95% HPD
s_w	0.9	(0.889,0.892)	0.5	(0.511,0.516)
C_w (S/m)	3.5	(3.267,3.351)	1	(1.056,1.065)
r_c (μm)	0.6	(0.603,0.635)	0.25	(0.254,0.260)
λ_c (S)	3×10^{-7}	(2.898×10^{-7}, 3.149×10^{-7})	2×10^{-7}	(2.076×10^{-7}, 2.136×10^{-7})
C_{sh} (S/m)	0.4	(0.407,0.417)	0.35	(0.327,0.338)
r_i (μm)	[redacted]		10	(10.159,10.880)

to the real values and the estimated water saturation, s_w, has only around 1% relative error in Synthetic Layer 1 and around 2.5% relative error for Synthetic Layer 2. The results for the synthetic layers indicate that the proposed MCMC-based stochastic-inversion method is reliable, accurate, and robust to noise.

3.2 Sensitivity of PPIP-SCAIP model to volume fractions of clay and pyrite

Before applying the newly developed MCMC-based stochastic inversion scheme coupled with PPIP-SCAIP model to process the broadband EM dispersion logs acquired in organic-rich shale formation, we evaluate the sensitivity of the PPIP-SCAIP model predictions of conductivity and permittivity to volume fractions of clay and pyrite. The high-volume fraction of clay minerals, low porosity, and presence of conductive pyrite significantly influence the relative dielectric permittivity and electrical conductivity of the formation under investigation (Anderson et al., 2006; Corley et al., 2010; Han and Misra, 2018b; Misra et al., 2016b). Fig. 8.5 shows the sensitivity of relative permittivity and electrical conductivity computed using the PPIP-SCAIP model to volumetric fraction of pyrite (V_i) and volumetric fraction of clay (V_c). We use the clay- and pyrite-rich Synthetic Layer 2 for the sensitivity analysis, such that V_i varies from 0 to 0.1 for V_c of 0.35, 0.45, and 0.55. The volumetric fraction of sand (nonconductive grains) changes with the alterations in the volumetric fractions of clay and pyrite while other petrophysical parameters, including total porosity and bulk conductivity of water, are kept constant. Fig. 8.5A and B show the variations in relative permittivity and electrical conductivity, respectively, with the variations in the volumetric

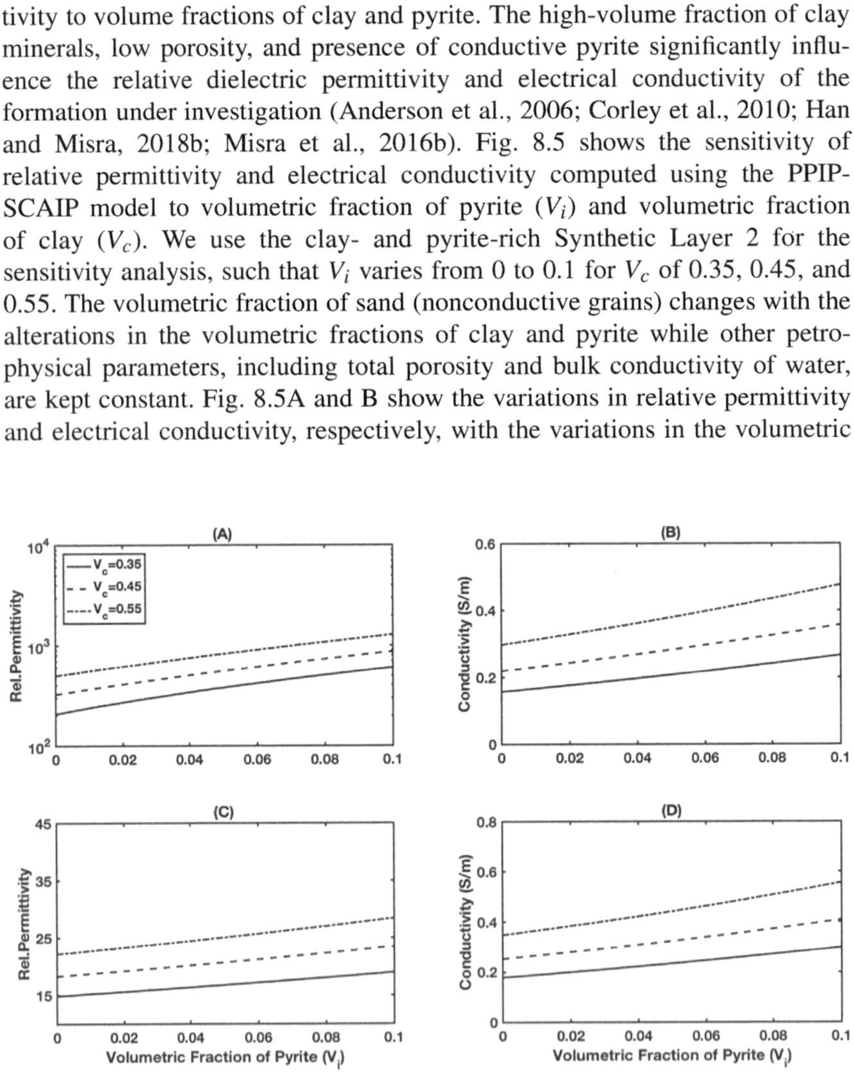

FIGURE 8.5 Sensitivity of the (A and C) relative permittivity and (B and D) electrical conductivity computed using the PPIP-SCAIP model to volumetric fraction of pyrite (V_i) and volumetric fraction of clay at 1 MHz and at 100 MHz, respectively (Han et al., 2019). *Credit: Adapted from Han et al. (2019)*

fractions of clay and pyrite at 1 MHz, which is close to the log-acquisition frequencies of LWD EM propagation tool. At the volumetric fraction of clay of 0.55 and frequency of 1 MHz, the relative permittivity increases from 500 to 1250 and electrical conductivity increases from 0.3 to 0.48 S/m with the increase in volumetric fraction of pyrite from 0 to 0.1. The relative permittivity equals to 670 and electrical conductivity equals to 0.35 S/m at volumetric fraction of pyrite equaling to 0.03, which is equal to that in the Synthetic Layer 2. Fig. 8.5C and D show the variations in relative permittivity and electrical conductivity, respectively, with the variations in volumetric fractions of clay and pyrite at 100 MHz, which is close to the log-acquisition frequencies of dielectric dispersion logging tool. When the volumetric fraction of clay equals to 0.55, the relative permittivity changes from 22.5 to 28.6 and conductivity changes from 0.35 to 0.56 S/m when increasing volume fraction of pyrite from 0 to 0.1. Relative permittivity equals to 24.1 and electrical conductivity equals to 0.4 S/m at volumetric fraction of pyrite of 0.03 and frequency of 100 MHz. These results illustrate that volumetric fractions of clay and pyrite minerals strongly influence the relative permittivity and electrical conductivity because of the frequency-dependent interfacial-polarization effects of clay and pyrite.

3.3 Application of the inversion-based method to process broadband electromagnetic dispersion logs acquired in the European organic-rich shale formation

The well under investigation was drilled in an organic-rich shale formation in Poland Lubelskie (Lublin Basin). Oil-based mud was used to reduce the invasion of drilling fluid into formation and the mud-filtrate invasion was negligible. EM induction resistivity, LWD propagation, and dielectric dispersion logging tools were deployed over a depth interval of 1500 m in this well (Wang and Poppitt, 2013). In doing so, for the very first time, continuous broadband EM dispersion measurements at seven EM log-acquisition frequencies ranging from kHz to GHz were acquired for purposes of robust water saturation estimation. Following that, effective relative permittivity and electrical conductivity at seven EM log-acquisition frequencies were derived from the broadband EM dispersion measurements (Wang and Poppitt, 2013). Finally, the newly developed stochastic inversion-based interpretation method processes the broadband EM dispersion logs (effective relative permittivity and electrical conductivity) acquired in the European organic-rich shale well. Three unknown petrophysical parameters, namely water saturation s_w, connate water conductivity C_w, and surface conductance of clay λ_c are continuously estimated along certain depth intervals. For all the depths of interest, we assume the average radius of clay r_c is 2 μm, average radius of pyrite grains r_i is 10 μm, and low-frequency conductivity in pure-clay zone C_{sh} is 0.25 S/m, which are typical values in organic-rich shale formations. By assuming r_i, r_c, and C_{sh} to be constant over certain depths, we mitigate the nonuniqueness problems of inversion results. The values of other petrophysical parameters are the same as those in Table 8.1 for the two synthetic layers. For the purpose of demonstrating the effectiveness of the stochastic inversion-based interpretation method, we use the data misfit between PPIP-SCAIP model predictions for the inversion-derived estimates and the measured broadband EM dispersion logs. In this study, we present results for two specific depths. The first chosen depth is X066 m, which is clay-rich and pyrite-bearing zone with low-total porosity ($\phi_T = 0.05$). At depth X066 m, the volumetric fraction of clay V_c is 0.47 and that of pyrite V_i is 0.01. The second chosen depth is X151 m which is clay-rich ($V_c = 0.57$) with higher total porosity ($\phi_T = 0.09$) and no pyrite content.

Fig. 8.6 shows histograms of inversion-derived estimates of petrophysical parameters, namely s_w, C_w, and λ_c, for the clay-rich and pyrite-bearing depth at X066 m. All the samples in the second half of each Markov chain are used to generate those histograms. Figs. 8.7 and 8.8 show the comparisons of the modeled relative permittivity and electrical conductivity dispersions computed using the clay-pyrite dispersion model based on the inversion-derived estimates against the measured broadband relative permittivity and electrical conductivity logs at depth X066m and depth X151m, respectively. The average relative errors between measured and modeled relative permittivity and electrical conductivity are 0.25 for the depth X066m (Fig. 8.7), and 0.13 for the depth X151m (Fig. 8.8). Fig. 8.7 indicates that the modeled and measured relative permittivity and electrical conductivity at depth X066 m exhibit good agreement except for relative permittivity at the LWD propagation

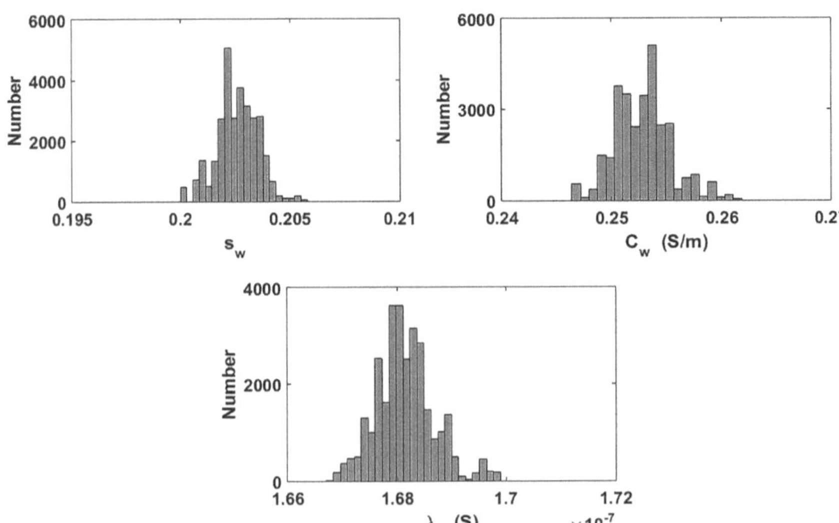

FIGURE 8.6 Histograms of estimated petrophysical parameters, s_w, C_w, and λ_c, of the clay-rich and pyrite-bearing depth at X066 m (Han et al., 2019). *Credit: Adapted from Han et al. (2019)*

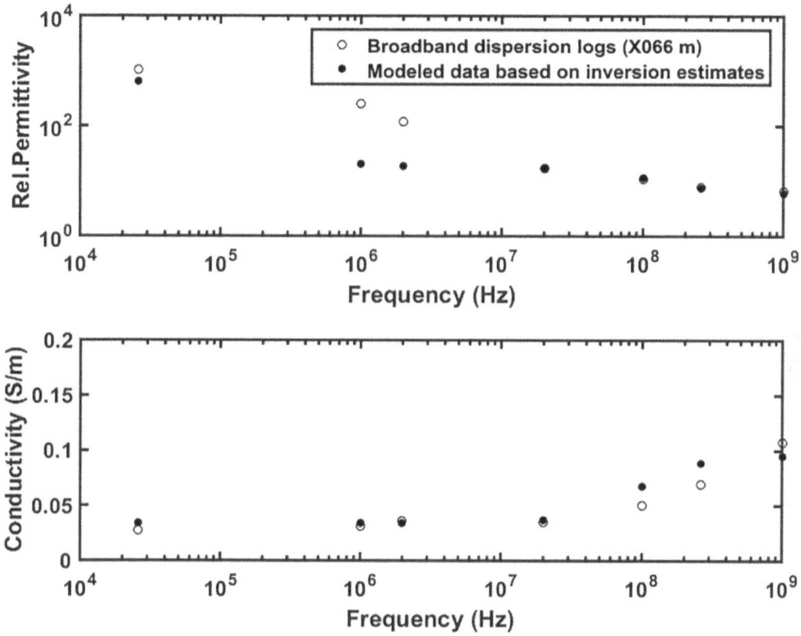

FIGURE 8.7 Comparisons of the modeled and measured relative permittivity and electrical conductivity of the clay-rich and pyrite-bearing depth at X066 m. Model predictions were generated using the PPIP-SCAIP model based on the inversion-derived estimates of s_w, C_w, and λ_c (Han et al., 2019). *Credit: Adapted from Han et al. (2019)*

log-acquisition frequencies (1 and 2 MHz). On the other hand, at depth X151m, the modeled and measured relative permittivity and electrical conductivity show good agreement over the entire frequency range. Table 8.4 lists the estimated 95% HPD and medians of s_w, C_w, and λ_c. The median of estimated s_w for depth X066m is 0.203 indicating hydrocarbon-bearing zone, and that for the depth X151m is 1 indicating water-bearing zone. Both the depths have similar surface conductance of clay, but the brine conductivity of X151m is higher than that of X066m. Table 8.4 presents the high certainty in the inversion-derived estimates.

SMD model can process dielectric dispersion logs acquired in clean sandstone formation for frequencies ranging from 20 MHz to 1 GHz. Unlike SMD model, the PPIP-SCAIP (PS) model considers the interfacial polarization arising from grains with surface charges or those from conductive/semi-metallic particles. Consequently, as compared to the SMD model, the PPIP-SCAIP model is more suitable for depths containing uniformly distributed clay and pyrite grains. For demonstrating the robustness and physical effectiveness of the PPIP-SCAIP model, we compare modeled and measured relative permittivity and electrical conductivity dispersions at depth of X070m

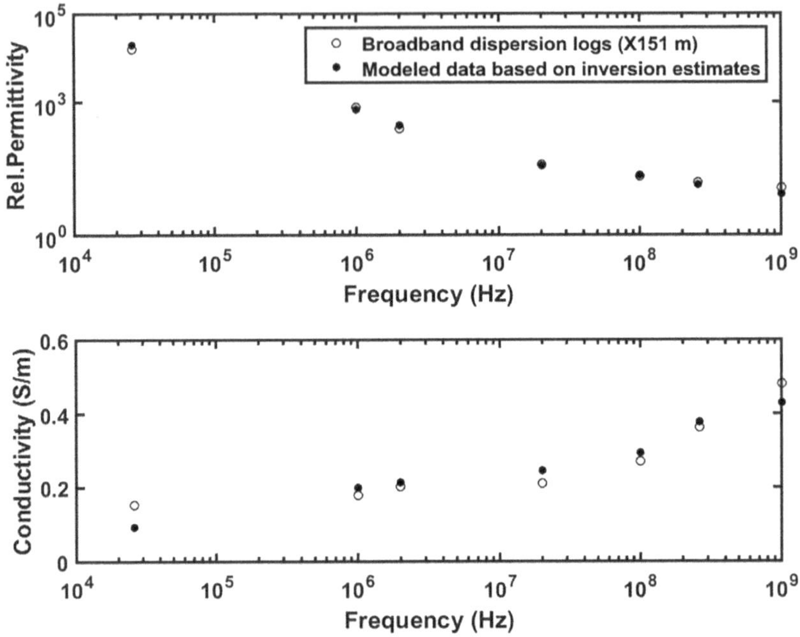

FIGURE 8.8 Comparisons of the modeled and measured relative permittivity and electrical conductivity of clay-rich depth at X151 m. Model predictions were generated using the PPIP-SCAIP model based on the inversion-derived estimates of s_w, C_w, and λ_c (Han et al., 2019). *Credit: Adapted from Han et al. (2019)*

TABLE 8.4 Estimated 95% HPD and medians of petrophysical parameters, namely s_w, C_w, and λ_c, in the clay-rich and pyrite-bearing depth of X066m and the clay-rich depth of X151m.

Model parameters	Depth X066 95% HPD	Depth X066 medians	Depth X151 95% HPD	Depth X151 medians
s_w	(0.201,0.204)	0.203	(1,1)	1
C_w (S/m)	(0.249,0.259)	0.253	(0.408,0.412)	0.410
λ_c (S)	$(1.671\times10^{-7}$, $1.691\times10^{-7})$	1.681×10^{-7}	$(5.888\times10^{-8}$, $5.953\times10^{-8})$	5.924×10^{-8}

for the dielectric-dispersion log-acquisition frequencies in Fig. 8.9. The depth X070 m is clay-rich ($V_c = 0.36$) with low-total porosity ($\phi_T = 0.04$). The relative permittivity and electrical conductivity data modeled using th PPIP-SCAIP model coupled with the MCMC-based stochastic

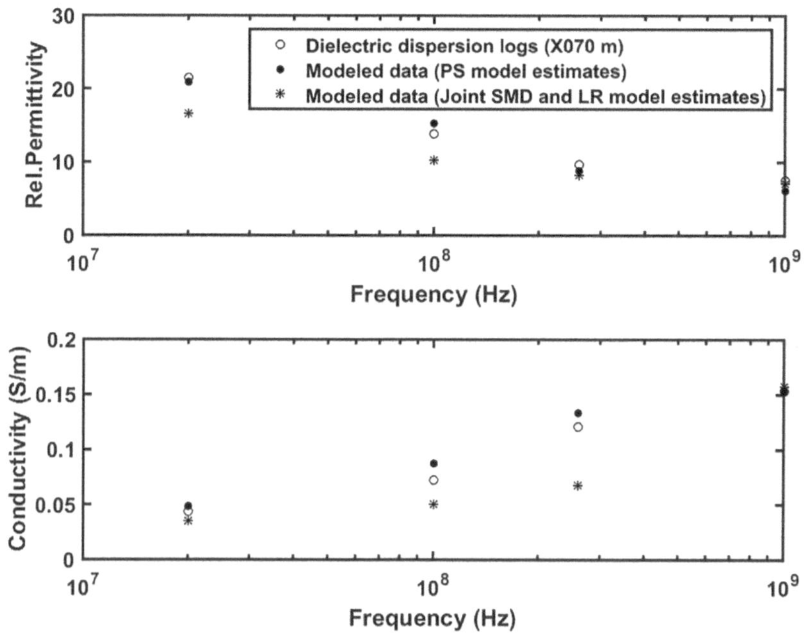

FIGURE 8.9 Comparisons of the measured relative permittivity and electrical conductivity against the PPIP-SCAIP (PS) model responses based on the stochastic inversion-derived estimates, and the joint SMD and LR model responses based on the deterministic inversion-derived estimates (Han et al., 2019). *Credit: Han et al. (2019)*

inversion-derived estimates are compared against those modeled using the joint SMD and LR model coupled with the Levenberg-Marquardt-based deterministic inversion-derived estimates. Chapters 6 and 9 describe in detail the application of deterministic inversion-based interpretation method. The data misfit in terms of average relative error is 0.10 when the modeled relative permittivity and electrical conductivity are computed using the inversion-derived estimates obtained with the PPIP-SCAIP model as the forward model coupled to the stochastic inversion scheme, whereas the average relative error is 0.21 when the modeled relative permittivity and electrical conductivity are computed using the deterministic inversion-derived estimates obtained by the joint SMD and LR model as the forward model. Implementation of PPIP-SCAIP model reduces the average relative error in fitting the dielectric dispersion logs at the depth of X070 m by 52%. The median value of the s_w estimates obtained when using the stochastic inversion coupled with the PPIP-SCAIP model is around 0.39, which indicates the presence of hydrocarbon. On the other hand, the s_w estimated computed using the deterministic inversion coupled with the joint SMD and LR model is 1, which indicates water-filled zone. The results indicate that we would overestimate s_w by using the conventional EM logs interpretation models when we ignore the clay and pyrite interfacial polarization effects, which is also reported in Chapter 9 and previous publication by Han and Misra (2018a) for dielectric dispersion logs interpretation in Bakken petroleum system.

3.4 Petrophysical interpretation and log analysis of the inversion-derived estimates in the European organic-rich shale formation

The newly proposed stochastic inversion-based interpretation method continuously processes the broadband EM dispersion logs acquired in the European organic-rich shale well for two depth intervals, namely X058m−X073m and X545m−X555m, which are clay and pyrite-rich, homogenous, and isotropic zones. Continuous estimations of S_w, C_w, and λ_c were obtained across the two intervals of interest in the European organic-rich shale formation. The proposed method assumes the formation to be homogeneous, isotropic, and rich in clay and conductive minerals. The EM logs being jointly processed in the proposed methodology are acquired at different frequencies and are sensitive to different formation volumes because of varying depths of investigation. Figs. 8.10 and 8.11 show the logs and inversion-derived estimates for the Formation Zone 1 (X058m−X073m) and Zone 2 (X545m−X555m), respectively. The layout of logs is the same for Figs. 8.10 and 8.11. The high gamma-ray values in both Zone 1 and Zone 2 are because of the high-volumetric content of clay and presence of kerogen, which are around 100−120 gAPI. The two intervals in the European organic-rich shale formation exhibit high resistivity indicated by the laterologs (>100 Ωm). Negligible effects of invasion can be observed and confirmed by the multiple DOI laterolog resistivity measurements (RLA1 - RLA5) presented in Track 3, and the caliper in Track 1 indicates that the borehole is regular along depth in the two zones of interest. Therefore, we can assume that the formation volumes sensed by EM induction, EM propagation, and dielectric dispersion logs have similar properties in absence of bed boundaries or thin bed layers. EM induction senses the largest formation volume, and with the increase in frequency and the change in transmitter-receiver spacing the formation volume sensed by the EM log decreases; consequently, dielectric log at 1 GHz is the shallowest sensing log. The bulk density in those two intervals is around 2.67 g/cm^3 and the photoelectric factor is around 3, which indicates shale formation and are illustrated in Track 4. ELAN multimineral analysis was performed to estimate total porosity and volumetric fractions of each mineral at continuous depth. The ELAN multimineral analysis is a joint inversion method combining deep and shallow resistivity, gamma-ray, photoelectric factor, bulk density, neutron porosity, sonic, and elemental spectroscopy logs. The TOC estimates obtained from Rock-Eval pyrolysis and weight fractions of minerals at some discrete depths obtained from X-ray powder diffraction (XRD) measurements were also included as additional constraints in the ELAN multimineral analysis. The volumetric fractions of various minerals

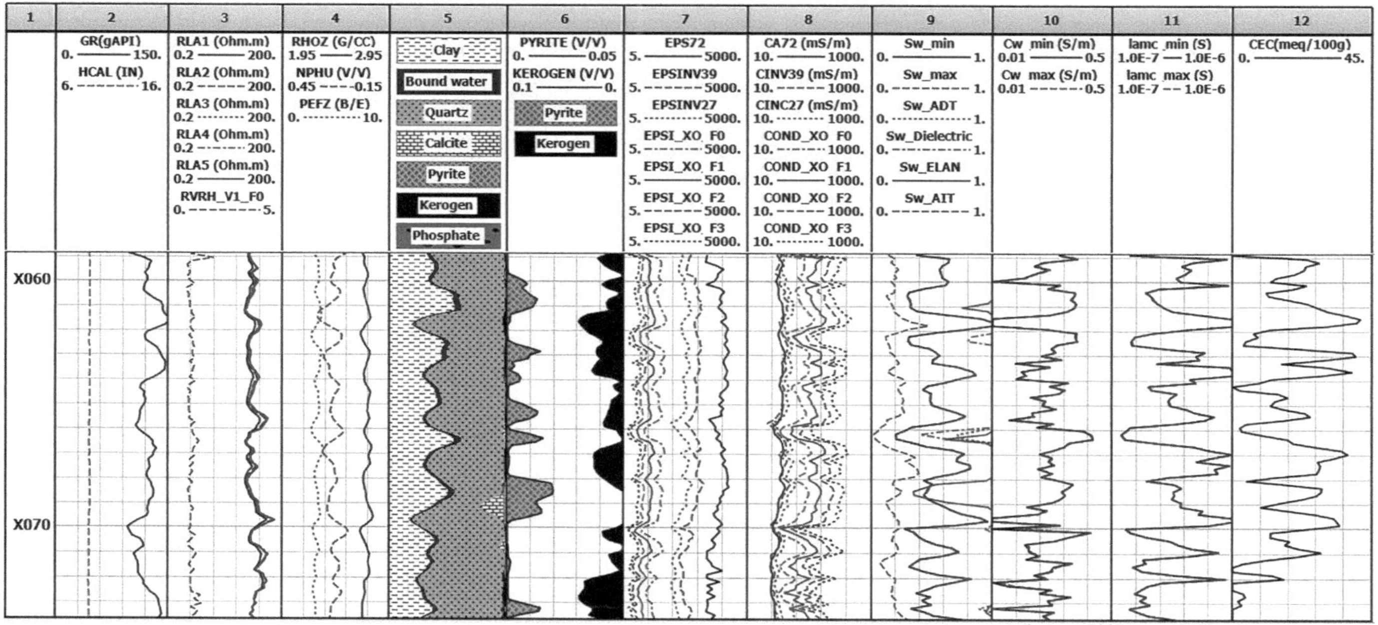

FIGURE 8.10 Logs for Zone 1 (X058 m-X073 m). Track 1 is depth, Track 2 contains gamma-ray (GR) and caliper (HCAL), Track 3 contains laterolog resistivity (RLA), Track 4 contains neutron porosity (NPHU), bulk density (RHOZ), and photoelectric factor (PEFZ), Track 5 contains volumetric fractions of various minerals: quartz, clay, kerogen, pyrite, calcite, and phosphate, Track 6 contains volumetric fractions of pyrite and kerogen, Track 7 contains relative permittivity logs at 7 frequencies, Track 8 contains electrical conductivity logs at 7 frequencies, Track 9 contains water saturation (S_w) estimated using various methods, Track 10 contains inversion-derived 95% HPD of water conductivity (C_w) estimates, Track 11 contains inversion-derived 95% HPD of surface conductance of clay (λ_c) estimates, and Track 12 contains formation CEC values interpreted by service company (Han et al., 2019). *Credit: Han et al. (2019)*

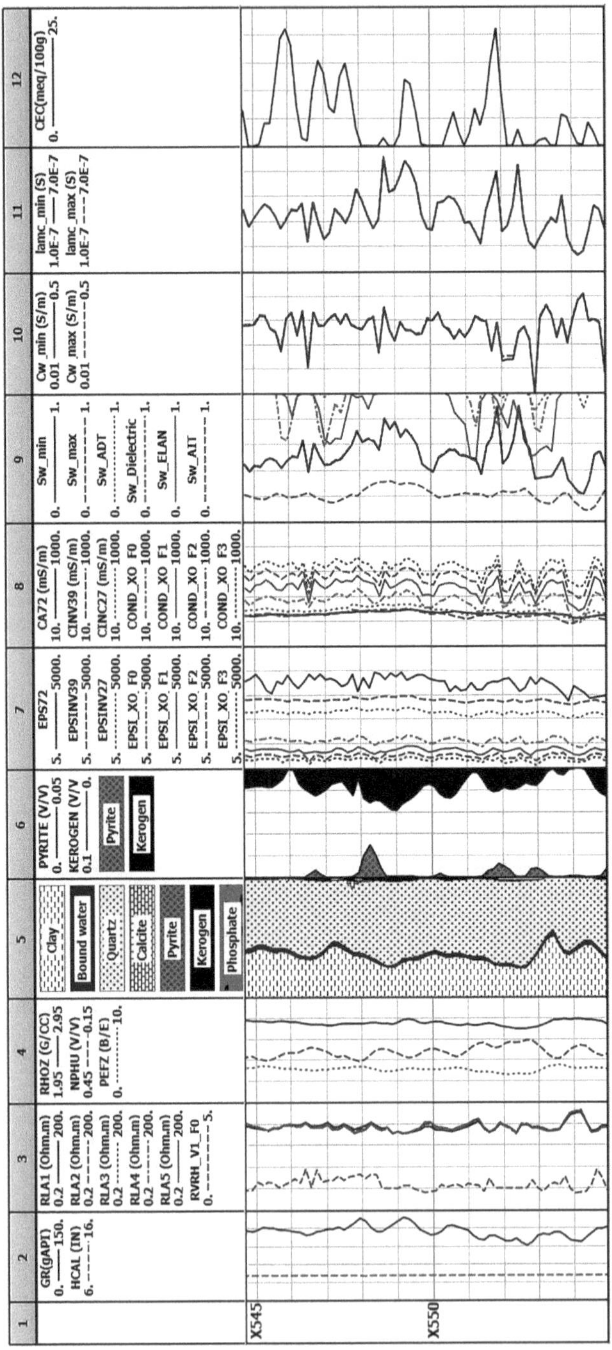

FIGURE 8.11 Logs for the Zone 2 (depth X545 m-X555 m), which is a clay-rich and pyrite-bearing interval. The logs and layout presented in this figure are similar to those in Fig. 8.10 (Han et al., 2019). *Credit: Han et al. (2019)*

including quartz, clay, kerogen, pyrite, calcite, and phosphate are illustrated in Track 5 indicating that the formation predominantly comprises clay and quartz. The total porosity in the two zones estimated by ELAN multimineral analysis is around 3%–7%, which is not included in the logs in Figs. 8.10 and 8.11. The volumetric fractions of special minerals including pyrite and kerogen are separately illustrated in Track 6. The relative permittivity and electrical conductivity logs at seven log-acquisition frequencies computed from EM measurements (Wang and Poppitt, 2013) are contained in Track 7 and Track 8, which varies from 5 to 1800, and 0.02 to 0.5 S/m, respectively. The ELAN multimineral analysis results indicate that the lithology is similar within the two zones. The volumetric fraction of clay in both zones are around 30%–50%. At depth X069m to X070m, there is 10%–20% volumetric fraction of calcite. The volumetric fractions of conductive pyrite is around 0%–2% and kerogen is around 1%–4%, which indicates a possibility of hydrocarbon in the two zones. The ratio of the horizontal resistivity (R_H) and vertical resistivity (R_V) which is approximate to 1 in both intervals (Track 3) indicates that the two zones are isotropic.

Tracks 9, 10, and 11 illustrate the inversion-derived estimates. Track 9 includes water saturation estimated using various methods, Track 10 contains inversion-derived 95% HPD of connate water conductivity estimates, and Track 11 contains inversion-derived 95% HPD of surface conductance of clay estimates. The ranges of estimated 95% HPD of all three parameters (e.g., S_w_min to S_w_max) are very narrow, as exhibited by the overlap of black continuous and blue (gray in printed version) dashed curves. S_w, C_w, and λ_c estimates in the two zones of interest in the European organic-rich shale formation are in the range of 0.2–0.9, 0.1 to 0.25 S/m, and 1×10^{-7} to 1×10^{-6} S, respectively. S_w estimates in the two zones computed using the stochastic inversion of broadband relative permittivity and electrical conductivity logs is between the S_w estimates derived from EM induction resistivity log interpretation and those derived from dielectric dispersion logs interpretation and multimineral analysis solver. The S_w estimates derived from the dielectric dispersion logs interpretation using deterministic inversion of the logs coupled with the joint SMD and LR model (S_w_ADT) (Han and Misra, 2018a) are close to those obtained by O&G service company's inversion of dielectric dispersion logs (S_w_Dielectric). S_w_ADT and S_w_Dielectric are all almost close to 1 and are typically 0.3 to 0.5 saturation unit higher than those obtained by the proposed stochastic-inversion methodology, namely S_w_min and S_w_max.

S_w_min and S_w_max obtained using the newly developed stochastic inversion is more accurate than the high values of S_w_ADT and S_w_Dielectric because the two zones contain 1%–4% volumetric fraction of kerogen, which tend to be oil-wet. The multimineral analysis solver also obtains higher S_w estimates (S_w_ELAN) than those obtained by the proposed method except for some intervals around X068-X069 m (in Zone 1) and X552-X553 m (in Zone 2) in which the two methods obtain similar values. The EM induction resistivity log was processed using the popular Simandoux model and the computed S_w_AIT is 0.2–0.4 saturation unit lower than those obtained using the proposed method in both zones. Frequency-independent Simandoux model accounts for the additional conductivity because of clay minerals but ignores the interfacial-polarization effects of clays and pyrite. We compare water saturation estimates by using various methods at two kerogen-rich (around 4%) depth: X072.8 m (in Zone 1) and X549 m (in Zone 2). The S_w estimates derived from EM induction resistivity log interpretation, the proposed stochastic inversion of broadband EM dispersion logs, and dielectric dispersion logs interpretation at depth X072.8 m is 0.2, 0.47, and 1, and at depth X549 m is 0.31, 0.77, and 1, respectively. The stochastic inversion-derived λ_c in Track 11 exhibits similar trend as the formation CEC values in Track 12 which are obtained by service company dielectric dispersion logs interpretation in both zones, especially in Zone 1. The variation of λ_c is between 1×10^{-7} S to 1×10^{-6} S in both zones, and this variation can be explained by the variation of the mineral content (illite, chlorite, and kaolinite), which can be confirmed and validated by the ELAN multimineral analysis and XRD data at some discrete depths.

The discrepancy in water saturation estimates obtained from various methods is primarily because of the model assumptions. Each of the three models is assuming different petrophysical conditions and accounts for different physical processes. Consequently, when these models are used to interpret the multifrequency electrical conductivity/relative permittivity measurements, we obtain different water saturation estimates. Following are few physical reasons giving rise to the discrepancies in the water saturation estimates:

1. The frequency-independent models such as Simandoux model or CRIM will always generate different water saturation estimates as compared to that obtained from the PPIP-SCAIP model, especially when frequency-independent model is applied on multifrequency electrical conductivity/relative permittivity at different frequencies.
2. CRIM is valid at high frequencies (approximately 1 GHz) and accounts for orientation/dipolar polarization of diploes in the mixture, whereas the Simandoux model is valid at low frequencies (around 1 kHz) and accounts for excess conductivity because of clay minerals.
3. Because of interfacial polarization phenomena, the relative permittivity and electrical conductivity exhibit dispersive behavior; consequently, electrical conductivity changes from 0.1 to 0.5 S/m for certain mixtures as frequency varies from 1 kHz to 1 GHz. The best model is the one that accounts for the interfacial polarization phenomena and any other processes that influence the electrical conductivity/relative permittivity dispersion.

4. In the presence of such dispersive behavior because of the interfacial polarization phenomena, water saturation estimates obtained by processing a single-frequency EM data will be different form that obtained by processing broadband-frequency EM data.
5. For clay-rich and pyrite-bearing shale formation, the interfacial polarization effects of clays and pyrite cannot be ignored. Both the Simandoux model and CRIM ignore the interfacial polarization effects of clays and pyrite; as a result, both generate inaccurate water saturation estimates.

4. Limitations and assumptions

The PPIP-SCAIP model assumes that the volume of formation under investigation rock is homogeneous and isotropic without Maxwell–Wagner polarization effects (Misra et al., 2016a). In addition, the proposed MCMC-based inversion interpretation method assumes that the volumes of formation detected by EM induction resistivity, LWD propagation, and dielectric dispersion logging tools at four frequencies have similar petrophysical properties and are free from laminations and fractures. In this chapter, MCMC-based inversion interpretation is applied to depth intervals with negligible mud-filtrate invasion. The clay-pyrite dispersion model or PPIP-SCAIP model, similar to certain log-scale or core-scale EM interpretation models such as CRIM and SMD model, only considers total porosity. Effective porosity, pore-network tortuosity, isolated pores, and connectivity of clay and pyrite grains are not considered in our approach. The information of pore-network or clay and pyrite-network can be obtained from pore-scale images of the rock or formation under investigation. Moreover, the advanced log-scale or core-scale EM interpretation model needs an upscaling method that includes the pore-scale network information for purposes of log-scale or core-scale modeling, which is beyond the scope of this chapter. We also assume that some petrophysical parameters, such as cementation exponent and saturation exponent, are constant in continuous depth intervals for the new inversion-based interpretation method, which should be different when lithology changes or there are lamination and fractures presence.

5. Conclusions

Downhole multifrequency EM logs were processed using a Markov Chain Monte Carlo (MCMC) sampling-based stochastic inversion method coupled with a new clay-pyrite EM dispersion model (referred as the PPIP-SCAIP) to estimate hydrocarbon saturation in an European organic-rich shale formations. The newly developed stochastic inversion method simultaneously processes the EM broadband dispersion logs acquired by EM induction tool (26 kHz), LWD propagation tool (1 and 2 MHz), and dielectric dispersion logs (20 MHz−1 GHz). The continuous estimations of formation hydrocarbon saturation, connate water conductivity, and surface conductance of clay with ranges of possible values (95% HPD) are obtained for two zones in the organic-rich shale formation. The proposed MCMC-sampling-based stochastic inversion method is robust to noise in the EM logs and suitable for clay and pyrite-rich, low-porosity shale formations. The MCMC-based stochastic inversion coupled with the new mechanistic PPIP-SCAIP model performs much better in comparison to the Levenberg-Marquardt-based deterministic inversion coupled to the traditional empirical models. The hydrocarbon saturation, connate water conductivity, and surface conductance of clay estimates in two clay-rich and pyrite-bearing, homogenous, and isotropic zones in the European shale formation are in the range of 0.2−0.9, 0.1 to 0.25 S/m, and 1×10^{-7} to 1×10^{-6} S, respectively. The hydrocarbon saturation estimates in the two zones of interest obtained using the proposed stochastic-inversion method are between the estimates derived using conventional interpretation that individually processes the EM induction resistivity log and those derived from deterministic-inversion of dielectric dispersion logs, conventional interpretation of dielectric permittivity log at 1 GHz, and multimineral analysis solver. Hydrocarbon saturation estimates based on the interpretations of dielectric dispersion or dielectric permittivity logs are close to 0 in the two zones of interest and are typically 0.3 to 0.5 saturation unit lower than those obtained by the newly proposed stochastic-inversion of broadband EM logs. The two zones of interest contain 1%−4% volume fraction of kerogen, which tend to be oil wet; consequently, the hydrocarbon saturation estimates derived using the newly proposed stochastic inversion are more physically consistent as compared to those derived using conventional dielectric permittivity/dispersion interpretation and inversion methods.

Nomenclature

Symbols and abbreviations

S_w Water Saturation
C_w Connate water Conductivity (S/m)
R_w Connate water resistivity ($\Omega \cdot$ m)
m Cementation exponent
n Saturation exponent
a Tortuosity factor
ϕ_T Total porosity of the formation
R_t Formation resistivity
V_{sh} Volumetric fraction of shale
ε_{rhy} Hydrocarbon relative permittivity
ε_0 Vacuum permittivity
ω Angular frequency
ε_{rw} Relative permittivity of water

V_i Volumetric fraction of pyrite grains
S_i Bulk conductivity of pyrite
ε_{ri} Relative permittivity of pyrite
D_i Diffusion coefficient of pyrite
r_i Radius of pyrite grains
V_c Volumetric fraction of clay
ε_{rc} Relative permittivity of clay
λ_c Surface conductance of clay
r_c Radius of spherical clay grains
C_{sh} Low frequency conductivity in pure clay formation
V_s Volumetric fraction of nonconductive grains
ε_{rs} Relative permittivity of nonconductive grains
λ_s Surface conductance of nonconductive grains
r_s Radius of nonconductive grains
D_w Diffusion coefficient of water
CEC Cation exchange capacity
CRIM Complex Refractive Index model
EM Electromagnetic
GHz GigaHertz
kHz KiloHertz
LR Lichteneker-Rother model
MCMC Markov-chain Monte Carlo
MHz MegaHertz
SMD Stroud-Milton-De
T Temperature (in $^{\circ}F$)
TOC Total organic carbon

References

Anderson, B., Barber, T., Luling, M., 2006. Observations of large dielectric effects on induction logs, or can source rocks be detected with induction measurements. SPWLA 47th Ann. Logg. Symp., SPWLA-2006-OOO.

Asquith, G.B., 1980. Log Analysis by Microcomputer, first ed. Petroleum Pub. Co., Tulsa.

Aster, R.C., Borchers, B., Thurber, C.H., 2005. Parameter Estimation and Inverse Problems, first ed. Elsevier Academic Press, Amsterdam.

Bérubé, C.L., Chouteau, M., Shamsipour, P., Enkin, R.J., Olivo, G.R., August 2017. Bayesian inference of spectral induced polarization parameters for laboratory complex resistivity measurements of rocks and soils. Comput. Geosci. 105, 51–64.

Botor, D., Golonka, J., Zając, J., Papiernik, B., Guzy, P., November 2019. Petroleum generation and expulsion in the Lower-Palaeozoic petroleum source rocks at the SW margin of the East European Craton (Poland). Ann. Soc. Geol. Pol. 89 (2), 153–174.

Brookst, S.P., April 1998. Markov chain Monte Carlo method and its application. J. Roy. Stat. Soc. 47 (1), 69–100.

Brovelli, A., Cassiani, G., August 2008. Effective permittivity of porous media: a critical analysis of the complex refractive index model. Geophys. Prospect. 56 (5), 715–727.

Chen, J., Kemna, A., Hubbard, S.S., November 2008. A comparison between Gauss-Newton and Markov-chain Monte Carlo-based methods for inverting spectral induced-polarization data for Cole-Cole parameters. Geophysics 73 (6), F247–F259.

Chib, S., Greenberg, E., November 1995. Understanding the metropolis-hastings algorithm. Am. Statistician 49 (4), 327–335.

Corley, B., Garcia, A., Maurer, H.M, Rabinovich, M.B., Zhou, Z., DuBois, P., Shaw, N., 2010. Study of unusual responses from multiple resistivity tools in the Bossier formation of the Haynesville shale play. SPE Ann. Tech. Confer. Exhibit., SPE-134494-MS.

Donadille, J.M., Leech, R., Pirie, I., 2016. Water salinity determination over an extended salinity range using a joint interpretation of dielectric and neutron cross-section measurements. SPE Ann. Tech. Confer. Exhibit., SPE-181366-MS.

Gelman, A., Rubin, D.B., November 1992. Inference from iterative simulation using multiple sequences. Stat. Sci. 7 (4), 457–472.

Han, Y., Misra, S., April 2018a. Bakken Petroleum System characterization using dielectric-dispersion logs. Petrophysics 59 (2), 201–217.

Han, Y., Misra, S., May 2018b. Joint petrophysical inversion of multifrequency conductivity and permittivity logs derived from subsurface galvanic, induction, propagation, and dielectric dispersion measurements. Geophysics 83 (3), D97–D112.

Han, Y., Misra, S., Wang, H., Toumelin, E., 2019. Hydrocarbon saturation in a Lower-Paleozoic organic-rich shale gas formation based on Markov-chain Monte Carlo stochastic inversion of broadband electromagnetic dispersion log. Fuel 243, 645–658.

Hastings, W.K., April 1970. Monte Carlo sampling methods using Markov chains and their applications. Biometrika 57 (1), 97–109.

Karcz, P., Janas, M., Dyrka, I., January 2013. Polish shale gas deposits in relation to selected shale gas prospective areas of Central and Eastern Europe. Przeglad Geol. 61 (7), 411–423.

Malinverno, A., November 2002. Parsimonious Bayesian Markov chain Monte Carlo inversion in a nonlinear geophysical problem. Geophys. J. Int. 151 (3), 675–688.

Metropolis, N., Rosenbluth, A.W., Rosenbluth, M.N., Teller, A.H., Teller, E., June 1953. Equation of state calculations by fast computing machines. J. Chem. Phys. 21 (6), 1087–1092.

Minsley, B.J., August 2011. A trans-dimensional Bayesian Markov chain Monte Carlo algorithm for model assessment using frequency-domain electromagnetic data. Geophys. J. Int. 187 (1), 252–272.

Misra, S., Torres-Verdín, C., Revil, A., Rasmus, J., Homan, D., March 2016a. Interfacial polarization of disseminated conductive minerals in absence of redox-active species—Part 2: effective electrical conductivity and dielectric permittivity. Geophysics 81 (2), E159–E176.

Misra, S., Lüling, M.G., Rasmus, J., Homan, D.M., Barber, T.D., June 25–29, 2016b. Dielectric effects in pyrite-rich clays on multifrequency induction logs and equivalent laboratory core measurements. In: SPWLA 57th Annual Logging Symposium.

Mosegaard, K., Tarantola, A., July 1995. Monte Carlo sampling of solutions to inverse problems. J. Geophys. Res. 100 (B7), 12431–12447.

Papiernik, B., Poprawa, P., Liana, B., Michna, M.A., Zabek, G., Machowski, G., Hadro, J., 2017. Comparative analysis of shale gas/oil resources in the Baltic and the Lublin basins (Poland) based on 3-D static modeling. AAPG Ann. Conv. Exhibit., 90291.

Tathed, P., Han, Y., Misra, S., December 2018a. Hydrocarbon saturation in Bakken Petroleum System based on joint inversion of resistivity and dielectric dispersion logs. Fuel 233, 45–55.

Tathed, P., Han, Y., Misra, S., May 2018b. Hydrocarbon saturation in upper Wolfcamp shale formation. Fuel 219, 375–388.

Wang, H., Poppitt, A., 2013. The broadband electromagnetic dispersion logging data in a gas shale formation: a case study. SPWLA 54th Ann. Logg. Symp., SPWLA-2013-AAA.

Wu, J., Mukerji, T., Journel, A.G., 2005. Improving water saturation prediction with 4D seismic. SPE Ann. Tech. Confer. Exhibit., SPE-95125-MS.

Zakri, T., Laurent, J.P., Vauclin, M., 1998. Theoretical evidence for 'Lichtenecker's mixture formulae' based on the effective medium theory. J. Phys. Appl. Phys. 31 (13), 1589–1594.

Chapter 9

Deterministic inversion based interpretation of multifrequency electromagnetic logs from wolfcamp and bakken shale formations

Pratiksha Tathed[1], Siddharth Misra[2,4], Yifu Han[3]

[1]*Petrophysicist, BP America Inc., Houston, TX, United States;* [2]*Harold Vance Department of Petroleum Engineering, College of Engineering, Texas A&M University, TX, United States;* [3]*Schlumberger Beijing Geoscience Center, Beijing, China;* [4]*Department of Geology and Geophysics, College of Geoscience, Texas A&M University, TX, United States*

This chapter applies deterministic inversion coupled with mechanistic model for interpreting multifrequency electromagnetic (EM) logs acquired in Wolfcamp and Bakken shale formations.

1. Introduction

Water saturation (S_w) and hydrocarbon saturation ($S_{hc} = 1 - S_w$) estimates in hydrocarbon-bearing formations are generally derived from Dean–Stark core measurements, nuclear magnetic resonance (NMR) log, and EM logs, such as induction log, galvanic resistivity (laterolog), or dielectric dispersion logs. In-situ estimation of hydrocarbon saturation in conventional reservoirs primarily rely on either deep-sensing or high-resolution EM logs. Water saturation estimation based on electrical and EM logs in organic-rich shale reservoirs is challenging because of effects of clay and conductive minerals. Conventional EM-log-interpretation models tend to break down for shale formations because they neglect the interfacial polarization effects and the dispersive behavior of EM properties. In unconventional reservoirs of Delaware basin, water saturation estimation is difficult because of complex mineralogy, higher clay content, low porosity, textural features, and high salinity. Similarly, Bakken Petroleum System (BPS) is composed of both

Multifrequency Electromagnetic Data Interpretation for Subsurface Characterization
https://doi.org/10.1016/B978-0-12-821439-8.00001-X

conventional and unconventional units, which exhibit significant variations in lithology, rock texture, clay content, total organic carbon (TOC), accompanied by high connate water salinity, presence of disseminated pyrite grains, and low values of porosity. These petrophysical attributes lead to disagreement between water saturation estimates obtained using different approaches, such as NMR, induction, dielectric, Dean–Stark core measurement method. A few of the above-mentioned challenges in water saturation estimation can be addressed by jointly processing the subsurface galvanic resistivity, induction, propagation, and dielectric dispersion logs using an integrated mechanistic model that accounts for various frequency-dependent interfacial polarization mechanisms. This chapter singificantly improves the interpretation and inversion methods proposed in Chapter 7.

Ramirez (Sanchez-Ramirez et al., 2010) implemented a probabilistic approach with Bayesian stochastic inversion to estimate porosity, water saturation, and volumetric shale concentration. They proposed a joint petrophysical inversion of gamma ray, density, and resistivity logs for the assessment of porosity and hydrocarbon saturation in thinly bedded formations. Glinskikh (Glinskikh et al., 2014) implemented two algorithms, Nelder-Mead nonlinear error minimization and pseudoinversion of sensitivity matrix, for the petrophysical inversion of high frequency induction logs acquired in a shaly sandstone formations of West and East Siberia to estimate water saturation, porosity, and clay fraction in the formation. The mixing model accounted for conductive dispersed clay assuming a three-component mixture of nonconductive quartz grains, conductive clay particles, and host fluid. Estimates of water saturation and porosity were of higher quality as compared to clay fraction.

At high salinity of about 70 ppk, the dielectric measurements lose sensitivity to salinity. Donadille et al. (2016) addressed the limitations in determining high connate-water salinity from dielectric dispersion logs. Joint inversion of neutron sigma measurements and dielectric dispersion logs showed excellent sensitivity to high-salinity values in Bakken shale formation. Chen and Heidari (2014) introduced analytical model combining conductivity and permittivity measurements for organic-rich source rocks with complex pore structure. They proposed joint interpretation of dielectric and resistivity measurements that significantly improves water-filled porosity and hydrocarbon saturation assessment. They suggested that spatial distribution and tortuosity of water, kerogen, and pyrite networks significantly affect dielectric permittivity and electrical resistivity.

Misra and Han (2016) carried out joint interpretation using conductivity and permittivity values obtained from EM induction at 26 kHz, EM propagation at 400 kHz and 2 MHz, and dielectric dispersion logs at 20 MHz, 100 MHz, 260 MHz, and 1 GHz to estimate the water saturation, bulk conductivity of brine, surface conductance of clay, and radius of spherical clay grains (Han and Misra, 2017). This work was further improved by Han and Misra (2018c) to develop robust inversion methodology for jointly processing galvanic, induction, propagation, and dielectric dispersion logs. Han and Misra (2018a) explained the complexity of analyzing dispersion measurements and introduced new inversion schemes and their sensitivity to various forms of noise. Han et al. (2017) proposed a joint-interpretation method using Modified Levenberg-Marquardt algorithm (LMA) for deriving accurate water-saturation estimates in pyrite- and clay rich-formations. A similar inversion scheme is used by Han and Misra (2018b) to process dispersive electrical conductivity and dielectric permittivity logs and compute range of water saturation estimates in BPS. They proposed a log-processing methodology by combining Lichtenecker–Rother model (LR), Stroud-Milton-De (SMD) model, and PS model, a mechanistic pyrite-clay dispersion model, for the estimation of water saturation, formation water salinity, homogeneity index, and cementation index in clay-lean and clay-rich units of Bakken shale. They carried their interpretation for a 300-feet depth interval in BPS and compared their estimates with Dean Stark core water saturation, NMR interpretation, and service company's inversion results.

Pirie et al. (2016) compared water saturation estimates obtained from triaxial resistivity induction log, NMR log, dielectric dispersion log, Techlog Quanti-ELAN, and Dean–Stark core measurements. S_w evaluated with the petrophysical model using R_h from induction resistivity logs at 2-feet vertical resolution compared very well with S_w obtained from dielectric measurements in Middle Bakken. The Dean–Stark core water saturation ranged from 40% to 60% in lower portion of Three Forks formation while dielectric measurements suggested 100% water saturation.

2. Proposed deterministic inversion–based interpretation method

This inversion technique processes multifrequency logs in the range of 1 kHz to 1 GHz to estimate water saturation (S_w), brine conductivity (C_w), textural index/cementation exponent (m), and saturation exponent (n) with associated uncertainties (Fig. 9.1). The estimates are generated at a vertical resolution of 2 inch. The proposed deterministic interpretation method (Tathed et al., 2018a) is similar to the modified LMA inversion scheme developed by Han and Misra (2018b), and it is coupled with an integrated mechanistic model, which combines the following three frequency-specific models (more explanation of these parameters and the models is provided in Chapters 1 and 2):

(1) Complex refractive index (CRI) model to process the conductivity and permittivity logs acquired at 1 GHz,
(2) SMD model to process the three conductivity-dispersion logs and three permittivity-dispersion logs in the frequency range of 1 –100 MHz, and
(3) Waxman–Smits (WS) model to process the single-frequency deep galvanic resistivity log(RLA5) acquired around 1 kHz or the single-frequency induction resistivity (AT90) log acquired around 20 kHz.

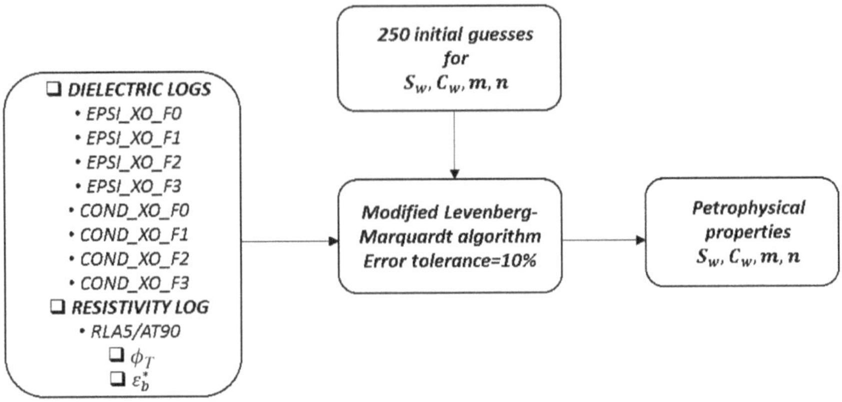

FIGURE 9.1 The proposed log-processing workflow for the deterministic inversion of multi-frequency electromagnetic logs, comprising glavanic/induction resistivity and dielectric dispersion logs. *Source: Authors.*

Measured data vector v comprises one induction/galvanic resistivity log($Res^{\text{meas,1kHz}}$) either at 1 kHz or 20 kHz, four permittivity logs ($Perm^{\text{meas,f}_i}$) in the range of 10 MHz to 1 GHz, and four conductivity logs ($Cond^{\text{meas,f}_i}$) in the range of 10 MHz to 1 GHz. The measured data vector is jointly inverted using the proposed log-processing scheme, illustrated in Fig. 9.1. $Res^{\text{meas,1kHz}}$ represents the true resistivity of the formation measured using the galvanic resistivity (laterolog) tool around 1 kHz or that measured using the induction resistivity tool around 20 kHz. $Perm^{\text{meas,f}_i}$ and $Cond^{\text{meas,f}_i}$ represent effective permittivity and conductivity of formation measured at four distinct dielectric-dispersion log-acquisition frequencies of 1 MHz $< f_0 < f_1 < f_2 < f_3 = 1$ GHz. The objective is to estimate petrophysical model parameter vector p comprising water saturation (S_w), formation water conductivity (C_w), homogeneity index (m), and saturation exponent (n) for each depth. A joint inversion is implemented to process the measured data vector v to estimate the model parameter vector p. The inversion scheme is coupled with an integrated mechanistic model, which combines CRI, SMD, and WS models to estimate the petrophysical parameters in the parameter vector p. The joint inversion scheme is initiated using 250 random initial guesses of the unknown model parameters in vector p, namely $S_w, C_w, m,$ and n. The three frequency-specific models, namely CRI, SMD, and WS models, have S_w and C_w in common. SMD and WS model share the m exponent, while n exponent is present only in WS model. More explanation of these parameters and the models is provided in Chapters 1 and 2.

The measured data vector v comprising galvanic resistivity, dielectric permittivity dispersions, and electrical conductivity dispersions are matched with modeled data vector m computed using the integrated CRI, SMD, and WS model. Various values of the vector m can be generated by varying the model parameter vector p. To that end, the modified LMA inversion algorithm (discussed in Chapter 6) iteratively computes the vector p for finding the best match between the vectors m and v that directs the convergence toward the optimum estimates for the unknown model parameters in vector p. This best estimate of vector p generates best fit between the measured data vector v and modeled data vector m, thus minimizing the data misfit. This inversion scheme requires minimization of cost function vector denoted as $F(p)$. Individual component of the vector $F(p^k)$ at the k-th iteration is denoted as $f_l(p^k)$, where the superscript k is the iteration counter for the inversion scheme, p^k is the model parameter vector computed at the k-th iteration of the inversion, and subscript l denotes one of the nine EM logs in the data vector v. $F(p^k)$ is divided by the standard deviation of the nine EM logs to improve the robustness of the parameter estimation. Cost function vector at k-th iteration $F(p^k)$ is expressed as

$$F(p^k) = [f_1(p^k), f_2(p^k), \ldots f_9(p^k)]^T \tag{9.1}$$

where $f_l(p^k)$ is expressed as

$$f_1(p^k) = \left[Res^{\text{mod,1kHz}}(p^k) - Res^{\text{meas,1kHz}}\right] \tag{9.2}$$

for $l = 1$, whereas

$$f_l(p^k) = \left[Perm^{\text{mod},f_i}(p^k) - Perm^{\text{meas},f_i}\right] \tag{9.3}$$

for $l = 2$, 3, 4, and 5 at the four distinct dielectric-dispersion log-acquisition frequencies of 1 MHz $< f_0 < f_1 < f_2 < f_3 = 1$ GHz, and

$$f_l(p^k) = \left[Cond^{\text{mod},f_i}(p^k) - Cond^{\text{meas},f_i}\right] \tag{9.4}$$

for $l = 6$, 7, 8, and 9 at four distinct dielectric-dispersion log-acquisition frequencies of 1 MHz $< f_0 < f_1 < f_2 < f_3 = 1$ GHz.

The Jacobian matrix of $F(p^k)$ is formulated as

$$J(p^k) = \begin{pmatrix} \dfrac{\partial f_1(p^k)}{\partial p_1^k} & \cdots & \dfrac{\partial f_1(p^k)}{\partial p_4^k} \\ \vdots & \ddots & \vdots \\ \dfrac{\partial f_9(p^k)}{\partial p_1^k} & \cdots & \dfrac{\partial f_9(p^k)}{\partial p_4^k} \end{pmatrix} \tag{9.5}$$

where $p^k = [p_1^k; p_2^k; p_3^k; p_4^k]$

The first order central-difference formula is implemented to approximate the derivative of $f_l(p^k)$ with respect to the j-th unknown parameters formulated as

$$\frac{\partial f_l(p^k)}{\partial p_j^k} \approx \frac{f_l\left(p_j^k + \Delta p_j^k\right) - f_l\left(p_j^k - \Delta p_j^k\right)}{2\Delta p_j^k} \tag{9.6}$$

such that $\Delta p_j^k = 0.001 p_j^k$, $1 \le j \le 4$ representing the index for model parameters to be estimated, and $1 \le l \le 9$ representing the index for the 9 EM logs in the data vector.

After convergence of the inversion-derived estimates, a line search method is applied to find the smallest error in the search space to select the estimates for the unknown model parameters, S_w, C_w, m, and n, in the vector p that generates the least data misfit. The search for global minimum among the 250 inversion searches is affected by the noise in the data and uncertainties in the inputs to the three frequency-specific mechanistic models, namely CRI, SMD, and WS models, coupled with the inversion scheme. Therefore, ranges of possible values for the unknown model parameters are generated rather than a single estimate. To that end, all possible values of the unknown model parameters, namely S_w, C_w, m, and n, are searched around global minimum that lead to data misfit within the predefined fitting error that does not exceed 1.1 times the data misfit at the global minimum. This generates minimum and maximum values for the four inversion-derived estimates. The total porosity used for water saturation estimation is computed using ELAN multimineral inversion of resistivity, gamma-ray, density, neutron, sonic, and elemental spectroscopy data (Ferraris et al., 2012). The computations are based on a probabilistic approach combining the above-mentioned logs with additional constraints of TOC, calcite transform, and pyrite transform.

3. Sensitivity analysis of the proposed joint inversion scheme

The comparison plots in this section explain the robustness and the sensitivity of the proposed deterministic inversion technique to jointly process the multifrequency resistivity and dielectric permittivity logs (Tathed, 2018). The sensitivity analysis replicates a material with low porosity, high-pyrite content, and high-clay content, representative of several organic-rich shale formations. Uncertainty of the inversion-derived estimates can be represented using the spread of local minima around the global minimum. A highly sensitive model will facilitate the error minimization and will lead to high certainty in the inversion-derived estimates. In subsequent figures from Figs. 9.2–9.9, the sensitivity of the model is denoted by sharp/fast change in the grayscale surface plot from black to white, e.g., Figs. 9.2B and 9.3B. A slow change in the grayscale surface plot from black to white is indicative of poor sensitivity, e.g., Figs. 9.3A and 9.4D. The darker regions have small relative change with respect to the global minimum, while the whiter regions have larger relative change with respect to the global minimum. A rapid change in the grayscale indicates higher sensitivity because of higher gradients. In this entire section, we lay emphasis on the sensitivity of the integrated model combining the CRI, SMD, and WS models at low-water saturation, which corresponds to high-hydrocarbon saturation.

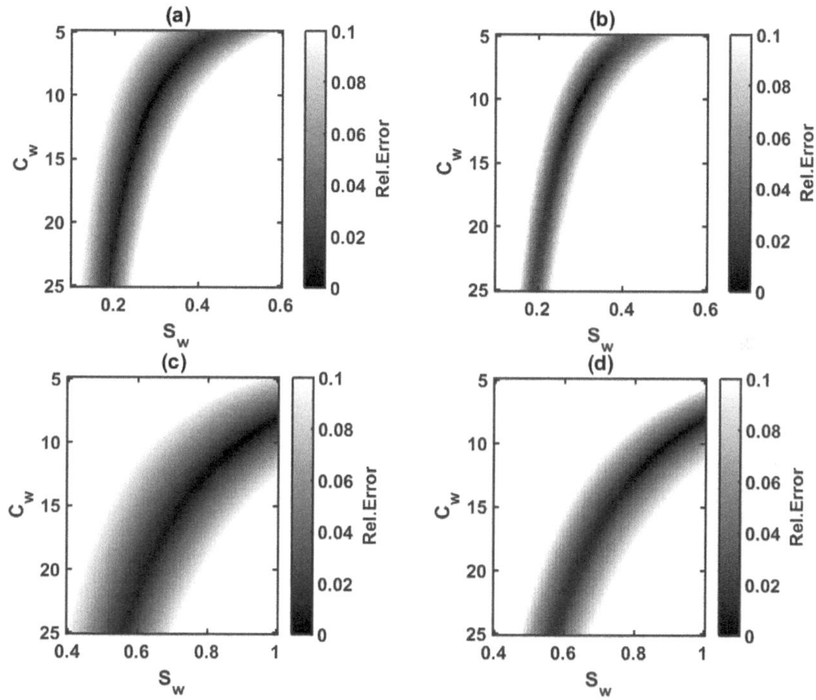

FIGURE 9.2 Comparison of (A and C) integrated model #2 combining CRIM and SMD models used for inverting only the dielectric dispersion logs against that of (B and D) integrated model #1 combining CRIM, SMD, and WS models used for the newly proposed joint inversion in terms of their sensitivities to the variations in S_w and C_w for the low water-saturation cases (A and B) around the default parameters: $m = 2$, $n = 2$, $C_w = 10$ S/m, $\phi_T = 0.02$, $BQ_v = 1.3$ mS/cc, and $S_w = 0.3$ s.u. and for the high water-saturation cases (C and D) around the default parameters: $m = 2$, $n = 2$, $C_w = 10$ S/m, $\phi_T = 0.02$, $BQ_v = 1.3$ mS/cc, and $S_w = 0.9$ s.u., respectively. *Source: Authors.*

3.1 Significance of the newly proposed joint inversion of galvanic/induction resistivity and dielectric dispersion logs

The proposed inversion workflow is coupled with the integrated WS, SMD, and CRI model to process galvanic/induction resistivity and dielectric dispersion logs. Conventional practice involves inversion of only the dielectric dispersion logs without using galvanic/induction resistivity log. We claim that the newly proposed joint inversion of galvanic/induction resistivity and dielectric dispersion logs is better than inversion of only the dielectric dispersion logs. To that end, in this section, we compare the sensitivity of the integrated model #1 combining WS, SMD, and CRI models (Fig. 9.2B and D) against that of the integrated model #2 combining SMD and CRI models, required to process only the dielectric dispersion logs (Fig. 9.2A and C).

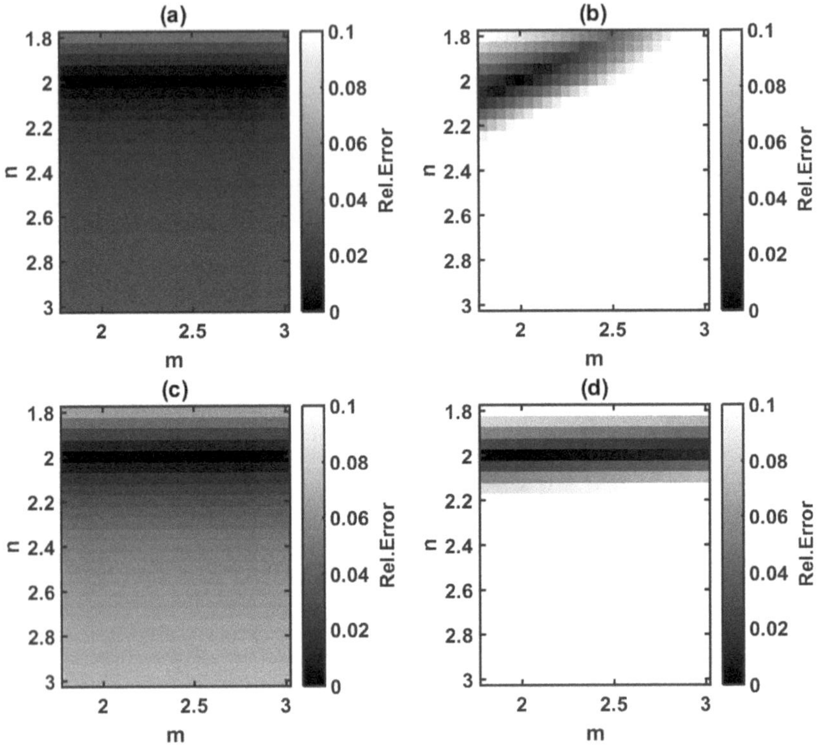

FIGURE 9.3 Comparison of (A and C) integrated model #2 combining CRIM and SMD models used for inverting only the dielectric dispersion logs against that of (B and D) integrated model #1 combining CRIM, SMD, and WS models used for the newly proposed joint inversion in terms of their sensitivities to the variations in n and m for the low water-saturation cases (A and B) around the default parameters: $m = 2$, $n = 2$, $C_w = 10$ S/m, $\phi_T = 0.05$, $BQ_v = 1.3$ mS/cc, and $S_w = 0.3$ s.u. and for the high water-saturation cases (C and D) around the default parameters: $m = 2$, $n = 2$, $C_w = 10$ S/m, $\phi_T = 0.05$, $BQ_v = 1.3$ mS/cc, and $S_w = 0.9$ s.u., respectively. *Source: Authors.*

A model exhibiting higher sensitivity will facilitate error minimization process during the inversion that will lead to more unique solutions. The proposed inversion is designed to estimate four unknown model parameters, namely S_w, C_w, m, and n. Consequently, to analyze the sensitivity of the two integrated models, the brine conductivity is varied in the range of 5 S/m to 25 S/m, water saturation in the range of 0.1–1 saturation unit (s.u.), m exponent in the range of 1.5–3.5, and n exponent in the range of 1.8–3 (Fig. 9.2).

When the unknown model parameters S_w, C_w, m, and n are slightly modified at each iteration during the inversion, the error minimization process is guided by the sensitivity of the integrated model. To compare the joint inversion of galvanic/induction resistivity and dielectric dispersion logs against the inversion of only the dielectric dispersion logs, we study the

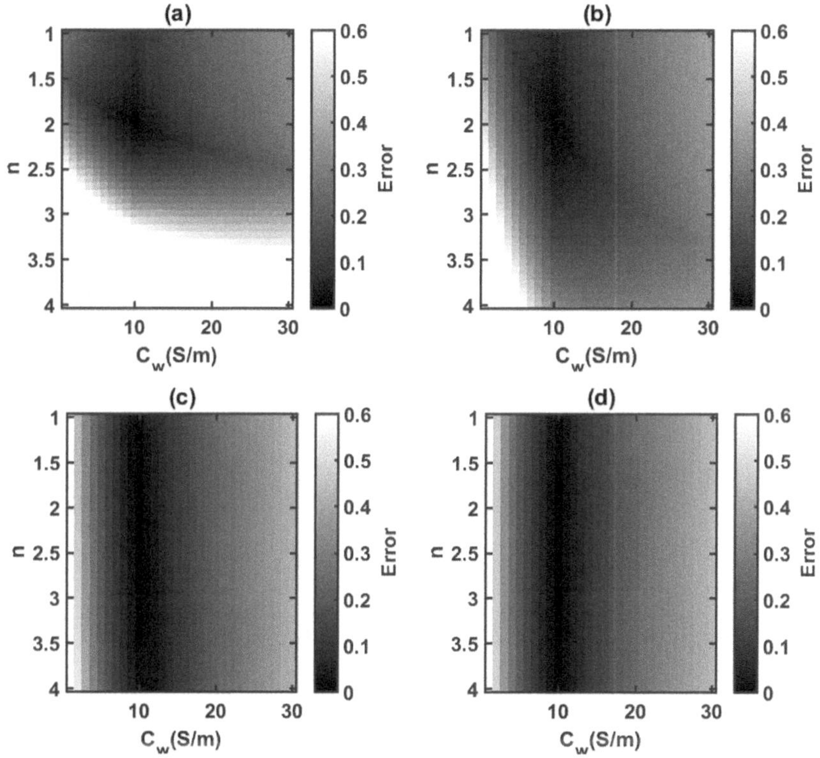

FIGURE 9.4 Comparison of sensitivity of the integrated CRIM, SMD and WS model to variations in n and C_w for (A) $S_w = 0.2$, (B) $S_w = 0.5$, (C) $S_w = 0.9$, and (D) $S_w = 0.99$ around default parameters: $n = 2$, $m = 2$, $\phi_T = 0.05$, $C_w = 10$ S/m, and $BQ_v = 1.3$ mS/cc. Comparison of Fig. 9.4A–D reveals the sensitivity to variations in n and C_w at various values of S_w. *Source: Authors.*

sensitivities of integrated model #1 and integrated model #2 to variations in S_w, C_w, m, and n around a global minimum. Figs. 9.2 and 9.3 show the relative-change contours (relative errors with respect to the global minimum) computed using the integrated models that ultimately direct the error minimization during the inversion.

Fig. 9.2A and B represent low-water saturation cases, where the global minimum is at $S_w = 0.3$ s.u. and $C_w = 10$ S/m and model sensitivity is studied for variations in S_w and C_w around $m = 2$, $n = 2$, $C_w = 10$ S/m, $\phi_T = 0.02$, and $S_w = 0.3$ s.u. Regions with high density of local minima around the global minimum are shown as the darker regions. The spread of such regions is indicative of the uncertainty in inversion-derived estimates for a specific case. At low-water saturation (Fig. 9.2A and B), the proposed joint inversion (Fig. 9.2B) shows steeper slope around global minima, as compared to the integrated SMD and CRI model (Fig. 9.2A). Fig. 9.2C and D denote the

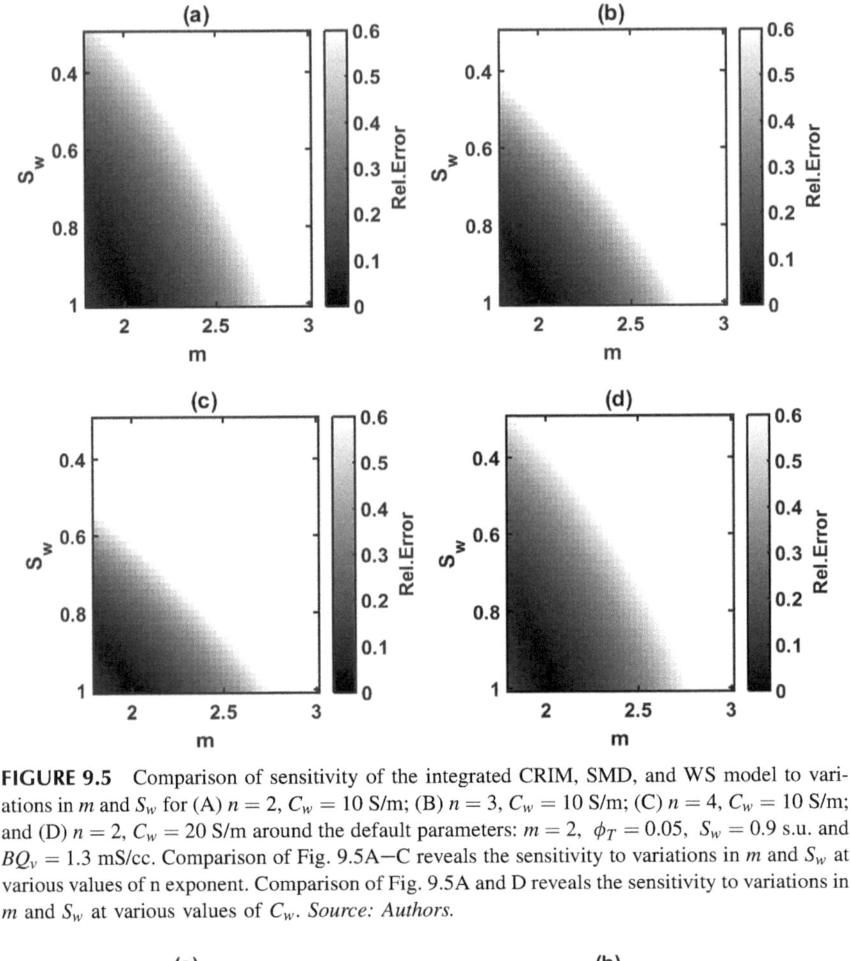

FIGURE 9.5 Comparison of sensitivity of the integrated CRIM, SMD, and WS model to variations in m and S_w for (A) $n = 2$, $C_w = 10$ S/m; (B) $n = 3$, $C_w = 10$ S/m; (C) $n = 4$, $C_w = 10$ S/m; and (D) $n = 2$, $C_w = 20$ S/m around the default parameters: $m = 2$, $\phi_T = 0.05$, $S_w = 0.9$ s.u. and $BQ_v = 1.3$ mS/cc. Comparison of Fig. 9.5A–C reveals the sensitivity to variations in m and S_w at various values of n exponent. Comparison of Fig. 9.5A and D reveals the sensitivity to variations in m and S_w at various values of C_w. *Source: Authors.*

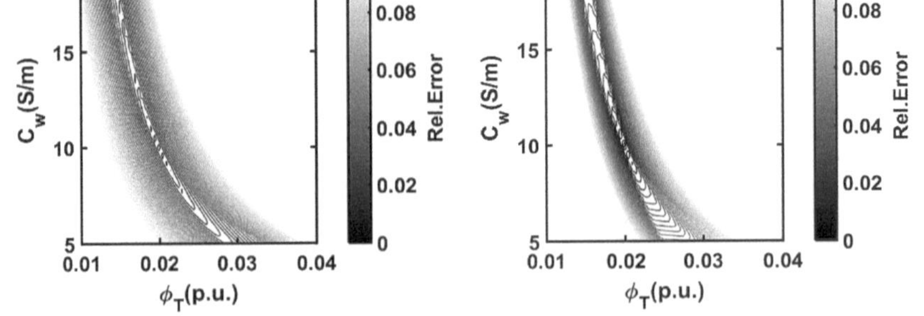

FIGURE 9.6 Comparison of sensitivity of (A) integrated CRIM and SMD model for dielectric dispersion inversion against that of (B) integrated CRIM, SMD, and WS model for the proposed joint inversion of galvanic/induction resistivity and dielectric dispersion logs to the variations in ϕ_T and C_w around default parameters: $n = 2$, $m = 2$, $\phi_T = 0.02$, $C_w = 10$ S/m and $S_w = 0.3$. The fast rate of change shown by the contours is indicative of high sensitivity. A model with higher sensitivity will exhibit sharp change in the grayscale from black to white. *Source: Authors.*

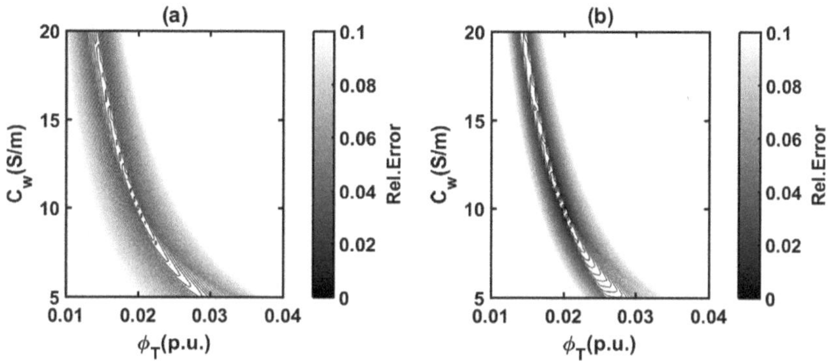

FIGURE 9.7 Comparison of sensitivity of (A) integrated CRIM and SMD model for dielectric dispersion inversion against that of (B) integrated CRIM, SMD, and WS model for the proposed joint inversion of galvanic/induction resistivity and dielectric dispersion logs to the variations in ϕ_T and C_w around default parameters: $n = 2$, $m = 2$, $\phi_T = 0.02$, $C_w = 10$ S/m and $S_w = 0.6$. A model with higher sensitivity will exhibit sharp change in the grayscale from black to white because of the faster change in the contours. *Source: Authors.*

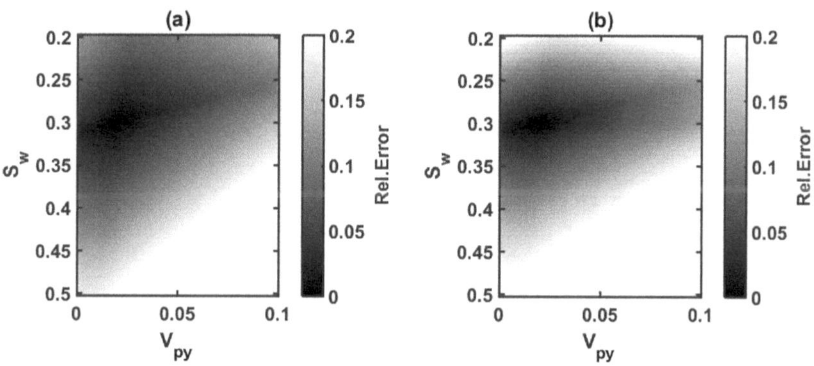

FIGURE 9.8 Comparison of sensitivity of (A) the integrated CRIM and SMD model used for dielectric dispersion inversion against that of (B) the integrated CRIM, SMD, and WS model used for the proposed joint inversion to variations S_w and V_{py} around the default parameters: $n = 2$, $m = 2$, $\phi_T = 0.02$, $C_w = 10$ S/m, $V_{py} = 0.02$, $BQ_v = 1.3$ mS/cc, and $S_w = 0.3$ s.u. Fig. 9.8 represents the model sensitivity at low water saturation and high volume fraction of pyrite. *Source: Authors.*

high water-saturation cases, where the global minimum is at $S_w = 0.9$ s.u. and $C_w = 10$ S/m keeping all the other parameters equal to the previous low water-saturation cases. At high-water saturation, the slopes around global minimum are relatively uniform and flatter compared to the lower water saturation cases. A comparison between Fig. 9.2A–D indicates that the proposed joint inversion of galvanic/induction resistivity and dielectric dispersion logs can generate the desired estimates more reliably at low-water saturations than at high-water saturations. Moreover, the joint inversion is much better than inversion of only the dielectric dispersion logs.

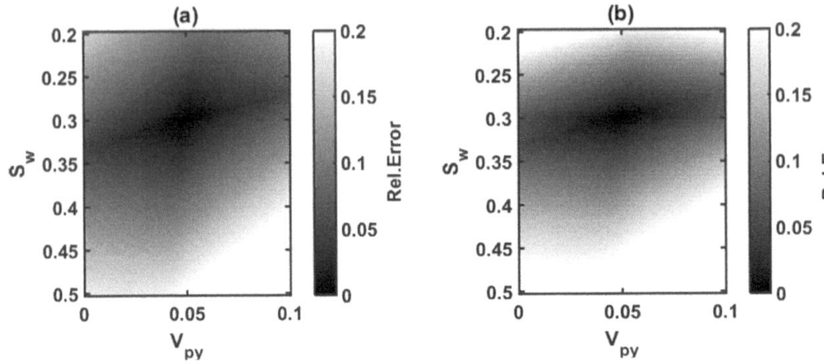

FIGURE 9.9 Comparison of sensitivity of (A) the integrated CRIM and SMD model used for dielectric dispersion inversion against that of (B) the integrated CRIM, SMD, and WS model used for the proposed joint inversion to the variations in S_w and V_{py} around default parameters: $n = 2$, $m = 2$, $\phi_T = 0.02$, $C_w = 10$ S/m, $V_{py} = 0.05$, $BQ_v = 1.3$ mS/cc, and $S_w = 0.6$ s.u. Compared to Fig. 9.8, Fig. 9.9 represents the model sensitivity at high water saturation and higher volume fraction of pyrite. *Source: Authors.*

For the purpose of comparing the sensitivities of integrated model #1 and integrated model #2, it is critical to understand the error minimization process when m and n change around the global minimum. To that end, we generate Fig. 9.3 similar to Fig. 9.2. The integrated model #1 combining WS, CRI, and SMD models for the proposed joint inversion generates well-defined error contours around the global minimum that will better direct the error minimization process during the joint inversion as compared to those obtained using the integrated model #2 combining CRI and SMD models. A comparison between Fig. 9.3A and B indicates the substantial improvement when the integrated WS, SMD, and CRI model is implemented instead of the integrated SMD and CRI model. In Fig. 9.3A, the spread of errors is very flat around the global minimum indicating that the error minimization process will not have enough resolution to generate unique solutions. Fig. 9.3A and B represent low water-saturation cases, where the global minimum is at $m = 2$ and $n = 2$ for $S_w = 0.3$ s.u. On the other hand, Fig. 9.3C and D represent high water-saturation cases with $S_w = 0.9$ s.u., while keeping other parameters of the global minimum similar to the low water-saturation cases. A comparison between Fig. 9.3B and D indicates that the proposed joint inversion can generate the desired estimates more reliably at low-water saturations and the inversion-derived estimates will tend to be nonunique at high-water saturations of around 0.9 s.u. This is similar to the observations for variations in S_w and C_w presented in the previous section. Similarly, in this case, with increase in water saturation, the sensitivity of the integrated WS, SMD, and CRI model to variation in m is significantly deteriorated while maintaining the sensitivity to variation in n. Deterioration of sensitivity of the integrated model to a specific model parameter indicate increased uncertainty in the corresponding estimates obtained from log processing. Overall, both Figs. 9.2 and 9.3 prove that error minimization during the inversion will be much easier when integrated model #1 combining WS, SMD, and CRI models is used for the proposed joint inversion of galvanic/induction resistivity and dielectric dispersion logs. This is evident from the comparison of Fig. 9.3A and B.

3.2 Uncertainties in inversion-derived estimates

After demonstrating that the newly proposed joint inversion will perform better compared to inverting only the dielectric dispersion logs, we test the sensitivity of the integrated CRIM, SMD, and WS model to variations in n and C_w for various S_w. Sensitivity of the integrated model to the petrophysical parameters, in this case n and C_w, is inversely correlated to the uncertainty in the parameter estimation when performing the joint inversion because the model sensitivity directly affects the error minimization process. A highly sensitive model will facilitate the error minimization. The sensitivity of the model is denoted by sharp change in the grayscale surface plot from black to white. In Fig. 9.4, global minimum is at default parameters: $n = 2$ and $C_w = 10$ S/m, identified as the dark region. Variations in n and C_w result in change in the relative error of the integrated model response with respect to the response obtained using the default parameters. It is evident from Fig. 9.4A–D that with increase in water saturation from 0.2 to 0.99 saturation units (s.u.), the sensitivity of the integrated model to n and C_w decreases. Consequently, the uncertainty in the estimations of n and C_w should increase with increase in water saturation. At $S_w = 0.9$, the joint inversion should generate unreliable nonunique estimates (Fig. 9.4C). At high-water saturation, the sensitivity of the integrated model decreases more drastically for n in comparison to C_w. Inversion-derived estimates for C_w will be slightly more reliable at higher water saturation as compared to lower water saturation. Inversion-derived estimates for C_w will be more reliable than that for n.

Fig. 9.5 illustrates the sensitivity of the integrated model to variations in m and S_w for various n and C_w. Global minimum is at default parameters: $m = 2$ and $S_w = 0.9$ s.u., identified as the dark region in Fig. 9.5. It is evident from Fig. 9.5A–C that with increase in n from 2 to 4, the sensitivity of the integrated model #1 (combining CRI, SMD, and WS models) to m and S_w increases; consequently, the uncertainty in the estimations of m and S_w should decrease. An increase in n relates to the increase in oil-wetness of the formation. The improvement in sensitivities for the two parameters with increase in n is relatively equal. With increase in C_w (Fig. 9.5D), there is not much change in the model sensitivity to m and S_w, as observed when comparing Fig. 9.5A and D. Inversion-derived estimates for m and S_w have relatively similar uncertainty.

In Fig. 9.6, the global minimum is at $\phi_T = 0.02$ and $C_w = 10$ S/m for $S_w = 0.3$. When ϕ_T and C_w are changed, the integrated WS, SMD, and CRI model for the proposed joint inversion generates well-defined error contours (Fig. 9.6B) around the global minimum that directs the error minimization during the joint processing of the nine EM logs. However, when using only SMD and CRI model to invert only the dielectric dispersion log, the error contour plots exhibit several local minima in Fig. 9.6A. The problem of uncertainty in the estimate increases with increase in water saturation as shown in Fig. 9.7. Fig. 9.7 shows the error contours when ϕ_t and C_w are changed around the global minimum of $\phi_T = 0.02$ and $C_w = 10$ S/m at $S_w = 0.6$. In Figs. 9.6A and 9.7A, the local minima become more prominent and widespread, especially in the case of Fig. 9.7A, which indicates a lower sensitivity of the integrated SMD and CRI model for inversion of only the dielectric dispersion logs.

3.3 Effect of pyrite in low-porosity, high-salinity formations

Errors in hydrocarbon saturation estimates from EM logs are largest for low-porosity, low-water-saturation, high-salinity formations (Seleznev et al., 2014). Presence of pyrite in the formation leads to the underestimation of hydrocarbon saturation (Clennell et al., 2010), especially when deriving them using dielectric dispersion logs. We analyze the effect of the presence of pyrite in low-porosity, high-salinity formation on the proposed joint inversion in comparison to that on the inversion of only dielectric dispersion log. To that end, we compare the sensitivity of the two integrated models, one for the proposed joint inversion and the other for the dielectric dispersion inversion, to the variations in S_w and volume fraction of pyrite (V_{py}) in the range of 0%−10%. In Figs. 9.8 and 9.9, the color of the map changes from white to dark indicating a reduction in the magnitude of the errors. Comparison of Fig. 9.8A and B indicates that the integrated CRIM, SMD, and WS model used for the proposed joint inversion is more sensitive to variation in volume fraction of pyrite compared to the integrated CRIM and SMD model used for dielectric dispersion inversion. Fig. 9.8A has a wider spread of dark region indicating lower sensitivity and higher uncertainty. Consequently, in the presence of pyrite, the proposed joint inversion is more reliable than the inversion of only dielectric dispersion logs. Based on Figs. 9.8B and 9.9B, when initializing the inversion scheme, faster convergence can be achieved by initializing the inversion with a low value of water saturation because the skew in the spread of the dark regions. Comparison of Figs. 9.8B and 9.9B indicates that the sensitivity of the integrated model for joint inversion is relatively invariant to the change in V_{py}.

3.4 Threshold of brine conductivity in Waxman–Smits, Stroud-Milton-De, and complex refractive index model

As stated earlier, at certain salinity, the dielectric measurements lose sensitivity to porosity and water saturation. The threshold of brine conductivity is temperature dependent and depends on the brine conductivity rather than salinity. We studied the sensitivity, or the rate of change of resistivity, permittivity and conductivity modeled using the integrated WS, SMD, and CRI model to brine conductivity for the nine EM log acquisition frequencies. We vary the brine conductivity in the range of 0 S/m to 100 S/m and normalize the derivative plot to generate the four permittivity dispersion logs, four conductivity dispersion logs, and one resistivity log. We analyze three distinct scenarios of water-rich layer of $S_w = 0.9$ s.u., oil-rich layer of $S_w = 0.1$ s.u., and clay-rich layer of $S_w = 0.9$ s.u. and volume of clay $V_c = 40\%$ at formation temperature of 230°F (Fig. 9.10). The permittivity at the three frequencies in the MHz range is the least sensitive to variation in C_w for C_w>40 S/m. Permittivity at 1 GHz shows high sensitivity to variations in C_w. It is evident that at high-water saturation (Fig. 9.10A), the derivative plot for conductivity decreases drastically till 20 S/m and then starts decreasing gradually till 40 S/m. After 40 S/m, the change in derivative plot for conductivity is negligible, which suggests that the integrated model #1 loses sensitivity to brine conductivity beyond this point. The derivative plot of resistivity decreases drastically till 100 S/m. Compared to water-rich interval, in the oil-rich interval, only resistivity and permittivity at 1 GHz are sensitive to C_w beyond 40 S/m (Fig. 9.10B). For clay-rich interval at high-water saturation in Fig. 9.10C, the results are relatively similar to the water-rich layer (Fig. 9.10A). With this study of sensitivity analysis, the upper bound of brine conductivity for purposes of our inversion was chosen to be 45 S/m.

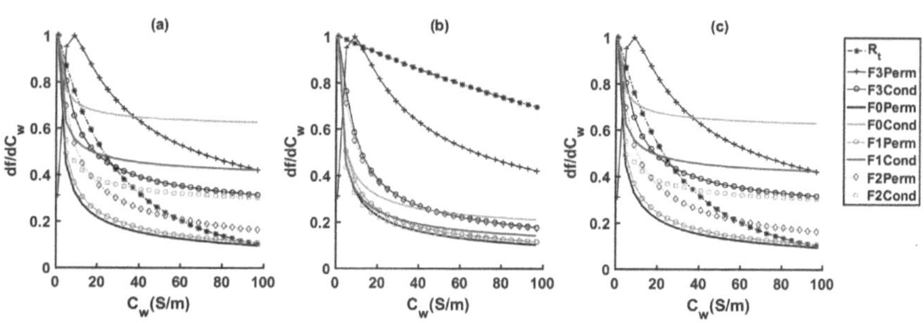

FIGURE 9.10 Comparison of sensitivity of resistivity at 20 kHz and permittivity and conductivity dispersions at frequencies: F0 (22 MHz), F1 (100 MHz), F2 (350 MHz), and F3 (1 GHz) to brine conductivity for (A) water rich layer of $S_w = 0.9$ s.u., (B) oil-rich layer of $S_w = 0.1$ s.u., and (C) clay-rich layer of $S_w = 0.9$ s.u. and $V_c = 40\%$ with default parameters $m = 2$, $n = 2$, $\phi_T = 0.07$, and $BQ_v = 75$ mS/cc. *Source: Authors.*

4. Application of the newly proposed joint inversion on synthetic electromagnetic data

Robustness of the proposed joint inversion could be assessed by inverting synthetic multifrequency EM data generated using the integrated model combining CRI, SMD, and WS models. After the evaluation on synthetic data, the proposed joint inversion is applied on multifrequency EM logs acquired in upper Wolfcamp shale formation and BPS.

The integrated model combining CRI, SMD, and WS models is used to generate synthetic EM data assuming porosity of 5% and brine conductivity of 10 S/m (approximating the upper Wolfcamp formation). Table 9.1 lists petrophysical properties assumed when generating the synthetic EM data, comprising 1 resistivity log at laterolog frequency and 8 permittivity/conductivity dispersion logs at the 4 dielectric-dispersion log frequencies. About 10% Gaussian noise was added to the synthetic EM data. These simulated conductivity, permittivity, and resistivity logs were processed using the proposed joint inversion coupled with the integrated CRIM, SMD, and WS model to estimate C_W, S_W, n, and m, similar to (Engle et al., 2016; Han and Misra, 2017). Inversion-derived estimates of the unknown model parameters, namely C_W, S_W, n, and m, match with the assumed values listed in Table 9.1.

Fig. 9.11 shows the histogram plot showing the frequency of occurrence of the four inversion-derived estimates for the 250 random initializations of the

TABLE 9.1 Values of petrophysical properties assumed for generating the synthetic multifrequency electromagnetic data used in Section 4. The last four rows are the four model parameters to be estimated using the proposed joint inversion.

Properties	Values
Total porosity	5%
Volume fraction of clay	25%
Cation exchange capacity per unit pore volume, Q_v	0.1 meq/cc
Equivalent conductance of sodium clay exchange cations, B	13 mS/meq
Relative permittivity of hydrocarbon	3.5
Relative permittivity of brine	80
Homogeneity index, α	0.5
Bulk conductivity of brine, C_W	10 S/m
Water saturation, S_W	30%
Cementation exponent, m	2
Saturation exponent, n	2

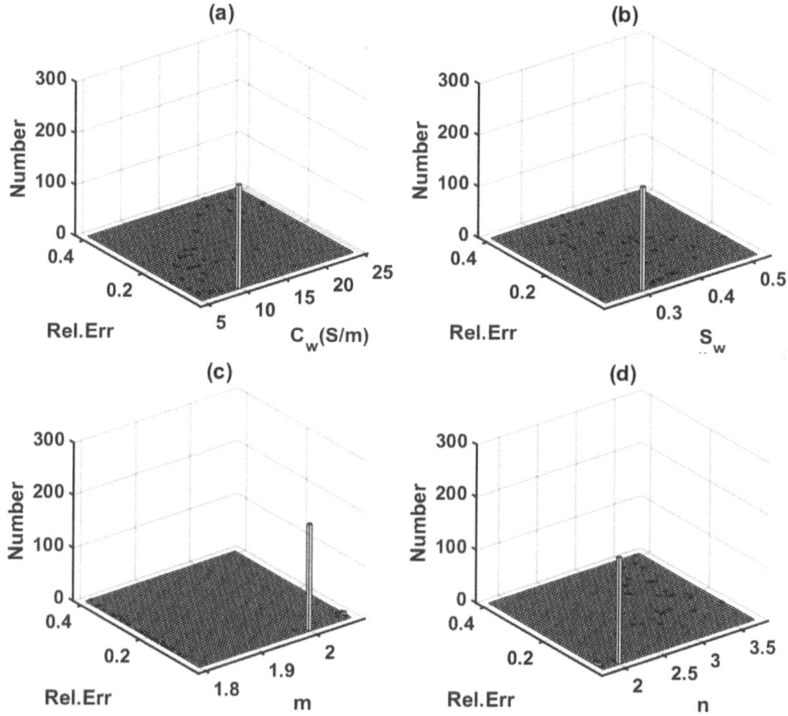

FIGURE 9.11 Frequency of occurrence of the inversion-derived estimates of (A) C_w, (B) S_w, (C) $m = 2$, and (D) $n = 2$ and the corresponding relative error in fitting the synthetic multi-frequency electromagnetic data as computed by the proposed joint inversion using 250 random initializations. Relative errors of the modes of inversion-derived estimates are close to zero. *Source: Authors.*

inversion scheme. For the synthetic case study, the inversion results show fast and robust convergence to desired values. Relative error in matching the synthetic data using the inversion-derived estimates is 0.01. The good agreement between the synthetic data and the integrated model predictions based on the inversion-derived estimates is illustrated in Fig. 9.12. For the 520-feet interval of upper Wolfcamp formation, no core data were available. The operating company did not retrieve any core samples in this interval. Core water saturation for other layers were available, but they were invaded with the invasion fluid. These water saturation values were very low because of the oil-mud invasion and cannot not be trusted (Tathed, 2018).

5. Application of the newly proposed joint inversion on multifrequency electromagnetic logs from Wolfcamp shale formation

The Delaware basin forms the western sub-division of Permian Basin of west Texas and southeast New Mexico extending over 10,000 square miles

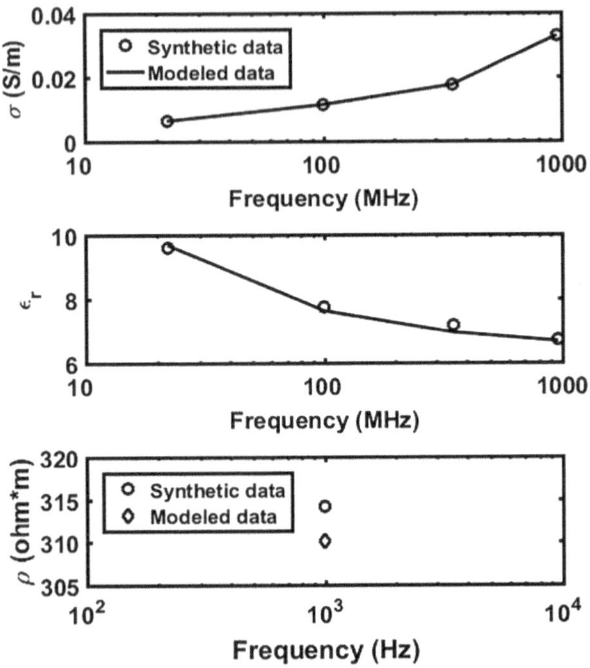

FIGURE 9.12 Comparison of synthetic effective conductivity (σ), synthetic relative permittivity (ε_r), and synthetic resistivity (ρ) data against the integrated model predictions based on the inversion-derived estimates. *Source: Authors.*

(Han and Misra, 2017). In the Delaware Basin, the Wolfcamp shale play forms one of the largest and complex unconventional reservoirs in the United Sates. Upper Wolfcamp formation is classified as a tight oil reservoir, with permeability usually in microdarcies and porosities ranging from 0.01 to 0.15 p.u. Single-frequency galvanic resistivity (laterolog) and multifrequency dielectric dispersion logs, comprising four conductivity and four dielectric permittivity logs measured at four distinct frequencies were acquired in few wellbores drilled in Wolfcamp formations. The joint inversion scheme proposed in Section 2 was applied to dielectric dispersion and laterolog resistivity logs acquired in a well in upper Wolfcamp shale formation. The proposed log-processing workflow uses an integrated mechanistic model, which combines CRI model to analyze the conductivity and permittivity logs acquired at 1 GHz, SMD model to analyze the three conductivity-dispersion logs and three permittivity-dispersion logs in the frequency range of 10 MHz to 0.3 GHz, and WS model to analyze the single-frequency deep galvanic resistivity log (RLA5) acquired at 1 kHz. We implemented this log-interpretation technique for the improved estimation of water saturation (S_w), brine conductivity (C_w), textural index/cementation exponent (m), and saturation exponent (n).

5.1 Geology of Wolfcamp shale formation

Upper Wolfcamp comprises mostly of shale, varying from almost black to gray and greenish gray with several interbedded limestone and calcareous sandstone layers. A continuous estimation of S_w, C_w, m, and n was carried out across the 520-ft depth interval in upper Wolfcamp formation. Unlike conventional interpretation methods, 250 random initializations were performed for each depth to find the global minimum and the range of uncertainties in the estimates. Estimated time required for the proposed inversion method to process the logs acquired from 520-feet depth interval was about 20 h.

Figs. 9.13–9.15 contain logs of inversion-derived estimates in the upper Wolfcamp formation. Fig. 9.13 represents top part of upper Wolfcamp formation from XX060-XX100 feet. This portion mainly comprises of 60%–80% quartz, 10%–20% clay, and calcite (Track 5) and negligible TOC. This interval of Wolfcamp formation has low resistivity on the laterologs (<10 Ω m). There are several laminations and beds exhibiting low porosity (Track 4). These low-porosity streaks are primarily calcite cementation (Track 5). Negligible effect of invasion was observed at different depths of investigation of laterolog resistivity measurements (Track 3). Therefore, we assume that the water saturations sensed by dielectric dispersion logs and resistivity log are equivalent in absence of bed boundaries. With an increase in depth deeper into the well, the upper Wolfcamp formation mainly consist of tight carbonates (Fig. 9.14, XX330–XX360 feet) as calcite content increases (Fig. 9.14, track 5). Formation microimages shows presence of healed, resistive fractures, and weak dissolution features in these tight carbonate formations. On triple combo logs, these intervals have low density (<2.55 gm/cc) and elevated resistivity (>30 Ω m). These depths exhibit small pores and tight formations as observed from the NMR T2 distribution (Track 13, Fig. 9.14). With further increase in depth (Fig. 9.15), there are thin laminations and interbeds of calcite and clays. TOC ranges from 2% to 3%. There are several streaks of high resistivity, low-porosity formations with high-calcite deposition. These depths exhibit wide range of pore size distribution as exhibited in Track 13 of Fig. 9.15.

5.2 Formation evaluation based on joint inversion of galvanic resistivity and dielectric dispersion logs

In the depth interval from XX060 to XX110 ft, S_w estimates obtained using the proposed inversion (Fig. 9.13, Track 7) are in good agreement with those estimated using the multimineral inversion performed by an O&G service company (Fig. 9.13). Inversion-derived S_w estimates are more than 20% higher than those estimated by the service company's multimineral inversion for depth intervals XX330–XX360 ft (Fig. 9.14) and XX430–XX460 ft (Fig. 9.15). Cementation exponent estimates computed using the inversion

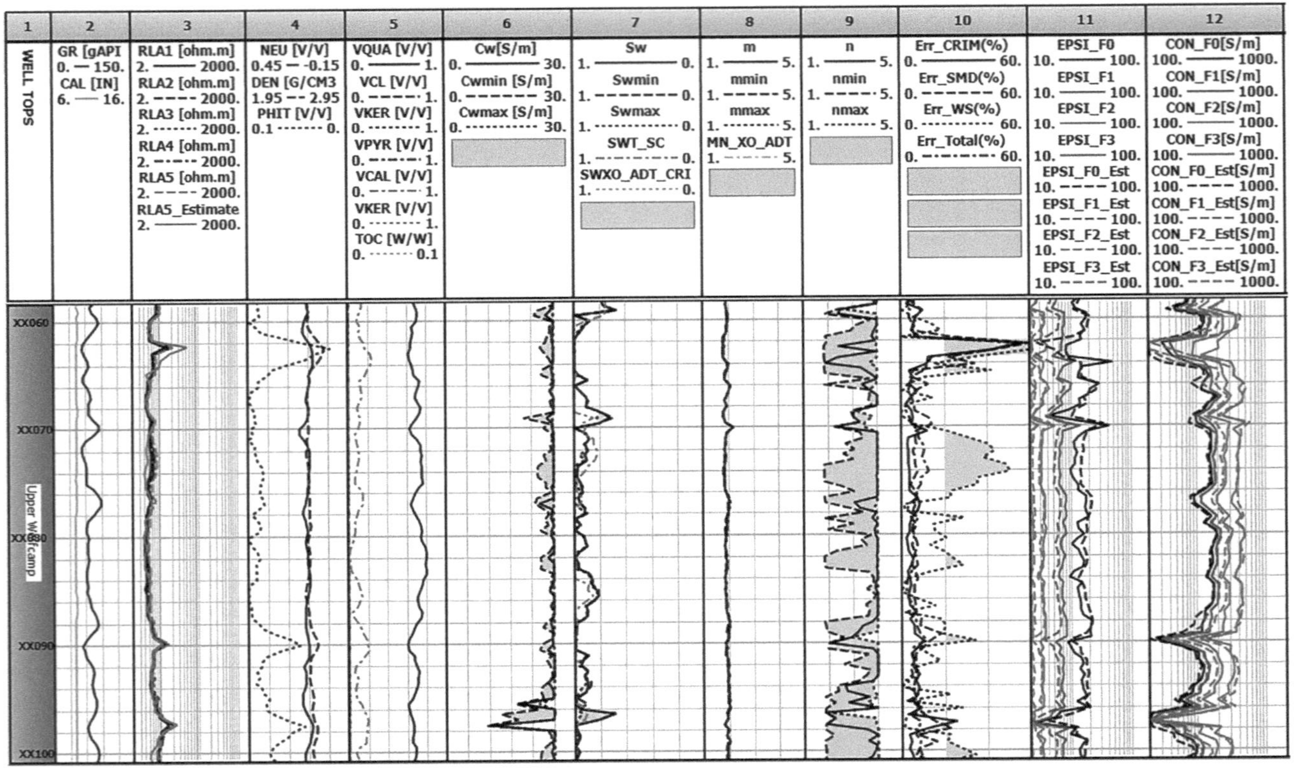

FIGURE 9.13 Track 1 is the depth (XX060–XX100) and formation type, Track 2 is the gamma ray (GR) and caliper (CAL), Track 3 is the laterolog resistivity (RLA), Track 4 contains the neutron porosity (NEU), bulk density (DEN), and total porosity (PHIT), Track 5 contains volume fractions of various minerals: quartz, chlorite, kerogen, pyrite, and calcite, Track 6 contains inversion-derived brine conductivity (C_w) estimate with a range, Track 7 contains water saturation estimated using various methods (S_w), Track 8 contains inversion-derived cementation exponent (m) estimate with a range, Track 9 contains inversion-derived saturation exponent (n) estimate with a range, Track 10 contains relative errors in the predictions of various models in the integrated model, Track 11 contains measured (bold) and modeled (dash) permittivity dispersion logs at four frequencies, and Track 12 contains measured (bold) and modeled (dash) conductivity logs at four frequencies. *Source: Authors.*

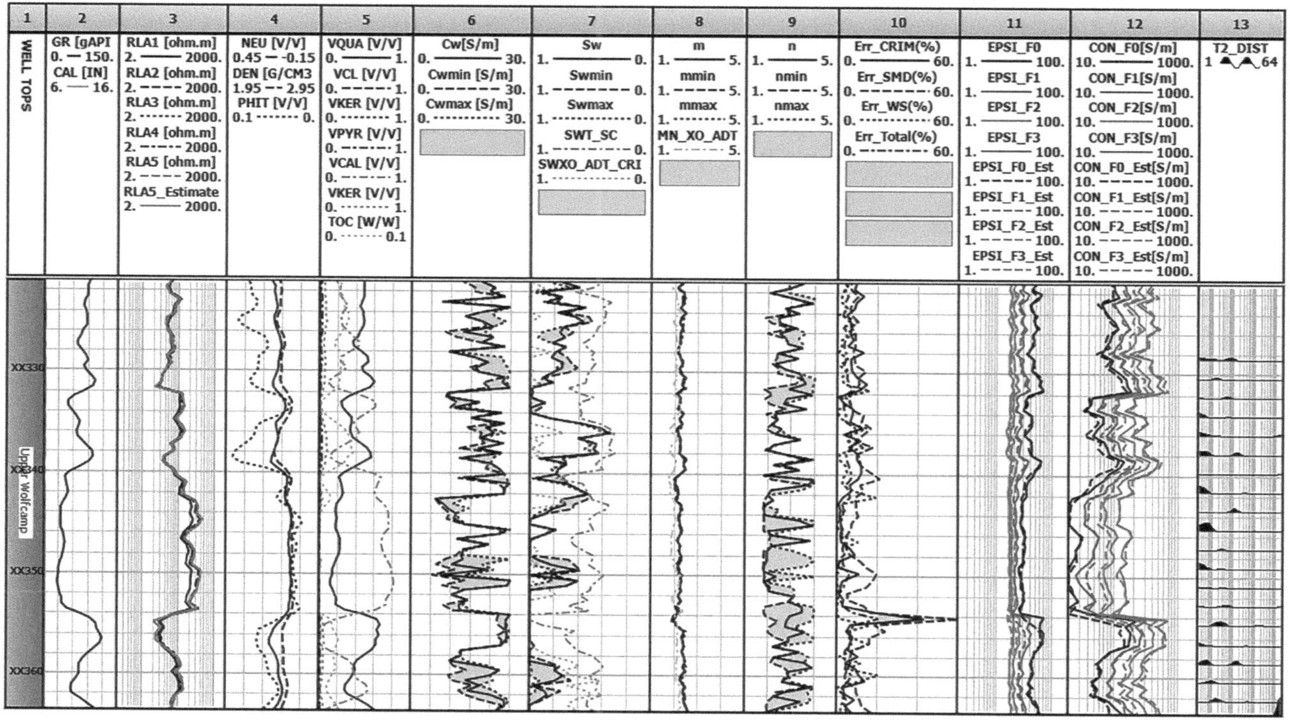

FIGURE 9.14 Logs for formation depth from XX322 to XX360. The logs and the layout presented in this figure are similar to those in Fig. 9.13. Track 13 contains T2 NMR distribution. *Source: Authors.*

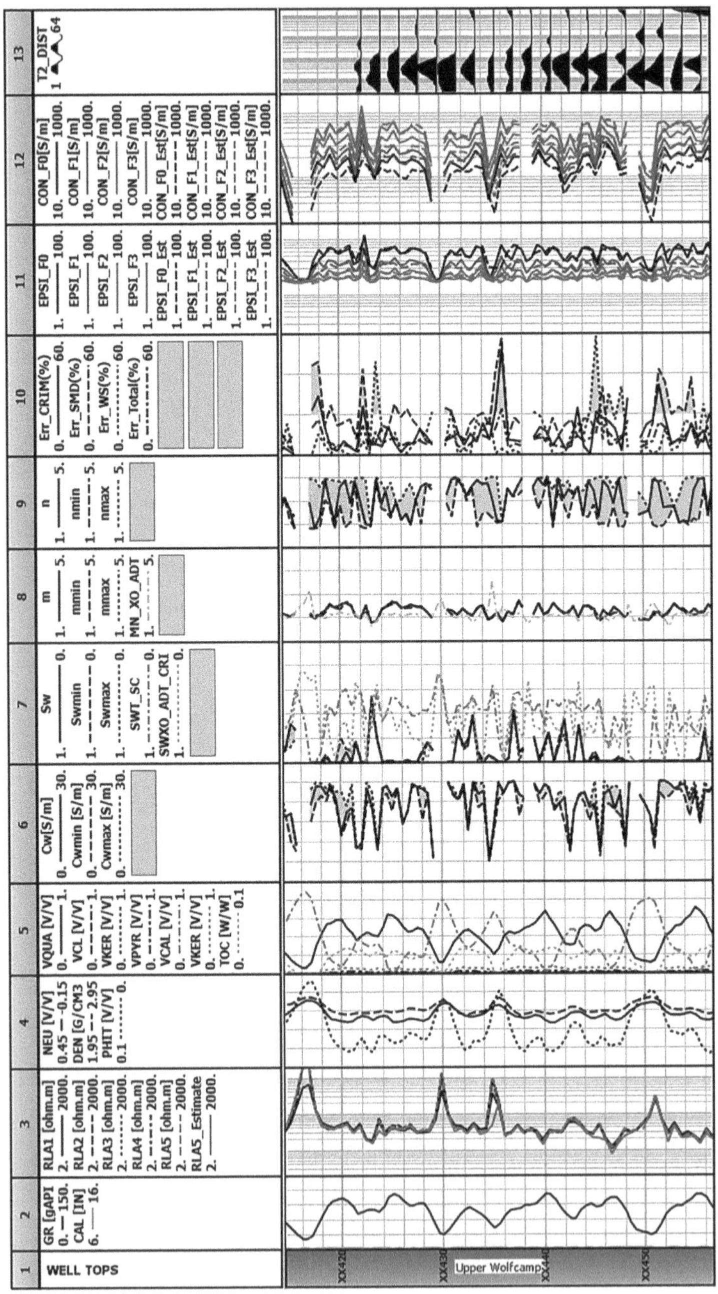

FIGURE 9.15 Logs for formation depth from XX412 to XX450. The logs and the layout presented in this figure are similar to those in Fig. 9.13. *Source: Authors.*

matches with the estimates obtained by the service company, as shown in Track 8 of Fig. 9.13. Estimates of cementation exponent obtained using the proposed inversion exhibits higher variations with the increase in depth, as shown in Track 10 of Fig. 9.15 in comparison to that of Fig. 9.13, which correlates with variations in clay content and porosity. Notably, based on the sensitivity analysis presented in Fig. 9.14, uncertainty in saturation-exponent estimates increase with increase in water saturation. Inversion-derived estimates of water saturation are maximum at depth XX098–XX104 feet, and in that depth interval the saturation exponent is the most uncertain. Similar to the sensitivity analysis performed in the earlier sections, water saturation and cementation exponent estimates have higher certainty in comparison to saturation exponent and brine conductivity estimates, as shown in Figs. 9.13–9.15.

In Fig. 9.14, at depth XX320–XX360 feet, high uncertainty observed in brine conductivity estimates is primarily because the integrated model is not able to account for the polarization mechanism that is resulting in large frequency-dependence of the effective conductivity measured by the dielectric dispersion tool (Track 12). Uncertainty in C_w estimates in the formation shown in Fig. 9.13 is low that correlates with the low-frequency dependence of the effective conductivity logs. Also, based on the sensitivity analysis, the proposed joint inversion shows unreliable estimates at high-water saturation and low-brine conductivity. From depth XX330–XX360 feet, uncertainty at high-water saturations can be observed. The proposed inversion achieves high certainty for the estimates of water saturation, brine conductivity, and cementation exponent from depth XX060–XX100 feet. This is a consequence of low-clay content and high porosity in that depth interval. Our estimates of brine conductivity and water saturation are uncertain in tight calcareous formations from XX330–XX360 feet. These unreliable estimates could be because of small pores and low-porosity beds in that depth interval.

5.3 Analysis of interpretation errors

Data misfit is calculated as the difference between the modeled and measured dielectric dispersion and resistivity logs. The logs are modeled based on inversion-derived estimates obtained using the proposed inversion scheme coupled with the integrated WS, SMD, and CRI model. Error in fitting the conductivity and permittivity logs acquired at 1 GHz using the CRI model predictions is referred as the $Error_{CRI}$. Error in matching the three conductivity and three permittivity dispersion logs acquired at 20, 100, and 260 MHz, respectively, using the SMD model predictions is referred as $Error_{SMD}$. Finally, the error in fitting the resistivity laterolog using WS model predictions is referred as $Error_{WS}$. Data misfit averaged over the one resistivity, four conductivity, and four permittivity logs is reported as the mean relative error, which is expressed as

$$\text{Error}_{\text{Total}} = \left[\frac{\text{Error}_{SMD} + \text{Error}_{CRI} + \text{Error}_{WS}}{5}\right], \tag{9.7}$$

such that

$$\text{Error}_{CRI} = \frac{\left[\dfrac{abs\left(Perm^{\text{meas,1GHz}} - Perm^{\text{mod,1GHz}}\right)}{Perm^{\text{meas,1GHz}}} + \dfrac{abs\left(Cond^{\text{meas,1GHz}} - Cond^{\text{mod,1GHz}}\right)}{Cond^{\text{meas,1GHz}}}\right]}{2} 100\% \tag{9.8}$$

$$\text{Error}_{SMD} = \sum_{i=0}^{2} \frac{\left[\dfrac{abs\left(Perm^{\text{meas},f_i} - Perm^{\text{mod},f_i}\right)}{Perm^{\text{meas},f_i}} + \dfrac{abs\left(Cond^{\text{meas},f_i} - Cond^{\text{mod},f_i}\right)}{Cond^{\text{meas},f_i}}\right]}{6} 100\% \tag{9.9}$$

$$\text{Error}_{WS} = \frac{abs\left(Res_{\text{mod,1kHz}} - Res_{\text{meas,1kHz}}\right)}{Res_{\text{meas,1kHz}}} 100\% \tag{9.10}$$

Track 10 in Figs. 9.13–9.15 contains the various data misfits in matching the corresponding logs with CRI model, SMD model, and WS model predictions. In Fig. 9.13, the depth interval XX060–XX080 exhibits large error because of data misfit in modeling the laterolog resistivity, which on an average range around 40%. This interval exhibits low resistivity and high-porosity formation with large volume fraction of quartz; these three characteristics correlate with zones where the Error_{WS} is large. Volume of clay content is low in these formations and high Error_{WS} may be a consequence of the assumed value of 1.3 for BQ_v for the entire formation. In Fig. 9.14, depth interval XX343–XX354 feet shows highest interval of low porosity at an average of 2.5 porosity units. Error in fitting laterolog resistivity log(Error_{WS}) and that in fitting the highest-frequency dielectric log(Error_{CRI}) is lowest in this interval. Error in fittingthe three lowest-frequency dielectric dispersion logs (Error_{SMD}) has the highest value in this interval. At the depth XX353.5 feet, Error_{SMD} is about 60% along with Error_{WS} attaining a large value of 50%, which might be a consequence of thin laminations and sand-shale bed boundary effects at that depth. High errors are also observed at XX418, XX436, and XX452 feet in Fig. 9.14 because of the errors in SMD and CRIM models. These high errors could be because the integrated model is not able to account for the polarization mechanism that gives rise to a large frequency-dependence of the effective conductivity measured by the dielectric dispersion tool.

6. Application of the newly proposed joint inversion on multifrequency electromagnetic logs from Bakken Petroleum System

The BPS, one of the largest tight oil deposits in the world, is situated in the Williston Basin that covers parts of North and South Dakota, Montana, and southern Canada (Hackley and Cardott, 2016). BPS consists of two layers of organic-rich shale (the Upper and Lower shales) sandwiching a dolomitic-siltstone interval of 45 ft thickness called the Middle Bakken. The Lodgepole formation overlying the Upper Bakken Shale mainly consists of limestone. Specifically, Scallion is the basal member of Lodgepole formation which overlies the Upper Bakken Shale. Included in the BPS below the Lower shale is the Three Forks formation, composed of laminations of dolostone and dolomitic-mudstone. There exist multifold complexities in formation evaluation, including hydrocarbon saturation estimation. Reservoirs in the Bakken Formation are characterized with low porosities and permeabilities, ranging from 5 to 15 pu and 1 to 20 mD, respectively (Kreis and Costa, 2005). Single-frequency induction resistivity and multifrequency dielectric dispersion logs, comprising four conductivity and four dielectric permittivity logs measured at four distinct frequencies were acquired in few wellbores drilled in Bakken shale formations. The similar joint inversion scheme proposed in Section 2 was applied to dielectric dispersion and induction resistivity logs acquired in a well in Bakken shale formation. The proposed log-processing workflow uses an integrated mechanistic model, which combines CRI model to analyze the conductivity and permittivity logs acquired at 1 GHz, SMD model to analyze the three conductivity-dispersion logs and three permittivity-dispersion logs in the frequency range of 10 MHz to 0.3 GHz, and WS model to analyze the single-frequency induction resistivity (AT90) log acquired at 20 kHz. We implemented this log-interpretation technique for the improved estimation of water saturation (S_w), brine conductivity (C_w), textural index/cementation exponent (m), and saturation exponent (n).

6.1 Geology of Bakken Petroleum System

The joint inversion scheme was applied to dielectric dispersion and induction resistivity logs acquired in a well in BPS for a continuous 350 ft interval (Tathed et al., 2018b). Unlike conventional interpretation methods, 250 random initializations were performed for each depth to find the global minimum and the ranges of uncertainties for the inversion-derived estimates. Estimated time required for the proposed inversion method to process the logs acquired from 350-feet depth interval was about 15 h.

Fig. 9.16 contains logs of inversion-derived estimates in the Lodgepole formation from XX610–XX670 feet. This portion mainly comprises of 70%–80% limestone with 2%–3% porosity (Track 5 and 6). These low-porosity streaks are primarily calcite cementation (Track 6). Negligible effect of invasion was observed at different depths of investigation of induction resistivity measurements (Track 4). Therefore, we assume that the water saturations sensed by dielectric dispersion logs and resistivity log are equivalent with the assumption that bed boundaries do not exist in the formation under investigation. Vertical resistivity R_v is higher than horizontal resistivity R_h and the resistivity measured by induction tool (Track 4). Lodgepole formation has anisotropy values ranging from six to nine calculated by R_t scanner (Track 7). These depths exhibit small pores and tight formations as observed from the NMR T_2 distribution (Track 15).

6.2 Formation evaluation based on joint inversion of induction resistivity and dielectric dispersion logs

Fig. 9.17 denotes complex reservoir interval of Middle Bakken from XX705.5–XX750 feet and Lower Bakken Shale from XX750–XX766 feet. Lower Bakken shale source rocks are characterized by high volume of clay about 30%–40% and pyrite content about 1%–3% (Fig. 9.17, Track 6). High gamma ray (Fig. 9.17, Track 3) is because of high TOC and high-kerogen content in this formation. Middle Bakken shows range of grain size and sorting from poorly sorted argillaceous siltstone to moderately well sorted fine-grained sandstone. Upper Middle Bakken contains well-sorted, fine grained sandstone that marks best reservoir quality (Pirie et al., 2016). The vertical resistivity R_v in the upper Middle Bakken (XX705.5–XX720 feet) has higher resistivity than horizontal resistivity, R_h and resistivity measured by induction tool (Track 4). Anisotropy value calculated by R_t scanner is around 4 (Track 7). Further in the lower Middle Bakken, from XX720–XX750 feet, negligible anisotropy is observed. Middle Bakken has highly conductive formations as compared to Lodgepole formation (Track 14). Wide range of pore size distribution is illustrated in Middle Bakken (Track 15). With an increase in depth deeper into the well, Three Forks formation mainly consist of dolostone (40%–60%), interbedded with clay-rich dolo-mudstone (Fig. 9.18, track 6). The uppermost sequence, Three Forks 1 (TF1) denotes the oil-bearing interval in Three Forks from XX775–XX820 feet. The petrophysical analysis of Three Forks formation is mainly complex because of presence of thin beds characterized by high-vertical resistivity (Track 4) and anisotropy values as high as 15 (Track 7). At the depth interval XX630.5 and XX654 feet, S_w estimates obtained using the proposed inversion (Fig. 9.16, Track 9) are in good agreement with those estimated using the multimineral inversion performed by an O&G service company and measured by dielectric ADT tool (Fig. 9.16). Inversion-derived S_w estimates are higher than those estimated by

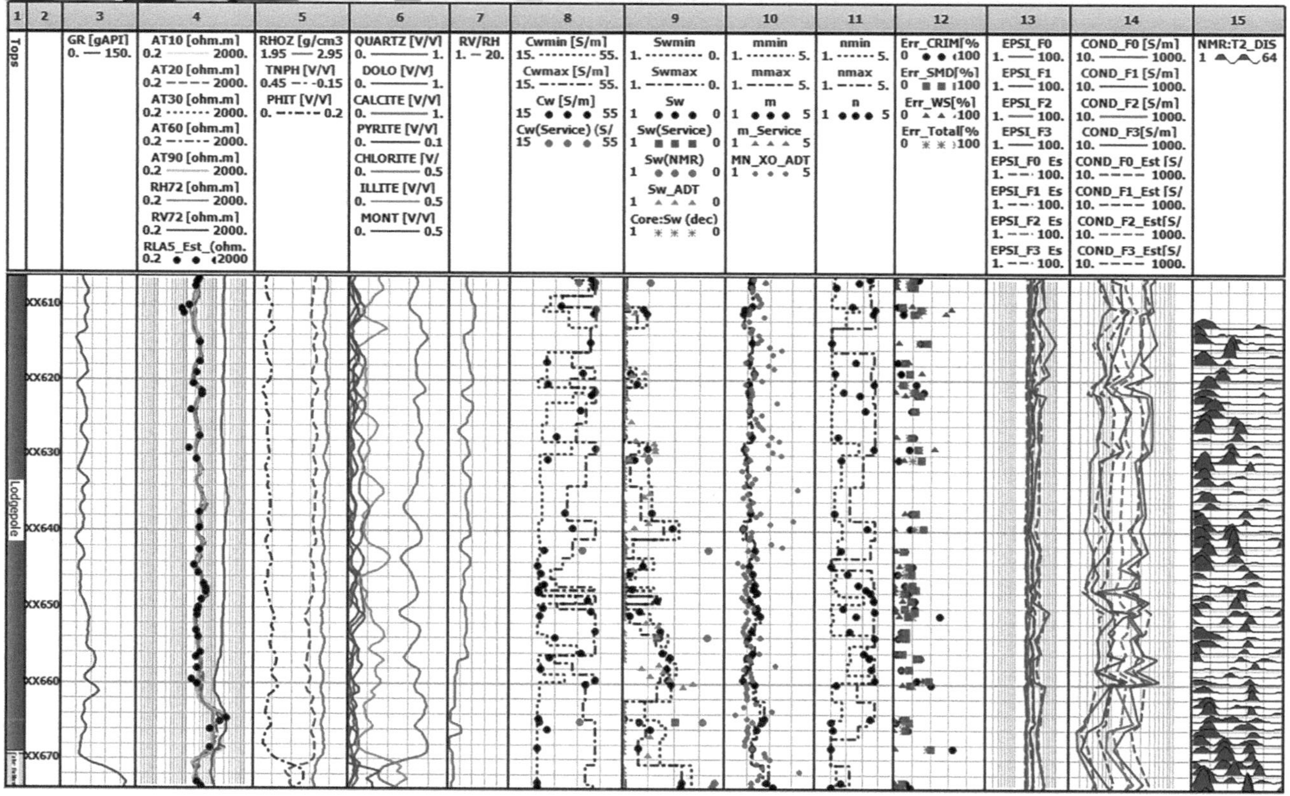

FIGURE 9.16 Track 1 is the formation type, Track 2 is the depth (XX610-XX670 ft), Track 3 is the gamma ray (GR), Track 4 contains the induction resistivity (AT), horizontal resistivity R_h, vertical resistivity R_v, and modeled resistivity, Track 5 contains the bulk density (RHOZ), neutron porosity (TNPH), and total porosity (PHIT), Track 6 contains volume fractions of various minerals: quartz, dolomite, calcite, pyrite, chlorite, illite, and montmorillonite, Track 7 contains anisotropy logs derived from R_t scanner, Track 8 contains inversion-derived brine conductivity (C_w) estimate with a range, Track 9 contains water saturation estimated using various methods (S_w), Track 10 contains inversion-derived cementation exponent estimate (m) estimate with a range, Track 11 contains inversion-derived saturation exponent estimate (n) estimate with a range, Track 12 contains relative errors in the predictions of various models in the integrated model, Track 13 contains measured (bold) and modeled (dash) permittivity dispersion logs at four frequencies, Track 14 contains measured (bold) and modeled (dash) conductivity logs at four frequencies, and Track 15 contains NMR T_2 distribution. *Source: Authors.*

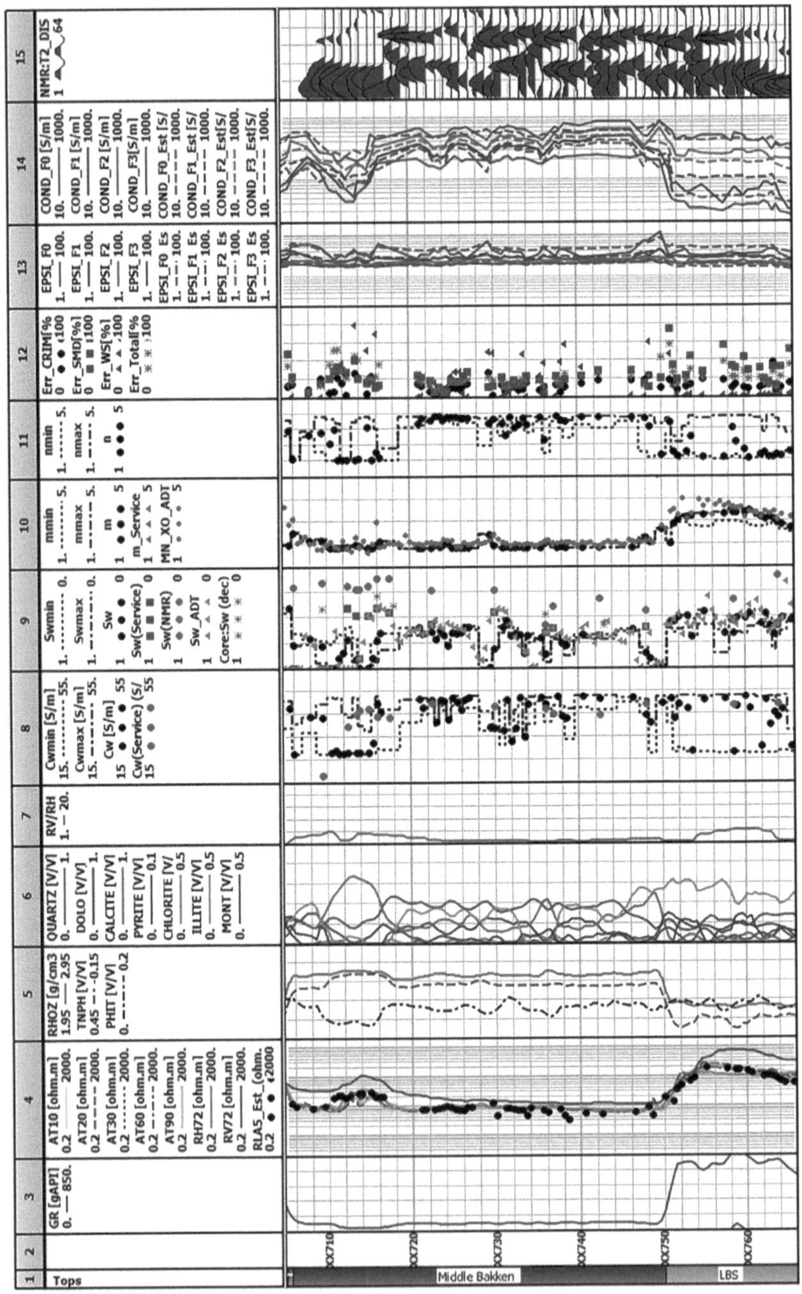

FIGURE 9.17 Logs for formation interval from the depth of XX710 to XX766 feet. The logs and the layout presented in this figure are similar to those in Fig. 9.16. *Source: Authors.*

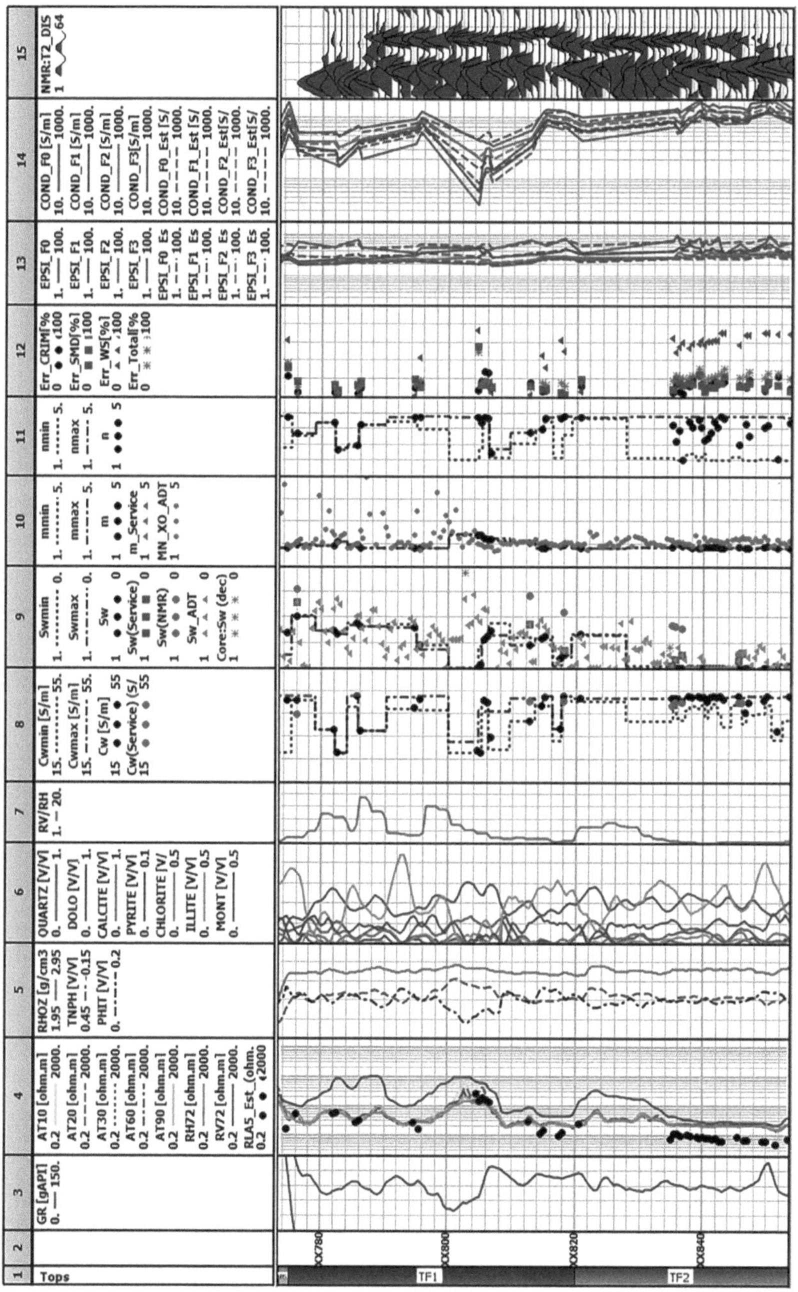

FIGURE 9.18 Logs for formation interval from the depth of XX780 to XX850 feet. The logs and the layout presented in this figure are similar to those in Fig. 9.16. *Source: Authors.*

the NMR log interpretation. Cementation exponent estimates computed using the inversion matches with the estimates obtained by the service company, as shown in Track 10 of Fig. 9.16.

In Fig. 9.16, at depth XX610-XX670 feet, high uncertainty observed in brine conductivity estimates (Track 8) is primarily because the integrated model is not able to account for the polarization mechanism that result in large frequency-dependence of the effective complex permittivity measured by the dielectric dispersion tool (Track 14). Uncertainty in saturation-exponent estimates increase with increase in water saturation from depth interval XX610–XX628 feet. These unreliable estimates could be because of small pores and low-porosity beds in that depth interval. Water saturation and cementation exponent estimates have higher certainty in comparison to saturation exponent and brine conductivity estimates, as shown in Figs. 9.16–9.18. In Fig. 9.17, from depth interval XX705.5–XX720 feet, uncertainties for water saturation and brine conductivity are large which could be because of low porosity (Track 5) and low salinity (Track 14) in this formation. Further in the Middle Bakken from depth XX720–XX750 feet, S_w estimates obtained using the joint inversion (Fig. 9.17, Track 9) gets better agreement with those estimated by service company and measured by dielectric ADT tool than Dean–Stark core water saturation (Fig. 9.17). Inversion-derived C_w estimates are also in good agreement than those estimated by the service company (Track 8). The proposed inversion achieves high certainty in this interval for the estimates of S_w, C_w, m, and n. This is a consequence of low-clay content and high porosity in that depth interval. The reduced uncertainty of the estimates could also be because of lower anisotropy in this depth interval. There is a good agreement between S_w estimates obtained using the proposed inversion (Fig. 9.18, Track 9) and those estimated using multimineral inversion performed by service company. There is high uncertainty in brine conductivity and saturation exponent in Three Forks 2 formation from XX820–XX850 feet (Fig. 9.18, Track 8 and Track 11). This could be the result of large frequency dependence of effective complex permittivity measured by the ADT tool.

6.3 Analysis of interpretation errors

As described in Section 5.3, data misfit is calculated as the difference between the modeled and measured dielectric dispersion and resistivity logs. Data misfit calculations are represented in Eqs. 9.7–9.10. Error in fitting the conductivity and permittivity logs acquired at 1 GHz using the CRI model predictions is referred as the Error_{CRI}. Error in matching the three conductivity and three permittivity dispersion logs acquired at 22, 100, and 350 MHz, respectively, using the SMD model predictions is referred as Error_{SMD}. Finally, the error in fitting the resistivity induction using WS model predictions is referred as Error_{WS}. Track 12 in Figs. 9.16–9.18 contains the various data misfits in matching the corresponding logs with the predictions of CRI model,

SMD model, and WS model. In Fig. 9.16, depth XX610 exhibits largest error because of data misfit in modeling induction resistivity, which is about 60%. The anisotropy value reported at this depth is about 9.67 (Track 7). Track four resistivity track also shows the presence of thin beds at this depth because of high-vertical resistivity at this depth. The depth interval XX835–XX853.5 in Fig. 9.18 exhibits highest error for modeling induction resistivity on an average of around 60%. This interval characterizes low resistivity and large volumes of illite. This high error could also be the result of calculated cationic exchange capacity values in this depth interval. Error in fitting the three low-frequency dielectric dispersion logs ($Error_{SMD}$) has the highest value in this interval. At the depth interval XX750–XX766.5 feet, $Error_{SMD}$ is about 70% along with $Error_{CRI}$ attaining a large value of 35%, which might be a consequence of high-pyrite content in that depth interval. SMD and CRIM model fails in high volume of pyrite. In Fig. 9.16, the depth interval XX720–XX730 feet exhibits lowest data misfit, in modeling the induction resistivity and dielectric dispersion, which is less than 20%. Such low errors could be because of negligible anisotropy and low-pyrite content in this depth interval.

7. Limitations of the proposed interpretation of multifrequency electromagnetic logs

The joint inversion assumes that the volumes sensed by the eight dielectric dispersion logs at the four frequencies and that sensed by the laterolog resistivity and induction resistivity have similar petrophysical properties. This assumption requires the formation to be free from fractures and beds/laminations thinner than 2 feet. Vertical resolution refers to the thinnest bed the tool can accurately sense without the bed-boundary effects. Dielectric tool generally has about 2-inch vertical resolution and 3-inch depth of investigation, whereas laterolog resistivity and induction resistivity generally have vertical resolution of 2 feet and depth of investigation can be as high as 5–8 feet. The pad-based microresistivity measurements in the 520-feet depth interval of the upper Wolfcamp shale indicate that there are few depths consisting of beds as thin as 0.1 feet. Notably, the use of total porosity for log-processing agrees with the physics of dielectric dispersion tool measurement; however, the use of total porosity will generate erroneous results for galvanic resistivity measurements when isolated pores are present. Another simplification required cationic exchange capacity to be calculated from the clay types and volumetric concentrations of clay minerals. Laboratory experimentation is necessary for more reliable values of cationic exchange capacity.

8. Conclusions

We successfully processed eight dielectric permittivity and conductivity dispersion logs (10 MHz–1 GHz) and one laterolog resistivity (~1 kHz) acquired in 520-feet depth interval of the clay- and carbonate-rich upper Wolfcamp formation. The proposed joint interpretation technique also processed eight dielectric permittivity and conductivity dispersion logs (22–960 MHz) and one induction resistivity (~20 kHz) logs acquired in 350-feet depth interval of the BPS. In doing so, a continuous estimation of water saturation, brine conductivity, cementation exponent (textural index), and saturation exponent with ranges of possible values were obtained for the upper Wolfcamp formation and BPS.

The proposed log processing technique is robust compared to processing only the dielectric dispersion logs, especially in pyrite-rich, low-porosity, and hydrocarbon-bearing formations. These estimates have higher certainty and better convergence at lower water saturation, which is desired feature of the processing technique. Compared to the estimates for saturation exponent and brine conductivity, water saturation and cementation exponent estimates exhibit higher certainty. Estimates of cementation exponent obtained using the proposed inversion exhibit higher variation with the increase in depth indicating an increase in heterogeneity/layering with depth. Hydrocarbon-bearing formation has connate water conductivity of 20 S/m. The saturation exponent estimates are close to 2.8 in the hydrocarbon-bearing layers and tend to exhibit large uncertainty in the nonreservoir sections. Sensitivity of resistivity, permittivity, and conductivity logs to brine conductivity is lost for brine conductivity greater than 45 S/m.

As stated earlier, the proposed inversion work assumes that formation should be free from beds less than two feet. These thin beds are bypassed because of the limitation of vertical resolution and hence could lead to significant under estimation of our results. For future work, we would recommend analyzing these thin beds. Thomas–Stieber approach could be used for thin bed analysis and calculating parameters like total porosity. For the experimental work, dielectric measurements could be conducted on shale samples over a range of frequencies to characterize electrical properties of shale and relate the responses to other petrophysical properties. Resistivity measurements could be further conducted at different saturation state and different salinity to develop a resistivity-saturation-porosity relationship for organic-rich shales.

Symbols and abbreviations

B Equivalent conductance of sodium clay exchange cations (mS/meq)
CEC Cation exchange capacity
CRI Complex Refractive Index
Cw Brine Conductivity(S/m)
EM Electromagnetic
$\boldsymbol{F(p)}$ Cost Function Vector
GHz GigaHertz
IP Interfacial polarization

kHz KiloHertz
LR Lichtenecker–Rother model
MHz MegaHertz
PHIT Total porosity
RHOZ Bulk density
RLA5 Resistivity Laterolog measured using galvanic resistivity tool at 1 kHz
SMD Stroud-Milton-De
TNPH Neutron porosity
TOC Total organic carbon
WS Waxman–Smits
S_w Water Saturation
m Cementation exponent
n Saturation exponent
Q_v Cation exchange capacity per unit pore volume (meq/cc)
p^k Model parameter vector computed at the k-th iteration of the inversion
$J(p^k)$ Jacobian matrix computed at the k-th iteration of the inversion
V_{py} Volume fraction of pyrite
Res_{mod} Modeled Resistivity
Res_{meas} Measured Resistivity
$\mathbf{Error}_{Total}$ Total Error
$\mathbf{Error}_{SMD}$ Error for permittivity and conductivity for SMD model
$\mathbf{Error}_{CRI}$ Error for permittivity and conductivity for CRI model
$\mathbf{Error}_{WS}$ Error for resistivity for Waxman Smits model
$Perm^{meas,1GHz}$ Measured dielectric permittivity at 1 GHz
$Perm^{mod,1GHz}$ Modeled dielectric permittivity at 1 GHz
$Cond^{meas,1GHz}$ Measured electrical conductivity at 1 GHz
$Cond^{mod,1GHz}$ Modeled electrical conductivity at 1 GHz
R_t True resistivity of the formation measured using galvanic resistivity tool at 1 kHz
ϕ_T Total porosity of the formation
ε_b^* Bulk complex permittivity of the formation
ε_m Matrix permittivity
ε_w^* Complex permittivity of water
ε_o Hydrocarbon permittivity
E_b Real value of water permittivity assumed as 80
E_o Vacuum permittivity with universal value of 8.85×10^{-12}
V_c Volume fraction of clay
R_h Horizontal resistivity
R_v Vertical resistivity

References

Chen, H., Heidari, Z., 2014. Assessment of hydrocarbon saturation in organic-rich source rocks using combined interpretation of dielectric and electrical resistivity measurements. In: SPE Annual Technical Conference and Exhibition, Society of Petroleum Engineers.

Clennell, M.B., Josh, M., Esteban, L., Piane, C.D., Schmid, S., Verrall, M., Hill, D., Woods, C., McMullan, B., 2010. The influence of pyrite on rock electrical properties: a case study from NW Australian gas reservoirs. In: SPWLA 51st Annual Logging Symposium, Society of Petrophysicists and Well-Log Analysts.

Donadille, J.M., Leech, R., Pirie, I., 2016. Water salinity determination over an extended salinity range using a joint interpretation of dielectric and neutron cross-section measurements. In: SPE Annual Technical Conference and Exhibition, Society of Petroleum Engineers.

Engle, M.A., Reyes, F.R., Varonka, M.S., Orem, W.H., Ma, L., Ianno, A.J., Schell, T.M., Xu, P., Carroll, K.C., 2016. Geochemistry of formation waters from the Wolfcamp and "Cline" shales: insights into brine origin, reservoir connectivity, and fluid flow in the Permian Basin, USA. Chem. Geol. 425, 76–92.

Ferraris, P., Borovskaya, I., Ribeiro, M., 2012. Advances in formation evaluation independent of conveyance method: state of the art logging while drilling & wireline petrophysical analysis in a carbonate reservoir offshore Brazil. In: SPWLA 53rd Annual Logging Symposium, Society of Petrophysicists and Well-Log Analysts.

Glinskikh, V.N., Nesterova, G.V., Epov, M.I., 2014. Forward modeling and inversion of induction logs from shaly sand reservoirs using petrophysical conductivity models. Russ. Geol. Geophys. 55 (5–6), 793–799 (1001–1010).

Hackley, P.C., Cardott, B.J., 2016. Application of organic petrography in North American shale petroleum systems: a review. Int. J. Coal Geol. 163, 8–51. https://doi.org/10.1016/j.coal.2016.06.010.

Han, Y., Misra, S., 2017. Improved Water-Saturation Estimates Derived from Inversion-Based Interpretation of Broadband Electromagnetic Dispersion Logs in Organic-Rich Shale Formations, SEG Technical Program Expanded Abstracts. Society of Exploration Geophysicists, pp. 3458–3462.

Han, Y., Misra, S., 2018a. A unified inversion scheme to process multifrequency measurements of various dispersive electromagnetic properties. J. Appl. Geophys. 151, 23–39.

Han, Y., Misra, S., 2018b. Bakken petroleum system characterization using dielectric-dispersion logs. Petrophysics 59 (02), 201–217.

Han, Y., Misra, S., 2018c. Joint petrophysical inversion of multifrequency conductivity and permittivity logs derived from subsurface galvanic, induction, propagation, and dielectric dispersion measurements. Geophysics 83 (3), D97–D112.

Han, Y., Misra, S., Simpson, G., 2017. Dielectric dispersion log interpretation in bakken petroleum system. In: SPWLA 58th Annual Logging Symposium, Society of Petrophysicists and Well-Log Analysts.

Kreis, L.K., Costa, A., 2005. Hydrocarbon potential of the bakken and torquay formations, southeastern saskatchewan;. In: Summary of Investigations 2005, Volume 1, Saskatchewan Geological Survey, Sask. Industry Resources, Misc. Rep. 2005-4.1, CD-ROM, Paper A-10, p. 10.

Misra, S., Han, Y., 2016. Petrophysical Interpretation of Multi-Frequency Electromagnetic Measurements in Clay- and Conductive-Mineral-Rich Mudrocks. American Association of Petroleum Geologists.

Pirie, I., Horkowitz, J., Simpson, G., Hohman, J., 2016. Advanced methods for the evaluation of a hybrid-type unconventional play: the Bakken petroleum system. Interpretation 4 (2), SF93–SF111.

Sanchez-Ramirez, J.A., Torres-Verdín, C., Wang, G.L., Mendoza, A., Wolf, D., Liu, Z., Schell, G., 2010. Field examples of the combined petrophysical inversion of gamma-ray, density, and resistivity logs acquired in thinly-bedded clastic rock formations. Petrophysics 51, 247–263.

Seleznev, N.V., Fellah, K., Philips, J., Zulkipli, S.N., Fournie, B., 2014. Matrix Permittivity Measurements for Rock Powder, SPE Annual Technical Conference and Exhibition. Society of Petroleum Engineers, Amsterdam, Netherland.

Tathed, P., Han, Y., Misra, S., 2018a. Hydrocarbon saturation in upper Wolfcamp shale formation. Fuel 219, 375–388.

Tathed, P., Han, Y., Misra, S., 2018b. Hydrocarbon saturation in Bakken petroleum system based on joint inversion of resistivity and dielectric dispersion logs. Fuel 233, 45–55.

Tathed, P., 2018. Interpretation of Galvanic Resistivity, Electromagnetic Induction, and Dielectric Dispersion Logs to Estimate Water Saturation in Organic Rich Shale. University of Oklahoma. Master's Thesis.

Section 5

Advanced inversion-based interpretation of multifrequency electromagnetic data

Chapter 10

Multifrequency electromagnetic data interpretation using stochastic Markov-chain Monte Carlo and simulated annealing methods

Yuteng Jin[4], Siddharth Misra[1,3], Yifu Han[2]
[1]*Harold Vance Department of Petroleum Engineering, College of Engineering, Texas A&M University, TX, United States;* [2]*Schlumberger Beijing Geoscience Center, Beijing, China;* [3]*Department of Geology and Geophysics, Texas A&M University, TX, United States;* [4]*Harold Vance Department of Petroleum Engineering, Texas A&M University, TX, United States*

1. Introduction

Engineers and scientists measure input-output (stimuli-response) data to better understand a system/process of interest. It is a common engineering/scientific practice to estimate the unknown factors/parameters that produced a set of observations/measurements. For example, it is of interest to know the porosity and permeability of the subsurface reservoir given the observed fluid production data. Another example involves the estimation of water saturation in the subsurface formations from the available subsurface electromagnetic (EM) measurements/logs.

Parameter estimation (also, referred as system identification or inverse modeling) is the general process of extracting information about a system/process from the measured input-output data. With an assumption of few candidate models that may represent the system/process, parameter estimation method performs computations on the measurements or the empirical data to estimate the unknown model parameters of the candidate models. In doing so, the newly estimated model can accurately and reliably represents the system/process of interest. The model can be static or dynamic, deterministic or stochastic, linear or nonlinear. Parameter estimation requires (1) experiment design for collecting high-quality data, (2) selection of a candidate model structure based on prior knowledge and trial and error, (3) calculation of the

Multifrequency Electromagnetic Data Interpretation for Subsurface Characterization
https://doi.org/10.1016/B978-0-12-821439-8.00004-5

best model within the selected structure based on certain criteria (also referred as objective function) of fitting the experimental data, and (4) rigorous evaluation of the properties of the best model to detect any inadequacies.

Parameter estimation, inverse modeling, or system identification solves an optimization problem to find the best model parameters within a permissible range (referred as solution space) by maximizing or minimizing an objective/cost function (for example, residual sum of squares). In most of the cases for parameter estimation, the optimization focusses on minimizing the data misfit (prediction error) while satisfying certain constraints on the model parameters. Parameter estimation algorithms can be broadly classified into deterministic algorithms and stochastic algorithms (Liberti and Kucherenko, 2005).

Deterministic inversion algorithms are designed to guarantee a single optimal solution that may fail on black-box formulations, extremely ill-behaved functions or complex large-scale problems. Stochastic inversion algorithms on the other hand compute a distribution of possible solutions, all plausible, which average to the deterministic solution. Stochastic inversion finds all the alternate solutions (referred as realizations) whose mean or average is the expected value of the underlying process/system. Compared to deterministic inversion, stochastic inversion provides an insight into the most probable solutions along with the uncertainty in the estimates that is suited for noisy data acquired in real-world environments. For most stochastic inversion, the probability of a specific solution tends to one given infinite computing time, but none guarantees a global optimal solution in finite computing time (Boender and Romeijn, 1995).

Nowadays, as the inversion modeling (parameter estimation) problems are becoming more and more complex along with the advances in computational technique, the use of stochastic inversion has gained popularity because it can find the global optimal solution along with estimation uncertainty. Compared to deterministic inversion, stochastic inversion takes more time for convergence, is more effective in global search, has lower risk of early convergence because of local minimum, and is easier to implement in parallel. Stochastic inversion is less effective for local search and simple problems.

Stochastic inversion algorithms include Markov-chain Monte Carlo (MCMC), simulated annealing (SA), swarm algorithms like particle swarm optimization (PSO) and ant colony optimization (ACO), and evolutionary algorithms like genetic algorithm, hill climbing, and tabu search. Deterministic inversion methods include neighborhood search, steepest ascent, gradient descent, branch and cut, dynamic programming, and branch and boundary, some of which have been discussed in detail in Chapters 6, 7, and 9. Going ahead, Chapters 10 and 11 focus on applying several popular stochastic inversion methods. MCMC and SA are discussed in Chapter 10, and PSO and ACO are discussed in Chapter 11.

1.1 Markov-chain Monte Carlo inversion

Markov chain describes a stochastic/random process with Markov property, where the probability of a random process transitioning to the next state is only dependent on the current state and it is independent of states that preceded the current state. In other words, Markov chain describes a memory-less process that solely depends on the current state/action. Several board games, such as snakes and ladders, can be modeled as Markov chain because the next state of the board depends entirely on the current state of the board and dice. However, a game of blackjack cannot be represented using Markov chain because each subsequent card depends on the cards that have already been dealt.

Monte Carlo method uses repeated random sampling to solve numerical problems, such as area inside a closed curve or the probability of simultaneously rolling two sixes using two dices. The repeated random sampling (ranging from thousands to millions of random samples) generates a likelihood of a range of possible outcomes of an uncertain event. Monte Carlo method is suited for complex numerical problem, when an alternate computational method is hard or impossible. Such situation arises in numerical integration, optimization, simulating physical system with coupled degrees of freedom, simulating a complex probabilistic phenomenon, long-term forecasting along with uncertainty quantification, and risk assessment. In simple terms, Monte Carlo method involves assigning multiple random values to an uncertain variable to generate multiple outcomes and then to average the multiple outcomes for obtaining a numerical estimate.

Based on Bayesian formulation, a stochastic inversion computes the posterior, which is the probability of model parameters given the data. The computed posterior is used to generate the probabilistic model of the process/system of interest. From a Bayesian standpoint, the posterior is the product of prior (assumed probability distribution of the model parameters before the availability of data) and the likelihood (probability distribution of data given a model parameter) divided by the evidence that the data were generated by the probabilistic model. Computation of the evidence (the denominator) in the Bayesian framework is hard; consequently, MCMC sampling is used to approximate the posterior without explicitly calculating the hard-to-compute evidence. MCMC sampling generates stochastic samples representative of the posterior while exploring the sample space using a Markov chain (Andrieu et al., 2003). In simple terms, MCMC is designed to construct a Markov chain of model parameters and then build a sequence of accepted model parameters that can be used for the Monte Carlo approximation of the posterior.

In our study, the MCMC sampling implements the Metropolis–Hastings algorithm. Metropolis–Hastings algorithm helps to iteratively select a sequence random samples having Markov property that on the long run approximates a target distribution from which direct sampling is difficult. In doing so, MCMC sampling does not draw from the posterior distribution of model parameters; instead, it draws from the unnormalized posterior distribution of the model parameters. The sequence generated by the Metropolis–Hastings algorithm is such that a candidate sample from the Markov chain is accepted based on a predefined acceptance ratio, which denotes how probable is the new sample (i.e., model parameter) representative of the target posterior distribution as compared to the current sample. Acceptance ratio governs the exploration of low-density regions in the model-parameter space. In this algorithm, when building the approximation of the posterior distribution of the model parameters, the hard-to-compute evidence is circumvented by taking the ratio of the product of prior and likelihood for the proposed sample to that of the current sample. Proposed model parameters are selected when they have higher posterior density than that of the current sample. Few proposed model parameters with lower posterior density are also selected randomly to allow the exploration of low-density regions of the model-parameter space. In summary, we can perform stochastic inversion using MCMC sampling based on Metropolis–Hastings algorithm.

MCMC-based stochastic inversion has a lot of successful applications in the oil and gas industry and other subsurface characterization tasks, including history matching (Dachanuwattana et al., 2018), reservoir description and forecasting (Li and Reynolds, 2017; Da Silva Moreira et al., 2019), and inverse problems (Han et al., 2019; Shen et al., 2018). MCMC-based inversion is suitable for sampling from complex, high-dimensional spaces that far exceeds the traditional grid-based search (Dunn and Shultis, 2011). However, there is no convergence diagnostics that is guaranteed to work for MCMC sampling (Sinharay, 2004).

1.2 Simulated annealing inversion

SA is a metaheuristic algorithm, i.e., it performs problem-independent local search for the solution to a parameter estimation (or optimization) problem. SA is inspired by the annealing process in metallurgy, which is used to alter physical and chemical properties of a material. In annealing, a solid is heated above its recrystallization temperature and then cooled down slowly, allowing sufficient time for the atoms in the material to move toward the defects and fill them. This minimizes Gibbs free energy. Annealing leads to the diffusion of atoms within the solid toward equilibrium state. Initial high-temperature phase of annealing increases the rate of diffusion of atoms by providing the energy needed to break bonds; and the free atoms can easily move to new positions and redistribute at higher temperatures. The subsequent cooling phase of the annealing process results in the stabilization of the atoms at the lowest energy state, thereby filling the defects in the solid.

SA tries to mimic the annealing process, where at high temperatures, atoms shift unpredictably leading to the elimination of impurities as the material cools down into pure crystal. The impurities/defect are analogous to the optimal solution. SA for parameter estimation starts searching the model-parameter space (i.e., solution space) for an optimal solution by assuming a high value for a search-parameter referred as temperature (analogous to an atom starting the search for low-energy defect locations during the initial high-temperature phase of annealing). The high temperature initiates the search for solution and promotes exploration of the model-parameter space. As the temperature gradually reduces based on a predefined temperature-reduction rule, the extent of exploration reduces and a locally optimal solution is obtained at each temperature level. For each temperature level, a neighboring solution is identified by slightly perturbing a current solution. The neighboring solution is accepted based on an acceptance criterion. When neighboring solution is better than current solution, the neighboring solution is unconditionally accepted. On the other hand, when neighboring solution is not better than current solution, the neighboring solution is accepted based on a probability that is directly proportional to the current temperature and inversely proportional to the extent of difference between current and neighboring solutions. These steps are performed at each temperature level. By occasionally accepting worse solutions during the estimation process, SA finds the optimal solution as the temperature tends to zero.

To avoid local solutions, SA uses Metropolis algorithm that accepts suboptimal trial solutions to explore more of the solution space. Suboptimal trials are selected based on certain criterion, such that, at higher temperature, more suboptimal/bad trials are accepted to allow the exploration of a large part of the model-parameter space. An important feature of SA is that it can reach to the optimum without the need of calculating derivatives. Toward the end of the SA process, there is still a possibility that a lower energy state (better solution) exists (Chibante, 2010). In engineering practice, an approximation to the exact optimum value is good enough. For the same reason, there is no need to select a very low value of final temperature for the successful implementation of SA inversion process. Moreover, the computational time of SA is highly dependent on the selection of a proper initial temperature as well as the dimensions of the objective function to be solved. For large values of initial temperature, the SA method may take too long to converge (Chibante, 2010). SA is a widely used technique in oil andand gas industry with many successful applications such as well placement optimization (Tukur et al., 2019), well log interpretation (Szucs and Civan, 1996), and reservoir characterization (Ouenes and Saad, 1993).

Compared to MCMC, simulated-annealing inversion uses a different acceptance criterion aiming to find a single best solution. SA is easier to code, and its computational requirements are predictable because the number of iterations can be predefined (Osman and Potts, 1989). Inversion based on MCMC sampling needs to obtain a stable distribution within the highest posterior density interval (HPDI). Consequently, it may take longer time to find the solution using MCMC as compared to SA. Moreover, in engineering practice, the inversion based on MCMC sampling can be a challenge because a large number of data samples is required to create the data covariance matrix. SA outshines traditional derivative-based methods when several local minima exist in the solution space.

In this study, we use 95% HPDI, which refers to threshold values of the posterior distribution that represent one or more intervals, such that the probability of the estimate to lie within the interval is 0.95, while the probability of the estimate to lie outside the interval is 0.05. For near-symmetric unimodal distribution, HPDI will have only one interval, while for bimodal distribution, HPDI may comprise two disjoint intervals.

1.3 Petrophysical estimation based on stochastic inversion of multifrequency electromagnetic data

In this chapter, multifrequency EM data acquired at frequencies corresponding to EM induction resistivity tool (operating at 10 and 20 kHz), logging while drilling propagation tool (operating at 1 and 2 MHz), and dielectric dispersion tool (operating at 20 MHz, 100 MHz, 260 MHz and 1 GHz) are jointly inverted to estimate four petrophysical properties at each depth of the formation. The four properties to be estimated include water saturation, water conductivity, surface conductance of clay, and contact angle of graphite. To that end, MCMC and SA-based stochastic-inversion methods are coupled with improved PPIP−SCAIP (PS) model to process the multifrequency EM data, comprising effective conductivity and effective relative permittivity at eight frequencies. In total, there are 16 data points at each depth that need to be inverted to obtain the four above-mentioned properties at the depth of interest. The improved PS model is discussed in detail in Chapter 5.

2. Method

2.1 The mechanistic electromagnetic model for subsurface characterization

In this chapter, the improved PS model (as discussed in Chapter 5) is used as the mechanistic EM model to guide the parameter estimation. The improved PS model considers the effect of wettability of conductive particles. In real world application, the improved PS model will need to be calibrated before inverting multifrequency EM data. MCMC-based and SA-based stochastic inversion methods are coupled with the improved PS model to process the multifrequency EM data. The inversion methods can be coupled to any other forward model to process any other multifrequency dataset.

2.2 The synthetic formation

The improved PS model is used to generate multifrequency EM data for a synthetic formation comprising 25 distinct layers. The synthetic formation is designed to study the efficacy of inversion methods to variations in wettability, fluid saturation, and volume fractions of conducive minerals and clays. Overall, the synthetic formation contains three zones, i.e., water-wet, intermediately wet, and oil-wet zones made of 11, 5, and 9 layers, respectively. The synthetic formation is illustrated in Fig. 10.1, and the properties of each layer are listed in Table 10.3. For a specific zone, each layer has distinct volume fractions of

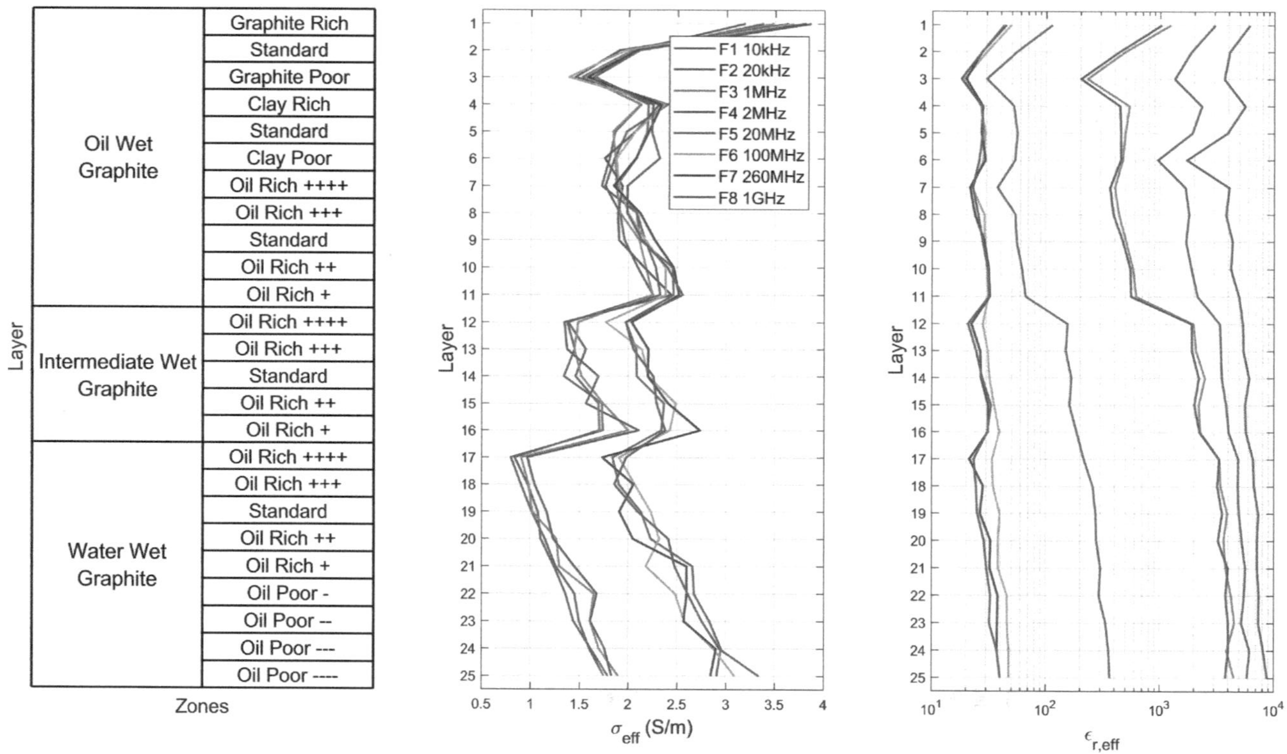

FIGURE 10.1 Multifrequency electromagnetic responses of the 25 layers in the synthetic formation along with qualitative descriptions of the layers. The multifrequency electromagnetic response comprise effective conductivity (σ_{eff}) and effective relative permittivity $(\varepsilon_{r,\text{eff}})$ at eight frequencies. *Source: Authors.*

conductive mineral, sand, clay, and oil. 21 of the 25 layers have the same porosity of 30%. MCMC-based and SA-based stochastic inversion methods are coupled with the improved PS model to process the synthetic multifrequency EM data acquired from the three zones comprising 25 synthetic layers in total. The petrophysical properties of inclusions and hosts assumed for the synthetic formation are listed in Tables 10.1 and 10.2, respectively. The petrophysical properties of each layer in the synthetic formation are listed in Table 10.3. For this synthetic formation, layers 1–11 contain oil-wet graphite, layers 12–16 contain intermediate-wet graphite, and layers 17–25 contain water-wet graphite.

Layers 1–3 are designed to show the effect of graphite content, layers 4–5 show the effect of clay content, and the other layers show the effect of oil saturation for the oil-wet case. For better comparison, the layers 2, 5, 9, 14, and 19 act as standard/base layers containing 10% graphite, 30% sand, and 30% clay with 30% porosity and 70% oil saturation. Improved PS model is used to generate the synthetic multifrequency EM data for the synthetic

TABLE 10.1 The petrophysical properties of inclusions (graphite, sand, clay, and oil) assumed for the synthetic formation comprising 25 layers, which have these properties in common. The value of parameter λ_c to be estimated is denoted using bold italicized font. a_i is the size of inclusion, $\varepsilon_{r,i}$ is the relative permittivity of inclusion, σ_i is the conductivity of inclusion, D_i is the diffusion coefficient of inclusion, and λ_i is the surface conductance of inclusion.

	a_i (μm)	$\varepsilon_{r,i}$	σ_i (S/m)	D_i (m²/s)	λ_i (S)
Graphite	10	12	500	5×10^{-5}	–
Sand	100	4	–	–	1×10^{-9}
Clay	10	8	–	–	$\boldsymbol{1 \times 10^{-5}}$
Oil	10	2	–	–	1×10^{-30}

TABLE 10.2 The petrophysical properties of hosts assumed for the synthetic formation comprising 25 layers, which have these properties in common. The value of parameter σ_w to be estimated is denoted using bold italicized font. ρ_h is the density of the host, and γ is the interfacial tension between water and oil.

	ρ_h (kg/m³)	$\varepsilon_{r,h}$	σ_h (S/m)	D_h (m²/s)	γ (N/m)
Water	1000	70	***5***	1×10^{-9}	0.05
Oil	800	2	1×10^{-30}	1×10^{-15}	

TABLE 10.3 The petrophysical properties of each of the 25 layers in the synthetic formation. The values of parameters θ_g and S_o to be estimated are denoted using bold italicized font. θ_g is the contact angle of graphite, ϕ_g is the volume fraction of graphite, ϕ_s is the volume fraction of sand, ϕ_c is the volume fraction of clay, ϕ_o is the volume fraction of oil, ϕ is the porosity, and S_o is the oil saturation, such that $\phi_o = \phi S_o$.

Layer	θ_g (°)	ϕ_g (%)	ϕ_s (%)	ϕ_c (%)	ϕ_o (%)	ϕ (%)	S_o (%)
1	***150***	20	30	30	14	20	***70***
2	***150***	10	30	30	21	30	***70***
3	***150***	5	30	30	24.5	35	***70***
4	***150***	10	30	40	14	20	***70***
5	***150***	10	30	30	21	30	***70***
6	***150***	10	30	10	35	50	***70***
7	***150***	10	30	30	27	30	***90***
8	***150***	10	30	30	24	30	***80***
9	***150***	10	30	30	21	30	***70***
10	***150***	10	30	30	18	30	***60***
11	***150***	10	30	30	15	30	***50***
12	***90***	10	30	30	27	30	***90***
13	***90***	10	30	30	24	30	***80***
14	***90***	10	30	30	21	30	***70***
15	***90***	10	30	30	18	30	***60***
16	***90***	10	30	30	15	30	***50***
17	***30***	10	30	30	27	30	***90***
18	***30***	10	30	30	24	30	***80***
19	***30***	10	30	30	21	30	***70***
20	***30***	10	30	30	18	30	***60***
21	***30***	10	30	30	15	30	***50***
22	***30***	10	30	30	12	30	***40***
23	***30***	10	30	30	9	30	***30***
24	***30***	10	30	30	6	30	***20***
25	***30***	10	30	30	3	30	***10***

formation. Notably, 5% Gaussian noise is added to the improved PS model predictions. The multifrequency EM response of the synthetic formation is shown in Fig. 10.1. In this parameter-estimation study, four petrophysical properties, namely surface conductance of clay λ_c, contact angle of graphite θ_g, water conductivity σ_w, and oil saturation S_o are to be inverted/estimated using the MCMC inversion and SA inversion. The known values for the four properties are denoted in bond italicized font in Tables 10.1–10.3.

2.3 Proposed inversion-based interpretation method

2.3.1 Markov-chain Monte Carlo inversion method

Similar to Chapter 8, MCMC method is used in this chapter for inversion of multifrequency EM data comprising conductivity and permittivity. To implement the MCMC method, the new solution vector $\boldsymbol{m}'$ is sampled from a multivariate normal distribution with mean $\boldsymbol{m}$, which is the current solution vector, and covariance matrix $\boldsymbol{C}_k$, which is defined in Chapter 8. If $\boldsymbol{m}'$ is out of the solution space S, the sampling should be performed again until $\boldsymbol{m}'$ is within S. Following that, $\boldsymbol{m}'$ is either accepted or rejected according to an acceptance probability defined in Chapter 8. The sampling should be done for multiple times till the convergence of the Markov chain. Unlike what is done in Chapter 8, instead of normalizing the data misfit vector $\boldsymbol{d}^{\text{meas}} - \boldsymbol{d}(\boldsymbol{m})$ by the standard deviation vector $\boldsymbol{\sigma}$ in the likelihood function, we use a different form of likelihood function, which is expressed as:

$$p(\boldsymbol{d}|\boldsymbol{m},\boldsymbol{I}) = \left[(2\pi)^N|\boldsymbol{C}_d|\right]^{-\frac{1}{2}} e^{\left[-\frac{1}{2}(\boldsymbol{d}^{\text{meas}}-\boldsymbol{d}(\boldsymbol{m}))^T\boldsymbol{C}_d^{-1}(\boldsymbol{d}^{\text{meas}}-\boldsymbol{d}(\boldsymbol{m}))\right]} \tag{10.1}$$

where $\boldsymbol{C}_d$ is the data covariance matrix which is a diagonal matrix with each diagonal element equal to the estimated data error variance for each measurement obtained at each frequency. For synthetic data, since we added 5% Gaussian noise for each measurement, the data error variance should be the square of 5% of the actual value. The definitions of other variables are the same as those previously defined in Chapter 8, the inversion procedure is the same as well.

2.3.2 Simulated annealing inversion method

In this chapter, the SA inversion method is evaluated and compared with the MCMC inversion method. A flowchart of the SA algorithm used in this chapter is shown in Fig. 10.2. The algorithm contains two loops: the inner loop and the outer loop. The outer loop is the annealing process, where temperature T decreases at each iteration, which is controlled by a constant temperature decline rate α. The parameter estimation starts at a high temperature to promote the exploration of the model-parameter space. The inner loop simulates the diffusion of atoms toward the thermal equilibrium. In the inner loop, Monte Carlo iterations are performed for *ite* times, aiming to find the best solution at each temperature level.

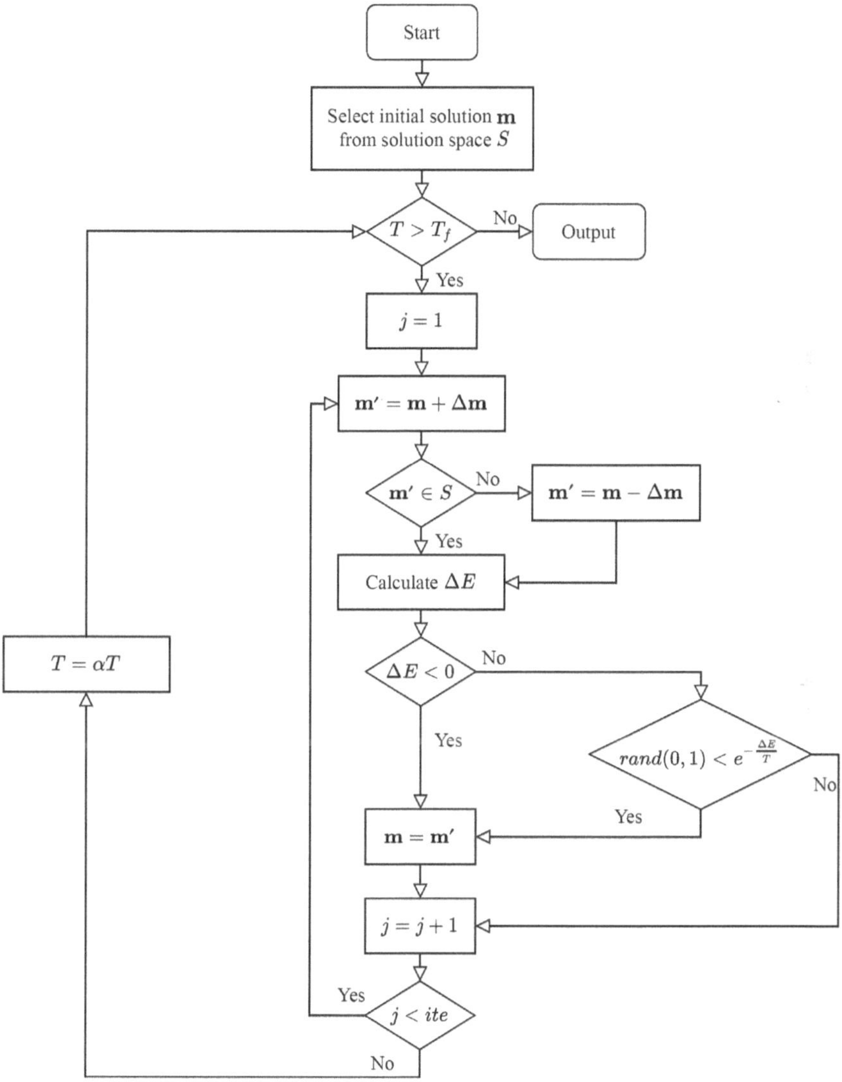

FIGURE 10.2 Flowchart of the simulated annealing inversion used to invert the multifrequency electromagnetic data to estimate the four petrophysical properties of each layer. *Source: Authors.*

For each temperature level, a neighboring solution $\boldsymbol{m}'$ is identified by slightly perturbing a current solution $\boldsymbol{m}$. Next, Metropolis algorithm is invoked to accept the neighboring solution based on an acceptance criterion. When neighboring solution is better than current solution (i.e., $\Delta E < 0$), the neighboring solution is unconditionally accepted $(\boldsymbol{m}' = \boldsymbol{m})$. On the other hand, when neighboring solution is not better than current solution, the neighboring solution is accepted based on a probability that is directly proportional to the current temperature and inversely proportional to the extent of difference between current and neighboring solutions. The annealing process continues until the final temperature T_f is reached. During the inversion process, the best solution generating the minimum energy state is recorded as the optimal solution of the inversion problem.

2.3.2.1 Method of generating the next solution

SA requires an initial solution obtained from the predefined solution space S as a starting point. The new solution vector $\boldsymbol{m}'$ is generated based on the current solution vector $\boldsymbol{m}$. In this chapter, the new solution is obtained by selecting a random value within the search window where the current solution is at the middle point. The length of the window is one 10th of the solution space. That means $\boldsymbol{m}'$ is obtained by moving a random length $\Delta\boldsymbol{m}$ from $\boldsymbol{m}$, where $\Delta\boldsymbol{m}$ is equal to one 10th of the length of solution space S times a random number between -0.5 and 0.5. Thus

$$\boldsymbol{m}' = \boldsymbol{m} + \Delta\boldsymbol{m} \tag{10.2}$$

If the new solution obtained by Eq. (10.2) is out of the solution space S, the following equation will be used instead of Eq. (10.2):

$$\boldsymbol{m}' = \boldsymbol{m} - \Delta\boldsymbol{m} \tag{10.3}$$

For example, if the solution space S is (0, 100), the current solution m is 98 and the random length Δm is selected to be 3. The new solution is then calculated to be 95 using Eq. (10.3) rather than Eq. (10.2) to make sure the new solution is within the solution space.

2.3.2.2 The cost function

In this chapter, the cost function is the energy in the SA process. Cost function is defined as

$$E = \sqrt{\sum \left[\log(\boldsymbol{d}^{\text{meas}}) - \log(\boldsymbol{d}(\boldsymbol{m}))\right]^2} \tag{10.4}$$

where $\boldsymbol{d}^{\text{meas}}$ is the measured data vector and $\boldsymbol{d}(\boldsymbol{m})$ is the modeled data vector, as defined in Chapter 8. Taking the common logarithm of the data vectors ensures the data vectors have the same orders of magnitude, such that both conductivity and permittivity are equally honored. The objective of the SA algorithm is to minimize the cost function E by selecting the optimal model parameter vector $\boldsymbol{m}$, such that the modeled data vector is closest to the measured data vector.

2.3.2.3 Acceptance criterion

The Metropolis algorithm is applied here to determine the acceptance criterion. The solution vector always moves from $\boldsymbol{m}$ to $\boldsymbol{m}'$ if the new energy state is lower than the current energy state (i.e., $\Delta E < 0$, where ΔE is the difference between the new energy state and the current energy state). However, if $\Delta E > 0$, the new solution vector $\boldsymbol{m}'$ is accepted based on probability P, which is expressed as

$$P = e^{-\frac{\Delta E}{T}} \tag{10.5}$$

As the temperature decreases, the probability of accepting the new solution becomes smaller, so the SA inversion reaches a stable solution.

3. Results and discussions

For both MCMC and SA inversion methods, we need to predefine the range of solution space and initial guesses for the unknown petrophysical properties to initiate the estimation. The predefined ranges and initial values of each of the three Markov chains for MCMC method are summarized in Table 10.4. As for SA inversion, the solution spaces are the same as the prior ranges used in MCMC method, and the initial values of the unknown properties are randomly selected from the solution spaces. The initial temperature T_o is 1, final temperature T_f is 0.00001, the temperature decline rate α is 0.99, and the iteration number *ite* for the inner loop is 50. Both MCMC and SA inversion methods are applied on all the 25 synthetic layers in the synthetic formation.

For Layer 2 in the synthetic formation, the inversion history of all the three Markov chains in MCMC method is shown in Fig. 10.3, and the histograms of the estimated petrophysical parameters based on MCMC inversion with the real values of the unknown petrophysical properties and the 95% HPDI is shown in Fig. 10.4. All the Markov chains with different starting points are stabilized around the same value after about 1000 iterations, and the MCMC inversion technique gives normal distributions of each estimated parameter. The real value of the properties are within the 95% HPDI, showing that the estimates are good approximations of the real values of these properties. For the Layer 2, compared to clay surface conductance λ_c, oil saturation S_o has a

TABLE 10.4 The prior ranges of the solution spaces and initial guesses of the unknown petrophysical parameters assumed for the three Markov chains of the proposed MCMC stochastic inversion.

Parameters	Prior ranges	Initial-1	Initial-2	Initial-3
λ_c (S)	$(1 \times 10^{-7}, 1 \times 10^{-4})$	1×10^{-5}	8×10^{-5}	2×10^{-7}
θ_g (°)	(0, 180)	60	90	120
σ_w (S/m)	(0.1, 10)	0.6	2	8
S_o (%)	(0.1, 99.9)	80	40	10

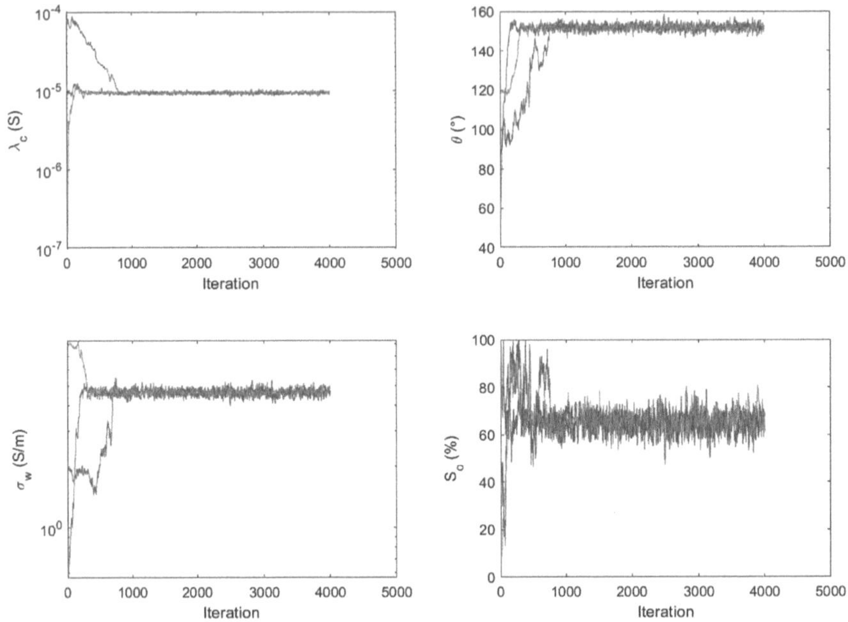

FIGURE 10.3 Markov-chain Monte Carlo inversion-derived estimates for the four unknown petrophysical properties, namely λ_c, θ_g, σ_w, and S_o, of all the three Markov chains for the Layer 2 in the synthetic formation, as shown in Fig. 10.1 and listed in Table 10.3. The chains are converged after about 1000 iterations. *Source: Authors.*

higher variation (as shown in Fig. 10.3); therefore, S_o estimate has a larger HPDI (as shown in Fig. 10.4). According to the improved PS model developed for mixtures containing conductive minerals of any wettability, the EM responses have the least sensitivity to S_o at high oil saturation as compared to clay surface conductance, volume fraction of conductive mineral, and the contact angle of the conductive mineral. In simpler terms, the variation in S_o at high oil saturation does not have much impact on the multifrequency EM response. Using the 95% HPDI corresponding to the four unknown model parameters, the average relative error in the S_o estimates over the 25 synthetic layers is 9.8%, as compared to an average relative error of 5% for each of the other three properties, namely λ_c, σ_w, and θ_g.

For the Layer 2 in the synthetic formation, Fig. 10.5 shows the history of the energy (cost function) using the SA method. Fig. 10.6 shows the estimates derived using the simulated-annealing based inversion as a function of number of iterations. The result converges after about 500 iterations, and the SA inversion technique gives a single estimation for each estimated parameter. In Fig. 10.5, the energy of the best solution is close to zero, showing the best result is a good approximation to the real parameters. This can also be concluded from the small separation between the red (dark gray in printed version) line (indicating the real values) and blue (gray in printed version) line (indicating the

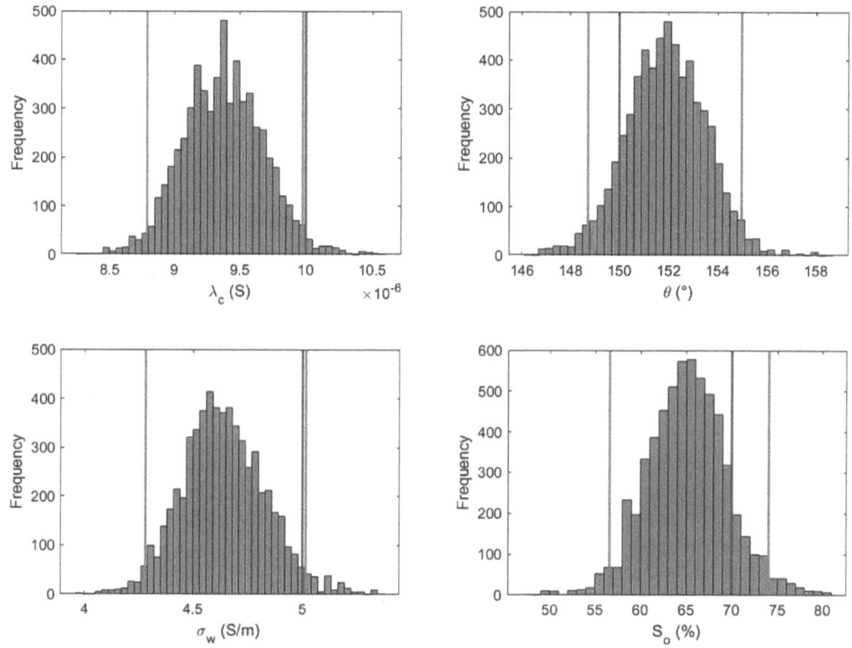

FIGURE 10.4 Histograms of the estimates for the four unknown petrophysical properties, namely λ_c, θ_g, σ_w, and S_o, obtained using the Markov-chain Monte Carlo sampling-based inversion for the Layer 2 in the synthetic formation, as shown in Fig. 10.1 and listed in Table 10.3. The *red line* (dark gray in print version) represents the real value of the unknown properties. The *two green lines* (gray in print version) represent the 95% highest posterior density interval. The estimates of λ_c, θ_g, and σ_w exhibit very low uncertainty, while the estimate of S_o exhibits a larger uncertainty ranging from 55% to 75%. *Source: Authors.*

estimates) in Fig. 10.6. Contact angle shows the best match, while oil saturation shows largest error in the estimate. Average relative error in the estimation of S_o over all the 25 synthetic layers is 10.9%, as compared to an average relative error of around 5% for the other three estimated properties.

Inversion-derived estimates computed using the MCMC sampling and SA for the whole synthetic formation is shown in Fig. 10.7. The real values of the unknown model parameters, i.e., the four petrophysical properties, namely λ_c, θ_g, σ_w, and S_o, are shown using the red (dark gray in printed version) line. The real values of clay surface conductance and the conductivity of the connate water stay constant across the 25 layers of the synthetic formation. The contact angle of graphite changes form oil wet at the top to intermediately wet in the model and water wet at the bottom of the synthetic formation. Errors in λ_c and σ_w estimates show high correlation. MCMC estimation is shown using the 95% HPDI. All the real values of the unknown properties are captured within the 95% HPDI and are close to the estimates derived using the simulated annealing-based inversion. For all the 25 synthetic layers, the average relative error of λ_c, θ_g, and σ_w is

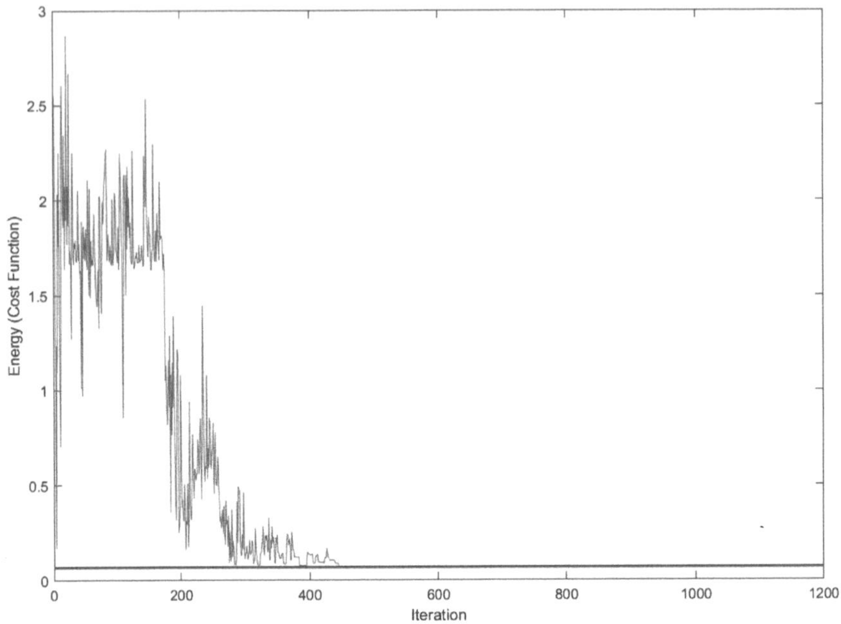

FIGURE 10.5 Energy at each temperature level as a function of number of iterations for the simulated annealing-based inversion to estimate the four unknown petrophysical properties, namely λ_c, θ_g, σ_w, and S_o, of the Layer 2 in the synthetic formation, as shown in Fig. 10.1 and listed in Table 10.3. The energy history stabilizes after about 500 iterations. The *blue line* (gray in print version) represents the energy of the optimal solution, which is close to zero. *Source: Authors.*

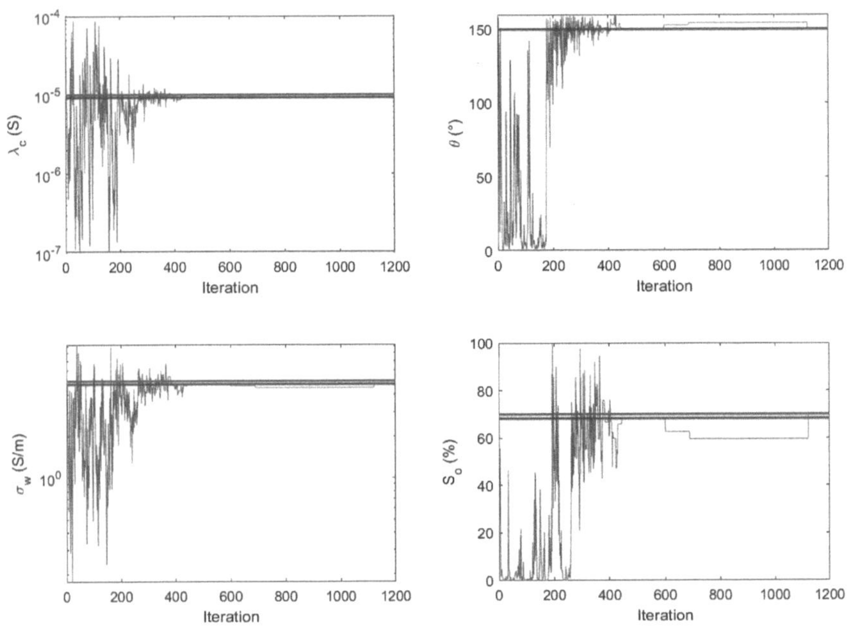

FIGURE 10.6 Simulated annealing-based inversion-derived estimates of the four unknown petrophysical parameters, namely λ_c, θ_g, σ_w, and S_o, for the layer 2 in the synthetic formation, as shown in Fig. 10.1 and listed in Table 10.3. The *red line* (dark gray in print version) represents the real value and the *blue line* (gray in print version) represents the optimal solution, which leads to the minimum energy state shown in Fig. 10.5. *Source: Authors.*

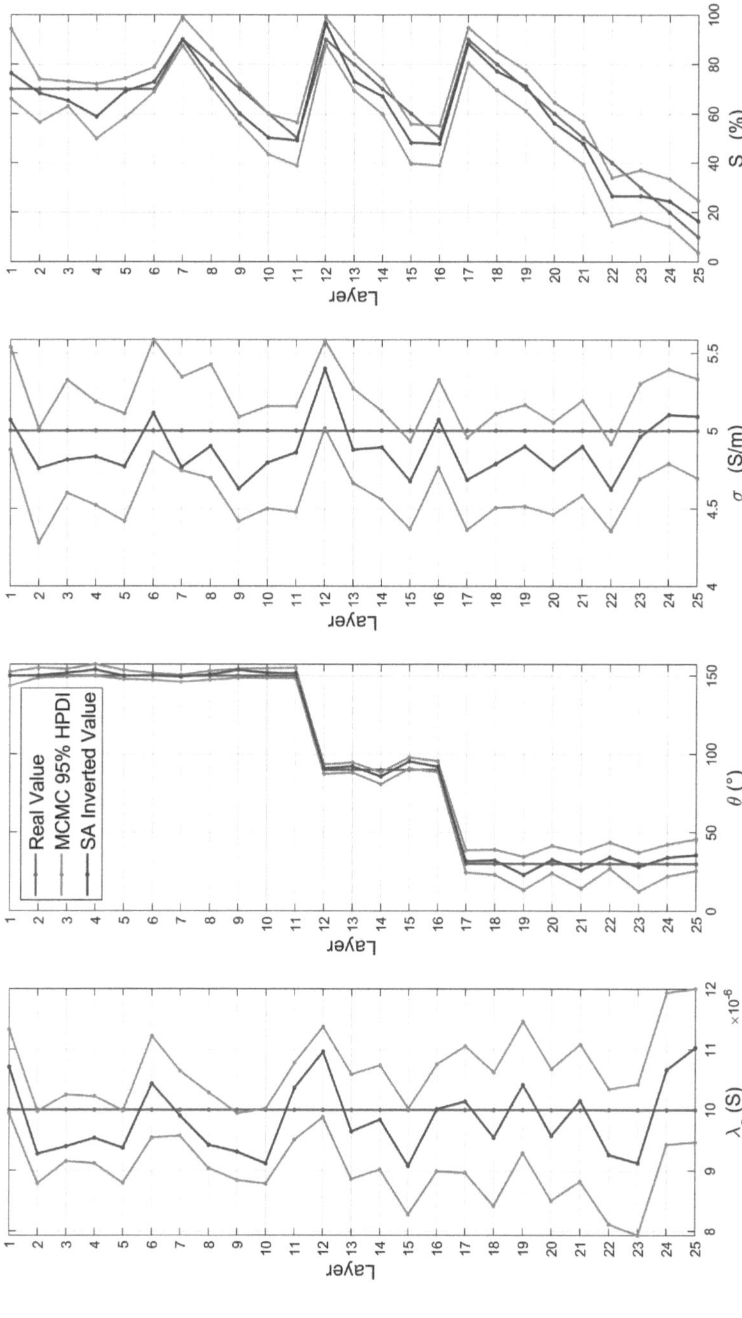

FIGURE 10.7 Inversion-derived estimates of λ_c, θ_g, σ_w, and S_o computed using the Markov-chain Monte Carlo (MCMC) sampling and simulated annealing (SA) for the whole synthetic formation, as shown in Fig. 10.1 and listed in Table 10.3. The *red line* (dark gray in print version) represents the real values of the four unknown properties, the *two green lines* (gray in print version) represent the 95% highest posterior density interval obtained using MCMC inversion method, and the *blue line* (black in print version) represents the estimates computed using the simulated annealing-based inversion. *Source: Authors.*

around 5% and that of S_o is around 10%. Moreover, to estimate the unknown parameters for the whole synthetic formation comprising the 25 synthetic layers, the MCMC inversion takes about 100 min, while SA inversion takes about 40 min. SA requires less iterations as compared to the MCMC, which incurs more iterations to generate the HPDI representing the uncertainty in the estimation. Moreover, it is not easy to determine a proper data covariance matrix $\boldsymbol{C}_d$ for MCMC inversion; consequently, the inversion may take longer time to achieve convergence. SA provides a good approximation in a shorter time; nonetheless, proper values of initial temperature and temperature decline rate must be selected to guarantee faster convergence.

4. Conclusions

In this chapter, water saturation, water conductivity, surface conductance of clay, and contact angle of graphite are simultaneously estimated by processing multifrequency EM measurements using the MCMC inversion and the SA inversion. MCMC inversion generates a range of values for the desired estimate denoting the uncertainty in the estimate. Multifrequency EM responses from 25 distinct layers in a synthetic formation were processed by the two inversion methodologies for purposes of comparing their estimation accuracies and computational times. The determination of data covariance matrix for the MCMC inversion can be a challenge. On the other hand, the SA inversion provides a single value of the desired estimate, which generally represents the optimum value of the estimate. Both the inversion methods provide reliable estimates. MCMC inversion takes about 100 min to perform inversion for all the 25 synthetic layers, while the SA inversion takes about 40 min. MCMC needs more iteration steps to ensure chain convergence as it explores the range of possible values of the desired estimate, representing the uncertainty in the estimation. The computation time of SA can be further reduced by selecting proper values of initial temperature and temperature-decline rate.

Nomenclature

Acronyms

ACO Ant Colony Optimization
EM electromagnetic
HPDI highest posterior density interval
MCMC Markov-chain Monte Carlo
MH Metropolis–Hastings
PPIP perfectly polarized interfacial polarization
PS PPIP–SCAIP
PSO Particle Swarm Optimization
SA Simulated Annealing
SCAIP surface-conductance-assisted interfacial polarization

Symbols

$\boldsymbol{a}$ characteristic length of inclusion phase (m)
α temperature decline rate
$\boldsymbol{C_d}$ data covariance matrix
$\boldsymbol{d}^{\text{meas}} - \boldsymbol{d}(\boldsymbol{m})$ data misfit vector
$\boldsymbol{d}^{\text{meas}}$ measured data vector
$\boldsymbol{d}(\boldsymbol{m})$ modeled data vector
$\boldsymbol{D}$ diffusion coefficient of charge carriers (m^2/s)
ΔE difference between the new energy state and the current energy state
$\Delta \boldsymbol{m}$ length between new solution vector and current solution vector
$\boldsymbol{E}$ cost function/energy
ε_r relative permittivity
γ interfacial tension between oil and water (N/m)
ite iteration number for the inner loop
λ surface conductance (S)
$\boldsymbol{m}'$ new solution vector
$\boldsymbol{m}$ current solution vector
$\boldsymbol{N}$ number of measurements
$\boldsymbol{p}(\boldsymbol{d}|\boldsymbol{m}, \boldsymbol{I})$ likelihood function
$\boldsymbol{P}$ probability of accepting the new solution
ϕ porosity of the porous media (%)
ϕ_i volume fraction of medium i in the mixture (%)
ρ density of host (kg/m^3)
$\boldsymbol{S}$ solution space
$\boldsymbol{S_o}$ oil saturation (%)
$\boldsymbol{\sigma}$ standard deviation vector
σ electrical conductivity (S/m)
$\boldsymbol{T}$ temperature of the SA process
$\boldsymbol{T_o}$ initial temperature of the SA process
$\boldsymbol{T_f}$ final temperature of the SA process
θ contact angle (°)

Subscripts

c clay
eff effective
g graphite
h host medium
i inclusion phase
o oil
r relative
s sand
w water

References

Andrieu, C., De Freitas, N., Doucet, A., Jordan, M.I., January 1, 2003. An introduction to MCMC for machine learning. Mach. Learn. 50 (1–2), 5–43.

Boender, C.G., Romeijn, H.E., 1995. Stochastic methods. In: Handbook of Global Optimization. Springer, Boston, MA, pp. 829–869.

Chibante, R. (Ed.), August 18, 2010. Simulated Annealing: Theory with Applications. BoD—Books on Demand.

Da Silva Moreira, P.H., Gomes da Silveira, T.M., Drexler, S., Couto, P., June 3, 2019. Application of MCMC optimization method to estimate relative permeability of carbonate rocks from unsteady-state core flood experiments. In: SPE Europec Featured at 81st EAGE Conference and Exhibition. Society of Petroleum Engineers.

Dachanuwattana, S., Xia, Z., Yu, W., Qu, L., Wang, P., Liu, W., Miao, J., Sepehrnoori, K., August 1, 2018. Application of proxy-based MCMC and EDFM to history match a shale gas condensate well. J. Petrol. Sci. Eng. 167, 486—497.

Dunn, W.L., Shultis, J.K., May 24, 2011. Markov chain Monte Carlo. In: Exploring Monte Carlo Methods. Elsevier.

Han, Y., Misra, S., Wang, H., Toumelin, E., May 1, 2019. Hydrocarbon saturation in a Lower-Paleozoic organic-rich shale gas formation based on Markov-chain Monte Carlo stochastic inversion of broadband electromagnetic dispersion logs. Fuel 243, 645—658.

Li, X., Reynolds, A.C., February 20, 2017. Generation of a proposal distribution for efficient MCMC characterization of uncertainty in reservoir description and forecasting. In: SPE Reservoir Simulation Conference. Society of Petroleum Engineers.

Liberti, L., Kucherenko, S., May 2005. Comparison of deterministic and stochastic approaches to global optimization. Int. Trans. Oper. Res. 12 (3), 263—285.

Osman, I.H., Potts, C.N., January 1, 1989. Simulated annealing for permutation flow-shop scheduling. Omega 17 (6), 551—557.

Ouenes, A., Saad, N., January 1, 1993. A new, fast parallel simulated annealing algorithm for reservoir characterization. In: SPE Annual Technical Conference and Exhibition. Society of Petroleum Engineers.

Shen, Q., Wu, X., Chen, J., Han, Z., Huang, Y., February 1, 2018. Solving geosteering inverse problems by stochastic Hybrid Monte Carlo method. J. Petrol. Sci. Eng. 161, 9—16.

Sinharay, S., December 2004. Experiences with Markov chain Monte Carlo convergence assessment in two psychometric examples. J. Educ. Behav. Stat. 29 (4), 461—488.

Szucs, P., Civan, F., May 1, 1996. Multi-layer well log interpretation using the simulated annealing method. J. Petrol. Sci. Eng. 14 (3—4), 209—220.

Tukur, A.D., Nzerem, P., Nsan, N., Okafor, I.S., Gimba, A., Ogolo, O., Oluwaseun, A., Andrew, O., August 5, 2019. Well placement optimization using simulated annealing and genetic algorithm. In: SPE Nigeria Annual International Conference and Exhibition. Society of Petroleum Engineers.

Chapter 11

Multifrequency electromagnetic data interpretation using particle swarm optimization and ant colony optimization methods

Siddharth Misra[1,2], Yuteng Jin[3]
[1]*Harold Vance Department of Petroleum Engineering, College of Engineering, Texas A&M University, TX, United States;* [2]*Department of Geology and Geophysics, Texas A&M University, TX, United States;* [3]*Harold Vance Department of Petroleum Engineering, Texas A&M University, TX, United States*

1. Introduction

This chapter adopts an outline similar to the previous chapter. Our objective is to apply and compare parameter estimation methods based on swarm intelligence and probabilistic multiagent algorithms for purposes of multifrequency electromagnetic (EM) data interpretation. To that end, particle swarm optimization (PSO) and ant colony optimization (ACO) will be used to process the multifrequency EM data. The underlying principle of swarm intelligence is as follows: a group of agents, each following simple rules without centralized control, can exhibit complexity, self-organization and creativity because a small change in one agent influences all the other agents, leading to the emergence of new behavior of the group as a whole. For instance, tiny ants forming intelligent cooperative colonies and bees building sophisticated hives.

PSO and ACO belong to the family of nature-inspired metaheuristics, also referred as metaphor-based metaheuristics, which is a heuristic (procedure) to find a heuristic that may find a sufficiently good solution to an optimization or parameter estimation problem. Other popular nature-inspired metaheuristics include evolutionary algorithm and artificial bee colony. Metaheuristics make few or no assumptions about the optimization problem while searching large spaces of candidate solutions. However, metaheuristics do not quantify the confidence in a solution, so quality of the solution is not known. Like Monte

Multifrequency Electromagnetic Data Interpretation for Subsurface Characterization
https://doi.org/10.1016/B978-0-12-821439-8.00009-4

Carlo Markov Chain (MCMC) sampling and simulated-annealing (SA) based inversion described in Chapter 10, PSO and ACO inversions do not require gradient calculations, which is especially useful when the gradient is impossible or too laborious to compute. ACO and PSO can be directly applied on the fitness measure or the objective function to be optimized without requiring gradient calculations. Similar to simulated annealing, both PSO and ACO are suited for global optimization.

1.1 Particle swarm optimization

PSO is a general-purpose method for parameter estimation, inversion modeling, and optimization that uses several agents, referred as particles, moving around in the model-parameter space (i.e., search phase) in search of candidate solutions (Kennedy and Eberhart, 1995). The location of a particle denotes a candidate solution. For each iteration, as particles move, the magnitude and direction of velocity of each particle is governed by three factors: (1) velocity of the particle in the previous iteration, (2) best candidate solution discovered thus far by the group as a whole, and (3) the best candidate solution discovered by the particle during the search process thus far. There are three parameters that determine the extent of each contribution; thereby determining the tradeoff between local search and global search. This algorithm is inspired by the behavior of birds during flocking.

Shi and Eberhart (1998) modified PSO by introducing the inertia weight, and Clerc (1999) modified PSO by introducing the constriction factor. Zhan et al. (2009) presented an adaptive PSO algorithm with higher efficiency and reliability. Shen et al. (2014) proposed a discrete PSO algorithm which is successfully applied on multicast routing problem in communication networks. Tasgetiren et al. (2004) illustrated a PSO algorithm based on variable neighborhood search strategy to solve a sequencing problem. PSO is also applied on fuzzy system control (Chiou et al., 2012), neural network training (Bashir and El-Hawary, 2009), and function optimization problems. The theory behind the PSO algorithm is simple, and it is easy to implement with small quantity of parameters to be adjusted. It can converge relatively quick with low-computational requirement. It takes advantage of both self-cognition and social information that helps bypass local optimum traps. However, the local search capability of PSO is poor leading to lower precision of the solution.

1.2 Ant colony optimization

Similar to the PSO, the ACO uses swarm intelligence (Dorigo, 1992). ACO finds an approximate solution for a combinatorial optimization problem by mimicking the behavior of real ant colonies. Combinatorial optimization aims to find an optimal object from a finite set of objects, when an exhaustive search is not computationally feasible. ACO is specially designed to find optimal paths. To apply ACO, the optimization problem needs to be transformed into a problem of finding the best path in a connected graph. ACO uses agents, referred as ants, which incrementally build probabilistic solutions by moving on the graph. ACO builds multiple probabilistic solutions by iteratively adding components to partial solutions by using heuristics and by using the continuously evolving search experience acquired by the ants. The stochastic component of ACO allows the ants to build a wide variety of solutions that facilitates the exploration of the model-parameter space.

At the start, ants leave the colony and randomly wander the model-parameter space (i.e., solution space) in search for food (i.e., candidate solution). When an ant finds the food (i.e., candidate solution), it takes a portion of the food and retraces the path back to the colony while leaving "markers" that highlight the path taken by the ant to get to the food. When other randomly wandering ants encounter markers, they follow the preidentified path to the food with a certain probability. These ants after reaching the left-over food will take a portion of the food and travel back to the colony by retracing their path from the colony to the food. As a predefined path is traveled by more ants, the path gets populated by more markers; subsequently, more ants will take the path that is getting densely populated with the markers. Consequently, few paths in the connected graph become denser than all other paths as more ants travel the specific few paths to the candidate solution. All this while, at any given time, several ants are always randomly scouting for shorter paths to candidate solutions and those that are closer to the colony. The underlying principle assumes that shorter paths to the food are more likely to get densely populated by markers; thereby, getting identified as the optimal path to the solution.

ACO algorithm has been successfully applied to a wide range of problems including traveling salesman problem (Gambardella and Dorigo, 1995), scheduling problem (Colorni et al., 1994), assignment problem (Stützle, 1997), vehicle routing problem (Donati et al., 2008), protein folding problem (Shmygelska et al., 2002), network routing problem (Okdem and Karaboga, 2009), graph coloring problem (Salari and Eshghi, 2005), and classification problem (Martens et al., 2007). The advantages of ACO algorithm include that it has a good performance in dynamic applications, where it can be run continuously and adapt to changes in real time. The ACO algorithm is suitable for parallel computing and efficient for computationally intractable problems, like traveling salesman problem. The positive feedback incorporated in the ACO algorithm helps to find the candidate solutions in a short time; however, a major disadvantage of the ACO algorithm is that it easily gets stuck with the local minima.

1.3 Petrophysical estimation based on stochastic inversion of multifrequency electromagnetic data

Similar to Chapter 10, in this chapter, multifrequency EM data acquired at frequencies corresponding to EM induction resistivity tool (operating at 10 and 20 kHz), logging while drilling propagation tool (operating at 1 and 2 MHz), and dielectric dispersion tool (operating at 20, 100, 260 MHz, and 1 GHz) are jointly inverted to estimate four petrophysical properties at each depth of the formation. The four petrophysical estimates include water saturation, water conductivity, surface conductance of clay, and contact angle of graphite in synthetic formation. To that end, the PSO and ACO methods are coupled with the improved PPIP–SCAIP (PS) model to process the multifrequency EM responses, comprising effective conductivity and effective relative permittivity at eight frequencies. In total, there are 16 data points at each depth that need to be inverted to obtain the four above-mentioned properties at the depth of interest.

2. Method

2.1 The mechanistic electromagnetic model for subsurface characterization

In this chapter, the improved PS model (as discussed in Chapter 5) is used as the mechanistic EM model to guide the parameter estimation. The improved PS model considers the effect of wettability of conductive particles. The improved PS model, which serves as the forward model, is discussed in detail in Chapter 5. In real world application, the improved PS model will need to be calibrated before inverting multifrequency EM data. The PSO and ACO methods are coupled with the improved PS model to process the multifrequency EM data. The inversion methods can be coupled to any other forward model to process any other multifrequency dataset.

2.2 The synthetic formation

The improved PS model is used to generate multifrequency EM responses for a synthetic formation comprising 25 distinct layers. There are 16 EM measurements, comprising eight conductivity and eight permittivity, for each depth. The synthetic formation is same as the one used in Chapter 10. The synthetic formation is designed to study the efficacy of inversion methods to variations in wettability, fluid saturation, and volume fractions of conducive minerals and clays. Overall, the synthetic formation contains three zones, i.e., water-wet, intermediately wet, and oil-wet zones made of 11, 5, and 9 layers, respectively. The synthetic formation is illustrated in Fig. 10.1, and the properties of each layer are listed in Table 10.3. For a specific zone, each layer has distinct volume fractions of conductive mineral, sand, clay, and oil. For more information about the synthetic formation, refer to Chapter 10. For this synthetic formation, layers 1–11 contain oil-wet graphite, layers 12–16 contain intermediate-wet graphite, and layers 17–25 contain water-wet graphite.

2.3 Proposed inversion-based interpretation method

2.3.1 *Particle swarm optimization based inversion*

The PSO algorithm is inspired by the model developed by Heppner and Grenander (1990) to describe the flocking behavior of birds. Each bird (particle) is assigned a location and velocity. The velocity is updated according to the current velocity, self-cognition (local information), and social information (global behavior of the flock/swarm). The location is then updated according to the velocity so that the birds move closer to the roost (optimal solution). The basic idea of PSO is that each particle in a swarm collaborates and shares information with each other to search for the best solution. Such interparticle interactions lead to the discovery of optimal solutions. The flowchart of the PSO algorithm used in this chapter is shown in Fig. 11.1.

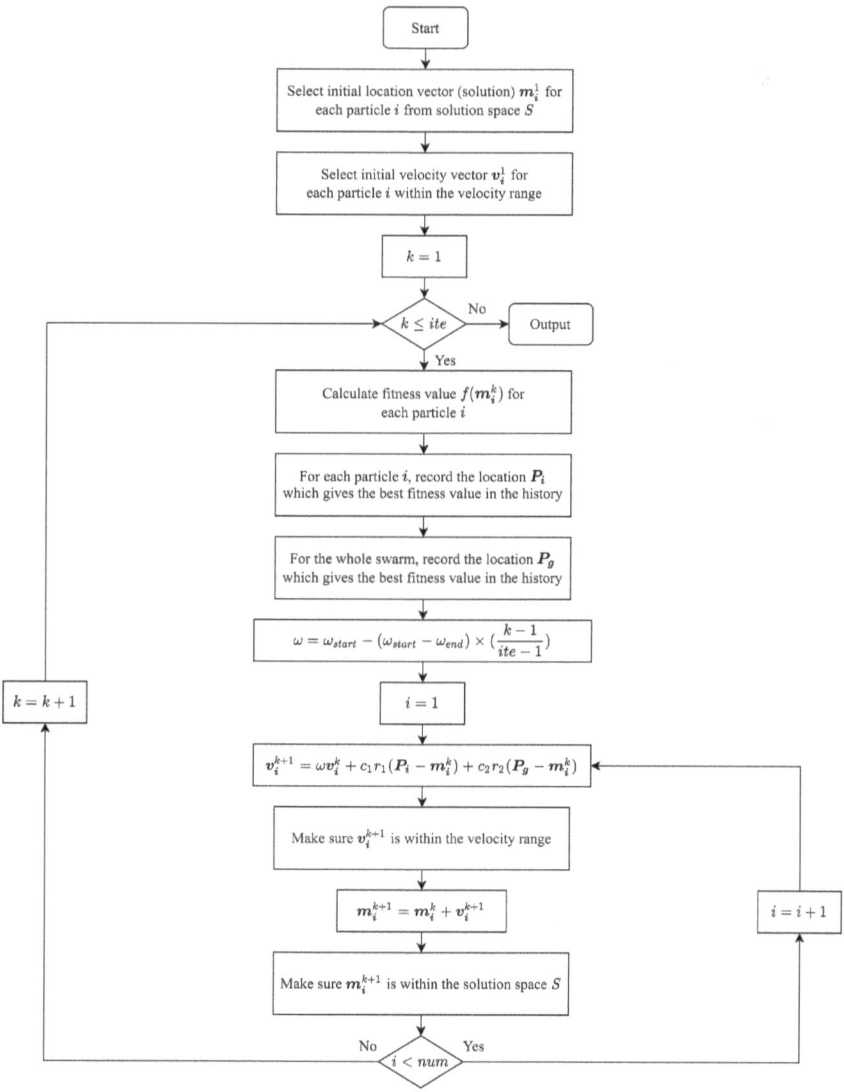

FIGURE 11.1 Flowchart of the particle swarm optimization-based inversion of the multifrequency electromagnetic data to estimate the four petrophysical properties of each layer. *Source: Authors.*

2.3.1.1 Update of particle velocity

In PSO, the initial location (solution) $\boldsymbol{m}_i^1$ to each particle i is randomly selected from the predefined solution space S. Each particle is also randomly assigned initial velocity $\boldsymbol{v}_i^1$ within the velocity range $[-\boldsymbol{v}_{max}, \boldsymbol{v}_{max}]$, where $\boldsymbol{v}_{max}$ is the maximum possible velocity of the particle, which equals to 10% of the range of the solution space S. In a standard PSO algorithm, the locations of the particles denote possible candidate solutions, and the swarm of particles is searching the solution space for the optimal solution. So, the PSO method focuses on changing the location of a particle by first computing the new velocity of the particle. As shown in Eq. (11.1), the velocity of a particle is updated based on three contributions: (1) influence of the current velocity of the particle, (2) influence of the local solution space around the particle, and (3) influence of all other particles. At each iteration $(k+1)$, the velocity of each particle is updated as

$$\boldsymbol{v}_i^{k+1} = \omega \boldsymbol{v}_i^k + c_1 r_1 \left(\boldsymbol{P}_i - \boldsymbol{m}_i^k\right) + c_2 r_2 \left(\boldsymbol{P}_g - \boldsymbol{m}_i^k\right) \tag{11.1}$$

where ω, c_1, and c_2 are predefined constants that govern the search for the solution, the superscript k denotes the iteration number, the subscript i represents the particle i, r_1 and r_2 are random numbers generated in the range $(0, 1)$, $\boldsymbol{P}_i$ is the location representing the best candidate solution identified by the particle i from the start to the $(k+1)$ iteration, and $\boldsymbol{P}_g$ is the location representing the best candidate solution identified by the whole swarm (all the particles) from the start to the $(k+1)$ iteration.

In Eq. (11.1), c_1 and c_2 are acceleration coefficients or learning factors, which controls the cognition influence (local information) and social influence (global trend), respectively. The two constants control the speed at which the particles move toward the best individual and the global candidate solutions ($\boldsymbol{P}_i$ and $\boldsymbol{P}_g$, respectively). Lower acceleration coefficients cause the particles to wander around exploring the solution space and slowly converge to the optimal solution, while higher acceleration coefficients cause the particles to bypass the optimal solution and never converge. If $c_1 = 0$, the algorithm becomes "social-only," where it quickly converges to a solution identified by the whole swarm, but it is much easier to fall into the local optimum trap. If $c_2 = 0$, the algorithm becomes "cognition-only," where each particle moves around without getting influenced and without influencing other particles; thus, the parameter estimation is hard to converge because most particles find distinct local-optimum solutions.

In Eq. (11.1), ω is inertia weight, which allows the particles to explore outside the neighborhood of optimum. In this chapter, we use a linearly decreasing inertia weight (LDIW) as proposed by Shi and Eberhart (1999). At the early stage of the algorithm, inertia weight is high, which helps the particles jump out of the local optimum trap; thereby promoting global search. As the algorithm iterates longer and approaches toward the end of iteration, the inertia weight decreases, which benefits local search to facilitate the convergence to the optimum solution. ω is updated at each iteration as

$$\omega = \omega_{start} - (\omega_{start} - \omega_{end}) \times \left(\frac{k-1}{ite-1}\right) \tag{11.2}$$

where k is an index representing iteration step, ite is the total iteration time, ω_{start} is the predefined initial inertia weight, and ω_{end} is the predefined final inertia weight. The velocity calculated at each iteration needs to be lower than the maximum velocity $\boldsymbol{v}_{max}$.

2.3.1.2 Update of particle location

The location of each particle is updated at each iteration as

$$\boldsymbol{m}_i^{k+1} = \boldsymbol{m}_i^k + \boldsymbol{v}_i^{k+1} \tag{11.3}$$

Once the particle reaches the boundary, it will continue to fly toward the opposite direction until the total distance traveled in that iteration satisfies the distance traveled due to the new particle velocity.

2.3.1.3 Fitness function

Parameter estimation, inverse modeling, or system identification solves an optimization problem to find the best model parameters within a permissible range (referred as solution space) by maximizing or minimizing an objective/cost/fitness function (for example, residual sum of squares). In this chapter, the fitness function to be minimized is defined as following:

$$f(\boldsymbol{m}) = \sqrt{\sum \left[\log(\boldsymbol{d}^{\text{meas}}) - \log(\boldsymbol{d}(\boldsymbol{m}))\right]^2} \tag{11.4}$$

where $\boldsymbol{d}^{\text{meas}}$ is the measured data vector and $\boldsymbol{d}(\boldsymbol{m})$ is the modeled data vector, as defined in Chapter 8. Taking the common logarithm of the data vectors ensures that the data vectors have the same orders of magnitude, such that both conductivity and permittivity are equally honored. Conductivity ranges from 0.001 to 10 S/m, while the relative permittivity ranges from 1 to 10^5. The logarithmic transformation of conductivity leads to values between -3 and 1 and that of relative permittivity leads to values between 0 and 5. Consequently, the logarithmic transformation ensures that the parameter estimation process assigns similar weightage to both permittivity and conductivity, which vary significantly in their scales. The logarithmic transformation serves as a scaling operation (more details are provided by Misra et al., 2019).

2.3.2 Ant colony optimization-based inversion

The ACO algorithm is inspired by the study that the ant colony can quickly find the shortest path from the nest to a food source by sensing each other's pheromone left as markers while exploring the environment for food. When an ant finds a food source, the ant walks back to nest with some food and releases pheromone while retracing its travel path to attract other ants to the path leading to the food. Over time, the pheromone on less traveled paths will evaporate to prevent ants from getting misled by the paths that do not lead to the food. Pheromone on longer paths to the food will also evaporate faster than the pheromone on shorter paths to the food because more ants can travel on the shorter path as compared to the longer path; thus, the pheromone density becomes higher on the shorter paths leading to food. As a result, the ants are more likely to follow the shorter path and the positive feedback eventually generates a single shortest path that leads to a single food source.

In this chapter, we use a max-min ant system (MMAS), which is developed by Stützle and Hoos (1996). Fig. 11.2 illustrates the workflow used in this chapter to achieve the desired ACO-based inversion. MMAS promotes the exploitation of the best solutions found during the search and prevents the early stagnation of the search. For the purpose of exploitation, only the global-best ant or iteration-best ant is allowed to add pheromone. Global-best ant is the one that has found the best solution from the start of search to the current iteration. Iteration-best ant is one that finds the best solution in the current iteration. This allows other ants to keep on searching without encountering pheromone of other ants; thereby, preventing the candidate paths to be found by a lot of ants. To avoid the stagnation of search, the amount of pheromone added by the iteration-best ant or the global-best ant is constrained within a specific range. This allows quicker evaporation of pheromone; thereby promoting exploration of the entire search space without getting biased by local optimum.

2.3.2.1 Initialization of ant location

In ACO, initial location (solution) $\boldsymbol{m}_i^1$ for each ant i is randomly selected from the predefined solution space S.

2.3.2.2 Objective function

Similar to the PSO method, Eq. (11.4) defines the objective function to be minimized in the ACO method.

2.3.2.3 Pheromone vector

In the traditional ACO, after an ant finds the food (i.e., candidate solution), the ant leaves pheromone (marker) on the path that led the ant to the food. Pheromone/maker on a path determines the likelihood of finding the solution when the path is chosen by other ants during the search. Pheromone acts as information on the path. In the Max-Min Ant System (MMAS), pheromone represents the attractive strength of an ant, which is different from the traditional ACO formulation. In MMAS, the pheromone τ is constrained

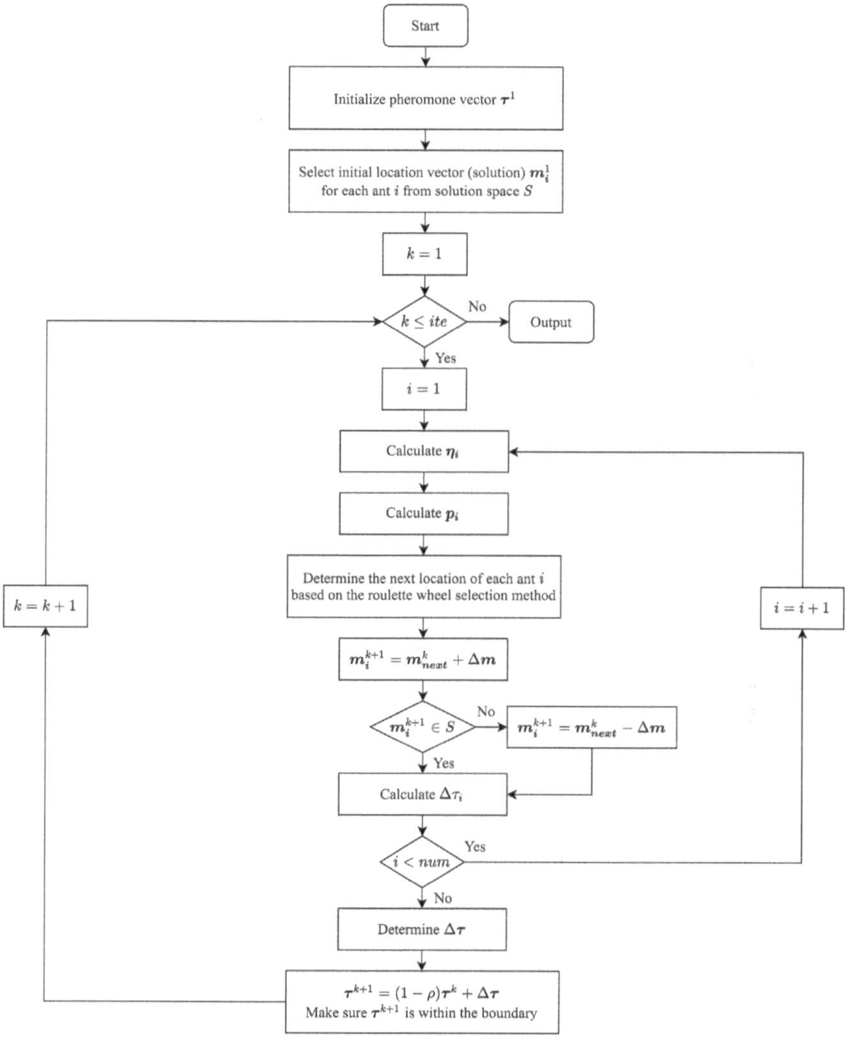

FIGURE 11.2 Flowchart of the ant colony optimization-based inversion of the multifrequency electromagnetic data to estimate the four petrophysical properties of each layer. *Source: Authors.*

within a specific range $[\tau_{min}, \tau_{max}]$ to avoid stagnation of the search. The pheromone is set to the boundary values when it exceeds the range. The initial pheromone of ant i at initial location is set as

$$\tau_i^1 = \tau_{max}\ (i = 1, 2, 3, \ldots, num) \tag{11.5}$$

where *num* is the number of ants, and subscript i represents an ant. Pheromone of each ant is combined into the pheromone vector $\boldsymbol{\tau}$.

2.3.2.4 Heuristic function

At each iteration k, the heuristic function $\eta_{i,j}$ of ant i evaluates the decrease of objective function if the ant moves from the current location to the location of ant j. $\eta_{i,j}=0$ when the move leads to an increase in objective function. When $\eta_{i,j}=0$, the movement of ant i toward ant j has zero possibility. $\eta_{i,j}$ is formulated as

$$\eta_{i,j}=\begin{cases} Q\left[f\left(\boldsymbol{m}_i^k\right)-f\left(\boldsymbol{m}_j^k\right)\right] & f\left(\boldsymbol{m}_i^k\right)>f\left(\boldsymbol{m}_j^k\right) \\ 0 & else \end{cases} \tag{11.6}$$

where Q is a constant controlling the magnitude of decrease, and $f(\boldsymbol{m})$ is the value of the objective function at location $\boldsymbol{m}$.

2.3.2.5 Possibility of ant movement

At each iteration k, ant i moves toward the ant *next* according to the possibility $p_{i,j}$, which determines the possibility of ant i moving toward ant j, is defined as

$$p_{i,j}=\begin{cases} \dfrac{\tau_j^\alpha \eta_{i,j}^\beta}{\sum_{j=1}^{num}\tau_j^\alpha \eta_{i,j}^\beta} & \eta_{i,j}>0 \\ 0 & \eta_{i,j}=0 \end{cases} \tag{11.7}$$

where α is pheromone importance factor, β is heuristic function importance factor, *num* is the number of ants, and subscripts i and j represent locations of any two ants. The selection of the ant *next* that ant i will move toward is based on the roulette wheel selection method. In this method for selecting ant *next*, first the cumulative sum of the possibility $p_{i,j}$ is calculated for each ant i with respect to each of the other ants, j. Then, a random number *rand* is selected from $(0,1)$. Following that, the lowest index of an ant for which the cumulative sum of $p_{i,j}$ is higher than *rand* is chosen as the index of the ant *next*. When no such index is found or when the cumulative sum of $p_{i,j}$ is zero, ant i will move around the current location of the ant and will not move toward any other ant in specific.

2.3.2.6 Update of ant location

After determining an ant *next* located at $\boldsymbol{m}_{next}^k$ toward which ant i will move, the new location of ant i is denoted by the $\boldsymbol{m}_i^{k+1}$, which can be calculated as

$$\boldsymbol{m}_i^{k+1}=\boldsymbol{m}_{next}^k+\Delta\boldsymbol{m} \tag{11.8}$$

where the superscript k represents the current iteration number, the subscript i or *next* represents the ant i or ant *next*, respectively, and $\boldsymbol{m}_{next}^k$ is the current location of ant *next* toward which ant i will move. $\Delta \boldsymbol{m}$ is referred as the disturbance that can be positive or negative and is formulated as

$$\Delta \boldsymbol{m} = (rand - 0.5)\frac{\boldsymbol{m}_{max} - \boldsymbol{m}_{min}}{2} r^k \tag{11.9}$$

where *rand* is a random number from $(0, 1)$, $\boldsymbol{m}_{max}$ and $\boldsymbol{m}_{min}$ is the upper and lower boundaries of the solution space S, respectively, r is the factor between 0 and 1 that determines the shrinkage of solution-search interval after each iteration, and k is the iteration number. As per Eq. (11.8), the disturbance $\Delta \boldsymbol{m}$ is added so that the ant i can explore new locations around the current location of ant *next*. As k increases and approaches the iteration time *ite*, r^k reduces; consequently, the ants will explore only around the optimal solutions. When the new solution obtained by Eq. 11.8 is out of the solution space S, the following equation will be used instead of Eq. 11.8 without recalculating $\Delta \boldsymbol{m}$:

$$\boldsymbol{m}_i^{k+1} = \boldsymbol{m}_{next}^k - \Delta \boldsymbol{m} \tag{11.10}$$

2.3.2.7 Incremental pheromone

In the MMAS, pheromone represents the attractive strength of an ant, which is a different formulation of pheromone as compared to the traditional ACO. The aim of the pheromone update is to increase the pheromone values associated with good solutions through selective pheromone increment, and to decrease those that are associated with bad solutions through pheromone evaporation. The incremental pheromone that an ant i can receive at the iteration k is expressed as

$$\Delta \tau_i = \begin{cases} Q\left[f\left(\boldsymbol{m}_i^k\right) - f\left(\boldsymbol{m}_{next}^k\right)\right] & f\left(\boldsymbol{m}_i^k\right) > f\left(\boldsymbol{m}_{next}^k\right) \\ 0 & \text{else} \end{cases} \tag{11.11}$$

where Q is a constant controlling the magnitude of incremental pheromone, and $f(\boldsymbol{m})$ is the value of the objective function at location $\boldsymbol{m}$. In MMAS, only the iteration-best ant or the global-best ant is allowed to add incremental pheromone $\Delta \tau$ at any iteration k, while the rest do not add any pheromone. At any given iteration, each ant moves toward respective ant *next*. The ant receiving the incremental pheromone is the one that experienced maximum change in the objective function among all ants because of the move from previous location to the new location. The amount of incremental pheromone $\Delta \tau_i$ is proportional to the reduction in the objective function when the ant i moved toward the ant *next*. In MMAS method, only the iteration-best ant receives the pheromone increment $\Delta \tau_i$; in other words, only the highest $\Delta \tau_i$ is used in the incremental pheromone vector $\Delta \boldsymbol{\tau}$.

The pheromone vector corresponding to all ants is updated after each iteration as follows

$$\tau^{k+1} = (1-\rho)\tau^{k} + \Delta\tau \tag{11.12}$$

where ρ is the pheromone volatilization factor. At any given iteration, all pre-existing pheromones will tend to evaporate and only the best ant adds incremental pheromone. Pheromone evaporation leads to forgetting candidate solutions so as to explore new areas in the search space.

3. Results and discussions

For both PSO and ACO-based inversion methods, we need to predefine the range of solution space and initial guesses for the unknown petrophysical properties to initiate the estimation. The solution spaces of the four unknown petrophysical properties are shown in Table 11.1. In this chapter, for PSO method, the total number of particles in the swarm (*num*) is 30, iteration time *ite* is 500, acceleration coefficient/learning factor 1 c_1 is 3, acceleration coefficient/learning factor 2 c_2 is 1.5, initial inertia weight ω_{start} is 0.8, and final inertia weight ω_{end} is 0.4. As for ACO method, the total number of ants in the colony (*num*) is 5, iteration time *ite* is 6000, pheromone importance factor α is 5, heuristic function importance factor β is 5, pheromone volatilization factor ρ is 0.5, solution search interval shrink factor r is 0.999, the constant Q is 1000, upper pheromone boundary τ_{max} is 1000, and lower pheromone boundary τ_{min} is 10. Both PSO and ACO inversion methods are applied on all the 25 synthetic layers in the synthetic formation.

For Layer 2, the inversion histories of the global best fitness valued in the PSO method and ACO method are shown in Figs. 11.3 and 11.5, respectively. The inversion histories of the four unknown petrophysical properties as estimated by the 30 particles in the PSO method and five ants in the ACO method are shown in Figs. 11.4 and 11.6. Global best fitness at any iteration corresponds to 1 of the 30 particle or 1 of the 5 ants that exhibits lowest value of the objective function. In Figs. 11.3 and 11.5, the blue (gray in printed version) lines represent the fitness value of the optimal solutions identified by PSO and ACO methods, respectively, which are close to zero. In Figs. 11.4 and 11.6, the blue (dark gray in printed version) lines identifying the petrophysical estimates are close to the red (gray in printed version) line identifying the real values of these unknown petrophysical properties. For both PSO and ACO-based inversion methods, we can conclude that the best estimates are

TABLE 11.1 The solution spaces of the unknown petrophysical properties for both particle swarm optimization and ant colony optimization-based inversion methods.

Parameters	λ_c (S)	θ_g (°)	σ_w (S/m)	S_o (%)
Solution space	$(1\times10^{-7}, 1\times10^{-4})$	(0, 180)	(0.1, 10)	(0.1, 99.9)

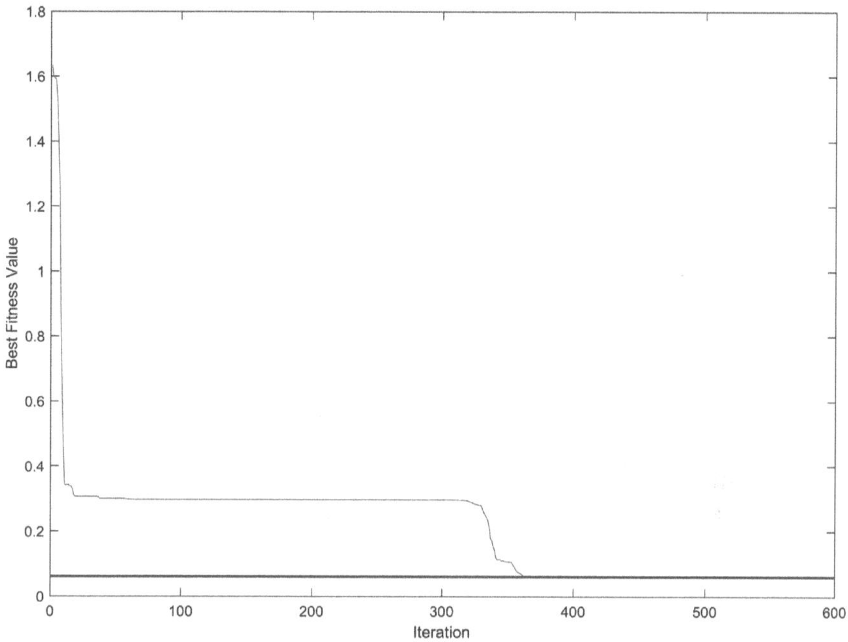

FIGURE 11.3 Global best fitness as a function of number of iterations for the particle swarm optimization-based inversion to estimate the four unknown petrophysical properties, namely λ_c, θ_g, σ_w, and S_o, of the Layer two in the synthetic formation, as shown in Fig. 10.1 and listed in Table 10.3. The blue (gray in printed version) line represents the fitness value of the optimal solution, which is close to zero. *Source: Authors.*

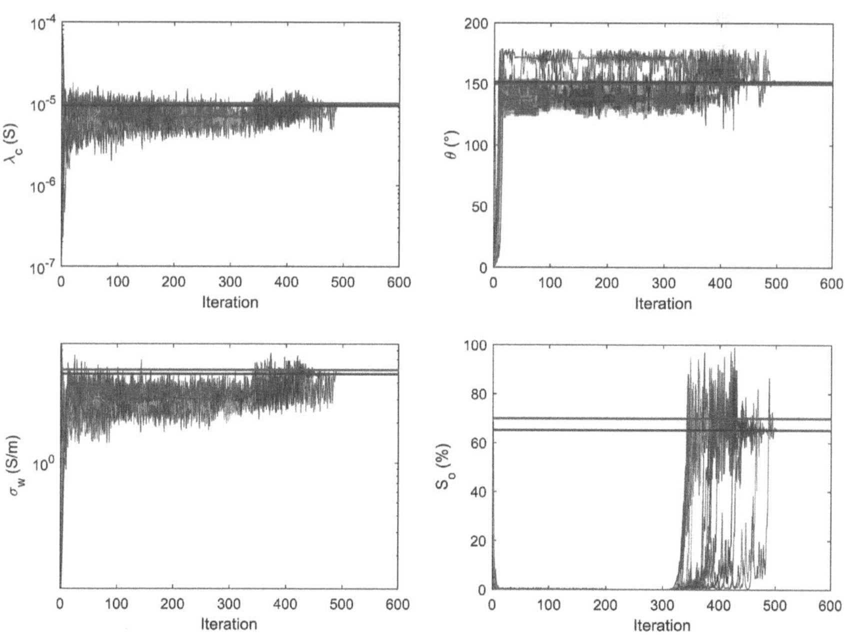

FIGURE 11.4 History of particle swarm optimization-based estimates for the four unknown petrophysical properties, namely λ_c, θ_g, σ_w, and S_o, as computed by the 30 particles in the swarm for the Layer two in the synthetic formation, as shown in Fig. 10.1 and listed in Table 10.3. The particles converged after about 500 iterations. The red (gray in printed version) line represents the real value, and the blue (dark gray in printed version) line represents the optimal solution that results in the lowest value of the fitness/objective function. *Source: Authors.*

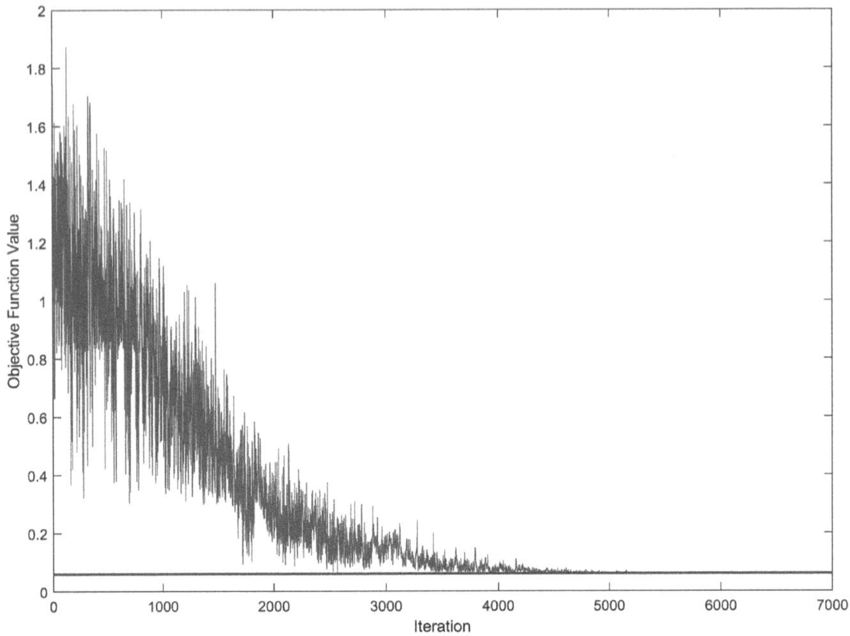

FIGURE 11.5 Global best fitness as a function of number of iterations for the ant colony optimization-based inversion to estimate the four unknown petrophysical properties, namely λ_c, θ_g, σ_w, and S_o, of the Layer two in the synthetic formation, as shown in Fig. 10.1 and listed in Table 10.3. The blue (gray in printed version) line represents the fitness value of the optimal solution, which is close to zero. *Source: Authors.*

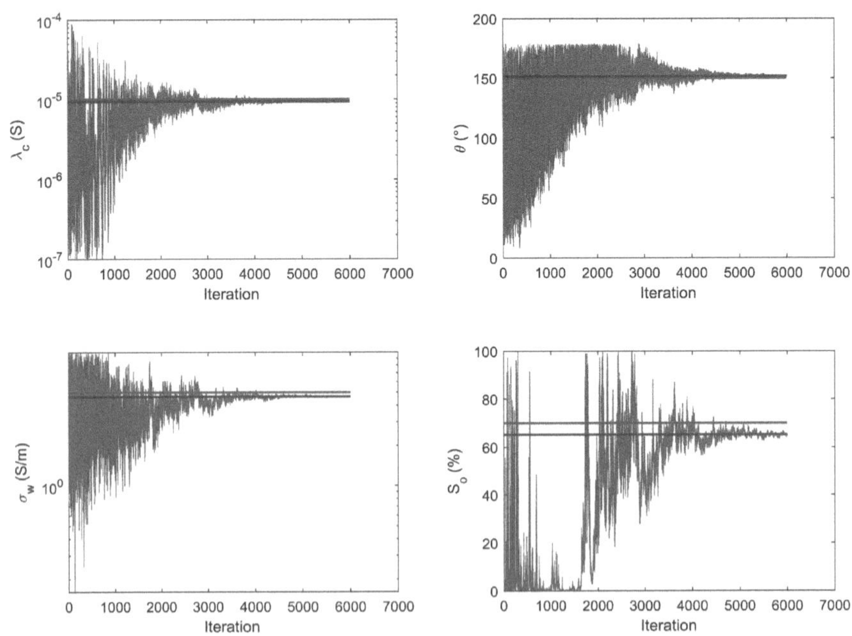

FIGURE 11.6 History of ant colony optimization-based estimates for the four unknown petrophysical properties, namely λ_c, θ_g, σ_w, and S_o, as computed by the five ants in the swarm for the Layer two in the synthetic formation, as shown in Fig. 10.1 and listed in Table 10.3. The particles converged after about 5000 iterations. The red (gray in printed version) line represents the real value and the blue (dark gray in printed version) line represents the optimal solution that results in the lowest value of the fitness/objective function. *Source: Authors.*

good approximations to the real values. Because of the low sensitivity of EM responses to oil saturation S_o at high-oil saturation, the estimation of S_o is the least accurate among the four petrophysical properties that were estimated. At high-oil saturation, estimates of oil saturation are less accurate as compared to the estimates of clay surface conductance, volume fraction of conductive mineral, and the contact angle of the conductive mineral. Chapter 10 also had similar results. Average relative error in the S_o estimates over the 25 synthetic layers is 10.3%, as compared to an average relative error of 5% for each of the other three properties, namely λ_c, σ_w, and θ_g. PSO method requires 500 iterations to stabilize, whereas the ACO method requires 5000 iterations.

Inversion-derived estimates computed using the PSO and ACO inversion results for the entire synthetic formation is shown in Fig. 11.7. The real value of the unknown model parameters, i.e., the four petrophysical properties, namely λ_c, θ_g, σ_w, and S_o, are shown using the red (dark gray in printed version) line. The real values of clay surface conductance and the conductivity of the connate water stay constant across the 25 layers of the synthetic formation. The contact angle of graphite, changes form oil wet at the top to intermediately wet in the model and water wet at the bottom of the synthetic formation. Errors in λ_c and σ_w estimates show high correlation. For all the 25 synthetic layers, the average relative errors of λ_c, θ_g, and σ_w is around 5% each and that of S_o is around 10%.

Moreover, to estimate the unknown parameters for the entire synthetic formation comprising the 25 synthetic layers, the PSO inversion takes about 12 min, while ACO inversion takes about 50 min. PSO is faster than ACO. In order of computational speed, PSO is the fastest followed by simulated annealing and then ACO with MCMC inversion being the slowest. PSO-based inversion is one order of magnitude faster than MCMC-based inversion. Simulated annealing and MCMC inversion methods are discussed in detail in Chapter 10.

4. Conclusions

In this chapter, water saturation, water conductivity, surface conductance of clay, and contact angle of graphite are simultaneously estimated by processing multifrequency EM measurements using the PSO and ACO-based inversion methods. PSO uses linearly decreasing inertia weight (LIDW) method and the ACO uses max-min ant system (MMAS). Multifrequency EM responses from 25 distinct layers in a synthetic formation were processed by the two inversion methodologies for purposes of comparing their estimation accuracies and computational times. Both the inversion methods provide reliable estimates. ACO-based inversion takes about 50 min to perform inversion for all the 25 synthetic layers, while the PSO-based inversion takes about 12 min. For all the 25 layers in the synthetic formation, the average relative errors of water conductivity, surface conductance of clay, and contact angle of graphite are around 5% each and that of water saturation is around 10%. Water saturation

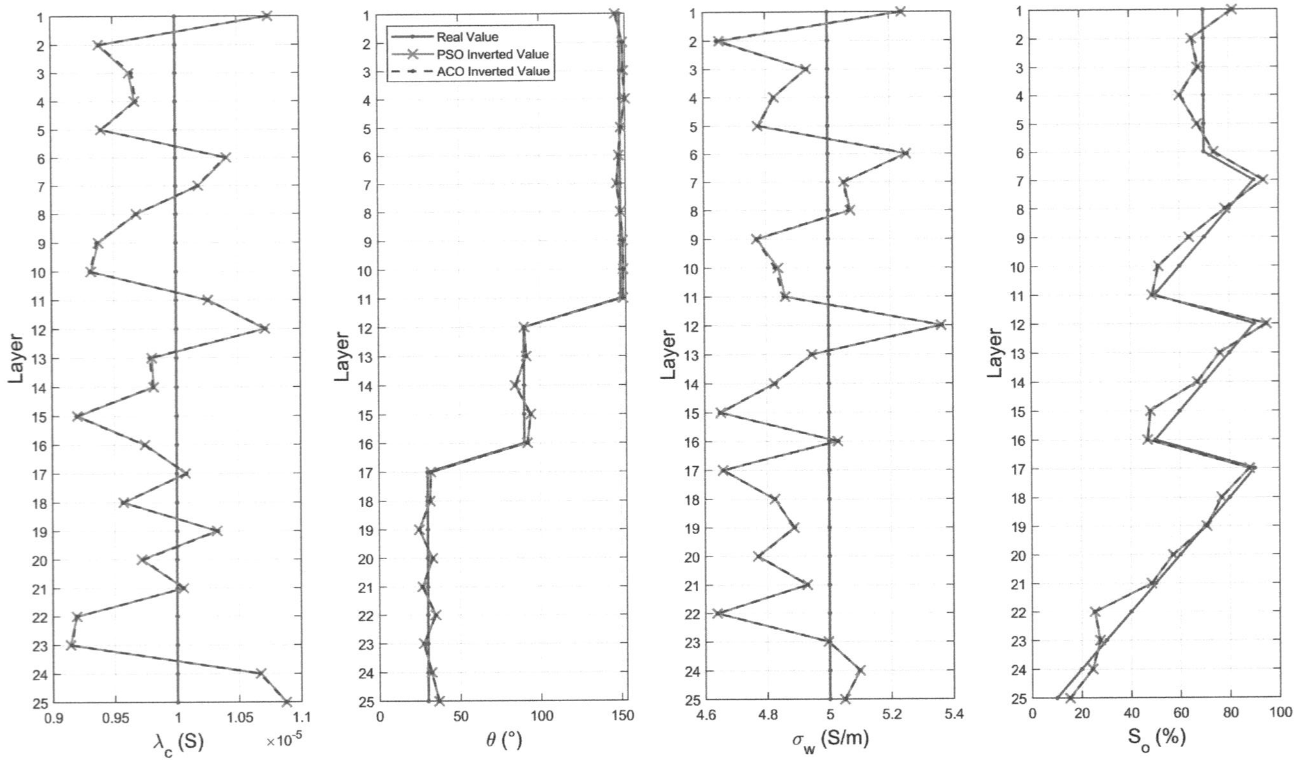

FIGURE 11.7 Inversion-derived estimates of λ_c, θ_g, σ_w, and S_o computed using the PSO and ACO-based inversion methods for the entire synthetic formation comprising 25 distinct layers, as shown in Fig. 10.1 and listed in Table 10.3. The red (dark gray in printed version) line represents the real values of the four unknown properties, the green (light gray in printed version) line represents the estimates computed using the PSO-based inversion, and the blue (gray in printed version) line represents the estimates computed using the ACO-based inversion. *Source: Authors.*

estimates are least accurate, while contact angle estimates are the most accurate when the oil saturation is high. In order of computational speed for the inversion of multifrequency EM data, PSO is the fastest followed by simulated annealing and then ACO with MCMC inversion being the slowest. PSO-based inversion is one order of magnitude faster than MCMC-based inversion. The parameters are relatively easy to tune in PSO method. We conclude that PSO inversion is the best for inverting multifrequency EM data.

Nomenclature

Acronyms

ACO Ant Colony Optimization
EM electromagnetic
LDIW linear decreasing inertia weight
MCMC Markov-chain Monte Carlo
MMAS Max-Min Ant System
PPIP perfectly polarized interfacial polarization
PS PPIP−SCAIP
PSO Particle Swarm Optimization
SA Simulated Annealing
SCAIP surface-conductance-assisted interfacial polarization

Symbols

$\boldsymbol{r_1}, \boldsymbol{r_2}$ **or** $\boldsymbol{rand}$ a random number from $(0, 1)$
α pheromone importance factor
β heuristic function importance factor
$\boldsymbol{c_1}$ acceleration coefficient/learning factor 1
$\boldsymbol{c_2}$ acceleration coefficient/learning factor 2
$\boldsymbol{d(m)}$ modeled data vector
$\boldsymbol{d^{meas}}$ measured data vector
$\boldsymbol{\Delta m}$ disturbance vector
$\Delta\tau$ incremental pheromone vector
$\eta_{i,j}$ heuristic function of ant i to ant j
$\boldsymbol{f(m)}$ objective function at location $\boldsymbol{m}$
$\boldsymbol{ite}$ iteration time for both PSO and ACO
$\boldsymbol{k}$ current iteration step
λ_c surface conductance of clay (S)
$\boldsymbol{m_i}$ location/solution vector of particle i for PSO or location/solution vector of ant i for ACO
$\boldsymbol{m_{max}}$ upper boundary vector of the solution space S
$\boldsymbol{m_{min}}$ lower boundary vector of the solution space S
$\boldsymbol{num}$ swarm size for PSO or ant quantity for ACO
ω inertia weight
ω_{start} initial inertia weight
ω_{end} final inertia weight
$\boldsymbol{P_g}$ the location vector which gives the best fitness value of the whole swarm in the history
$\boldsymbol{P_i}$ the location vector which gives the best fitness value of particle i in the history
$\boldsymbol{p_{i,j}}$ possibility of ant i moving toward ant j

Q a constant controlling the magnitude of objective function
r solution search interval shrink factor
ρ pheromone volatilization factor
S solution space
S_o oil saturation (%)
σ_w water conductivity (S/m)
τ pheromone vector
τ_{max} upper pheromone boundary
τ_{min} lower pheromone boundary
θ_g contact angle of graphite (°)
v_i velocity vector of particle i
v_{max} maximum possible velocity vector of the particle

Superscripts

1 initial
k iteration number

Subscripts

i ant i
j ant j
next ant *next*

References

Bashir, Z.A., El-Hawary, M.E., January 13, 2009. Applying wavelets to short-term load forecasting using PSO-based neural networks. IEEE Trans. Power Syst. 24 (1), 20–27.

Chiou, J.S., Tsai, S.H., Liu, M.T., August 1, 2012. A PSO-based adaptive fuzzy PID-controllers. Simulat. Model. Pract. Theor. 26, 49–59.

Clerc, M., July 6, 1999. The swarm and the queen: towards a deterministic and adaptive particle swarm optimization. In: Proceedings of the 1999 Congress on Evolutionary Computation-CEC99 (Cat. No. 99TH8406), vol. 3. IEEE, pp. 1951–1957.

Colorni, A., Dorigo, M., Maniezzo, V., Trubian, M., 1994. Ant system for job-shop scheduling. Belgian J. Oper. Res. 34, 39–53.

Donati, A.V., Montemanni, R., Casagrande, N., Rizzoli, A.E., Gambardella, L.M., March 16, 2008. Time dependent vehicle routing problem with a multi ant colony system. Eur. J. Oper. Res. 185 (3), 1174–1191.

Dorigo, M., 1992. Optimization, Learning and Natural Algorithms (Ph. D. thesis). Politecnico di, Milano, Italy.

Gambardella, L.M., Dorigo, M., January 1, 1995. Ant-Q: a reinforcement learning approach to the traveling salesman problem. In: Machine Learning Proceedings 1995, pp. 252–260 (Morgan Kaufmann).

Heppner, F., Grenander, U., 1990. A stochastic nonlinear model for coordinated bird flocks. In: The ubiquity of chaos, vol. 233, p. 238.

Kennedy, J., Eberhart, R., November 27, 1995. Particle swarm optimization. In: Proceedings of ICNN'95-International Conference on Neural Networks, vol. 4. IEEE, pp. 1942–1948.

Martens, D., De Backer, M., Haesen, R., Vanthienen, J., Snoeck, M., Baesens, B., October 1, 2007. Classification with ant colony optimization. IEEE Trans. Evol. Comput. 11 (5), 651–665.

Misra, S., Osogba, O., Powers, M., 2019. Unsupervised outlier detection techniques for well logs and geophysical data. Machine Learn. Subsurface Characterization 1.

Okdem, S., Karaboga, D., February 2009. Routing in wireless sensor networks using an ant colony optimization (ACO) router chip. Sensors 9 (2), 909–921.

Salari, E., Eshghi, K., December 15, 2005. An ACO algorithm for graph coloring problem. In: 2005 ICSC Congress on Computational Intelligence Methods and Applications. IEEE, p. 5.

Shen, M., Zhan, Z.H., Chen, W.N., Gong, Y.J., Zhang, J., Li, Y., March 27, 2014. Bi-velocity discrete particle swarm optimization and its application to multicast routing problem in communication networks. IEEE Trans. Ind. Electron. 61 (12), 7141–7151.

Shi, Y., Eberhart, R., May 4, 1998. A modified particle swarm optimizer. In: 1998 IEEE International Conference on Evolutionary Computation Proceedings. IEEE World Congress on Computational Intelligence (Cat. No. 98TH8360). IEEE, pp. 69–73.

Shi, Y., Eberhart, R., July 6, 1999. Empirical study of particle swarm optimization. In: Proceedings of the 1999 Congress on Evolutionary Computation-CEC99 (Cat. No. 99TH8406), vol. 3. IEEE, pp. 1945–1950.

Shmygelska, A., Aguirre-Hernandez, R., Hoos, H.H., September 12, 2002. An ant colony optimization algorithm for the 2D HP protein folding problem. In: International Workshop on Ant Algorithms. Springer, Berlin, Heidelberg, pp. 40–52.

Stützle, T., Hoos, H.H., August 1996. Improving the Ant System: A Detailed Report on the MAX–MIN Ant System. FG Intellektik, FB Informatik, TU Darmstadt, Germany. Tech. Rep. AIDA–96–12.

Stützle, T., 1997. MAX-MIN Ant System for Quadratic Assignment Problems.

Tasgetiren, M.F., Sevkli, M., Liang, Y.C., Gencyilmaz, G., September 5, 2004. Particle swarm optimization algorithm for permutation flowshop sequencing problem. In: International Workshop on Ant Colony Optimization and Swarm Intelligence. Springer, Berlin, Heidelberg, pp. 382–389.

Zhan, Z.H., Zhang, J., Li, Y., Chung, H.S., April 7, 2009. Adaptive particle swarm optimization. IEEE Trans. Syst. Man Cybernet. B (Cybernetics) 39 (6), 1362–1381.

Index

Note: 'Page numbers followed by "*f*" indicate figures and "*t*" indicate tables.'

F

G

H

I

M

T

U

V

W

Y

Z

FSC
www.fsc.org
MIX
Papier | Fördert gute Waldnutzung
FSC® C083411

Kolibri 360 GmbH
Ferdinand-Jühlke-Straße 7
99095 Erfurt, Deutschland
produktsicherheit@kolibri360.de